Lecture Notes in Mathematics

A collection of informal reports and seminars
Edited by A. Dold, Heidelberg and B. Eckmann, Zürich

Series: Mathematics Institute, University of Warwick
Adviser: D. B. A. Epstein

294

Stability of Stochastic Dynamical Systems

Proceedings of the International Symposium

Organized by "The Control Theory Centre",
University of Warwick, July 10–14, 1972

Sponsored by the "International Union of Theoretical and Applied Mechanics"

Edited by Ruth F. Curtain
University of Warwick, Coventry Warwickshire/England

Springer-Verlag
Berlin · Heidelberg · New York 1972

AMS Subject Classifications (1970): 34F05, 34H05, 60H10, 93-02, 93D99, 93E05, 93E10, 93E15, 93E99

ISBN 3-540-06050-2 Springer-Verlag Berlin · Heidelberg · New York
ISBN 0-387-06050-2 Springer-Verlag New York · Heidelberg · Berlin

Offsetdruck: Julius Beltz, Hemsbach/Bergstr.

INTRODUCTION

The symposium on the "Stability of Stochastic Dynamical Systems", held at Warwick University, July 10 - 14th, 1972, was sponsored by the International Union of Theoretical and Applied Mechanics, (IUTAM), whose support we appreciate. Following IUTAM policy, participation was by invitation and was limited to around 60 so as to encourage lively discussion. A full list of participants is given.

The main theme of the symposium was the stability and other properties of differential equations with stochastic coefficients. Both the general mathematical aspects and applications were discussed.

The contents have been arranged according to the daily sessions, when a "key-note" speaker gave a one hour address to set the daily "theme". Other lectures were of twenty minutes duration. As each contributor was asked to submit a complete typed version of his address for the symposium, these proceedings are in general more detailed than the actual talks and are in no need of further introduction.

Ruth F. Curtain, 1972

International Scientific Committee

Prof. J.A. SHERCLIFF (U.K.) Chairman
Prof. P. BLAQUIERE (France)
Dr. P. BRUNOVSKY (Czechoslovakia)
Prof. K. ITO (Japan)
Prof. R.E KALMAN (U.S.A.)
Prof. L. MARKUS (U.K. and U.S.A.)
Prof. H. ROSENBROCK (U.K.)
Prof. V.V. SOLODOVNIKOV (U.S.S.R.)
Prof. H.R. SCHWARZ (W. Germany) (IFAC Representative)

List of Participants

Prof. S.T. ARIARATNAM	Canada
Dr. L. ARNOLD	West Germany (FDR)
Prof. K.J. ÅSTRÖM	Sweden
Dr. M.C. AUMASSON	France
Dr. J.F. BARRETT	U.K.
Dr. A. BENSOUSSAN	France
Prof. P. BLAQUIERE	France
Prof. R. BROCKETT	U.S.A.
Dr. P. BRUNOVSKY	Czechoslovakia
Dr. HELGA BUNKE	DDR
Dr. P. CAINES	U.K.
Dr. J.M.C. CLARK	U.K.
Dr. RUTH CURTAIN	U.K.
Dr. M. DAVIS	U.K.
Prof. J.L. DOUCE	U.K.
Prof. T. DUNCAN	U.S.A.
Dr. R. ELLIOTT	U.K.
Prof. A. FRIEDMAN	U.S.A.
Dr. A.T. FULLER	U.K.
Dr. C.J. HARRIS	U.K.
Prof. P.J. HARRISON	U.K.
Prof. U. HAUSSMANN	Canada
Mr. D.B. HERNANDEZ-CASTANO	U.K.
Dr. M. HUGHES	U.K.
Prof. K. ITO	U.S.A.

Dr. J.G. JAMES	U.K.
Prof. F. KOZIN	U.S.A.
Prof. H.J. KUSHNER	U.S.A.
Prof. J.A. LEPORE	U.S.A.
Dr. P. MANDL	Czechoslovakia
Prof. L. MARKUS	U.S.A./U.K.
Prof. D. MAYNE	U.K.
Prof. R. MONOPOLI	U.S.A.
Dr. T. MOROZAN	Rumania
Dr. L.A. MYSAK	Canada
Dr. T. NAKAMIZO	Japan
Dr. P.C. PARKS	U.K.
Dr. A. PRITCHARD	U.K.
Prof. P. SAGIROW	West Germany (FDR)
Prof. P. SETHNA	U.S.A.
Prof. J.A. SHERCLIFF	U.K.
Dr. Ing. W. WEDIG	West Germany
Prof. P. WHITTLE	U.K.
Prof. J.C. WILLEMS	U.S.A.
Prof. J.L. WILLEMS	Belgium
Dr. D. WISHART	U.K.
Dr. J.A. ZEMAN	Austria

Local Organizing Committee (University of Warwick)

Prof. J.A. SHERCLIFF (Chairman)

Prof. L. MARKUS

Dr. RUTH CURTAIN

Dr. P.C. PARKS

CONTENTS

* These were key-note lectures

STOCHASTIC DIFFERENTIALS OF CONTINUOUS LOCAL QUASI-MARTINGALES*

KIYOSI ITÔ

Cornell University

The purpose of this paper is to unify the known results on stochastic differentials in terms of differential quasi-martingales so that we can understand the intuitive meaning more easily.

1. NOTATIONS AND DEFINITIONS

Let $\{\mathcal{F}_t, 0 \leq t \leq 1\}$ be a right continuous increasing family of σ-algebras. The conditional expectation relative to $\mathcal{F}_t$ is denoted by E_t. We consider the following classes of stochastic processes.

$\mathfrak{A}$ is the class of all stochastic processes adapted to $\{\mathcal{F}_t\}$ and having continuous sample paths.

$\mathfrak{M}$ is the class of all continuous local martingales relative to $\{\mathcal{F}_t\}$, namely all processes $X \in \mathfrak{A}$ such that there exists a sequence of martingales $X_n \in \mathfrak{A}$ satisfying

$$P(X_n(t) = X(t) \text{ for every } t) \to 1.$$

$\mathcal{L}$ is the class of all $X \in \mathfrak{A}$ whose sample paths are of bounded variation almost surely.

$\mathcal{Q}$ is the class of all continuous local quasi-martingales, namely all X such that

$$X = M + L, \ M \in \mathfrak{M}, \ L \in \mathcal{L};$$

see D.L. Fisk [2], S. Orey [7] and K.M. Rao [8].

$d\mathcal{Q}$ is the class of all random interval functions dQ (stochastic differentials) induced by Q as follows:

$$dQ(s,t) = Q(t)-Q(s), \ (s < t).$$

Similarly for $d\mathfrak{M}$, $d\mathcal{L}$ and $d\mathcal{B}$.

---

*This work was supported by NSF GP-28109.

2. STOCHASTIC DIFFERENTIALS

Let us introduce three basic operations on stochastic differentials.

DEFINITION 1. $dQ_2 = edQ_1$ <u>if and only if</u>

$$dQ_2(s,t) \equiv Q_2(t) - Q_2(s) = \int_s^t E_\theta \, dQ_1(\theta) \quad (s < t).$$

The integral is defined in the same way as stochastic integrals. If $Q_1 = M_1 + L_1$, such that both $M_1(t)$ and $L_1(t)$ are integrable and that $M_1 \in \mathcal{M}$ and $L_1 \in \mathcal{L}$, then

$$\int_s^t E_\theta \, dQ_1(\theta) = \lim_{|\Delta|\to 0} \sum_{i=1}^{n} E_{\theta_{i-1}}(Q_1(\theta_i) - Q_1(\theta_{i-1})) = L_1(t) - L_1(s)$$

where $\Delta = \{0 = \theta_1 < \theta_1 < \cdots < \theta_n = 1\}$ is an arbitrary division of the interval $[0,1]$ and $|\Delta| = \max_i |\theta_i - \theta_{i-1}|$. The existence of this limit can be easily verified and this limit defines a unique $dQ_2 \in d\mathcal{Q}$. The extension to the general $dQ_1 \in d\mathcal{Q}$ is obvious.

DEFINITION 2. $dQ_2 = \alpha dQ_1$ $(\alpha \in \mathcal{O})$ <u>if and only if</u>

$$dQ_2(s,t) \equiv Q_2(t) - Q_2(s) = \int_s^t \alpha(\theta) dQ_1(\theta) \quad (s < t)$$

The integral can be defined in the same way as the usual stochastic integral. If we write Q_1 as $Q_1 = M_1 + L_1$, $M_1 \in \mathcal{M}$ and $L_1 \in \mathcal{L}$, then the above integral is written as

$$\int_s^t \alpha dQ_1 = \int_s^t \alpha dM_1 + \int_s^t \alpha dL_1 .$$

The first integral on the right hand side is the martingale stochastic integral introduced by Ph. Courrège [1]; see also H. Kunita and S. Watanabe [4] and P. Meyer [6]. The second integral is the usual Stieltjes integral defined for each path.

DEFINITION 3. $dQ_3 = dQ_1 \cdot dQ_2$ <u>if and only if</u>

$$dQ_3(s,t) = Q_3(t) - Q_3(s) = \int_s^t dQ_1(\theta)\, dQ_2(\theta).$$

The integral is defined as follows:

$$= \lim_{|\Delta|\to 0} \sum_{i=1}^{n} (Q_1(\theta_i) - Q_1(\theta_{i-1}))(Q_2(\theta_i) - Q_2(\theta_{i-1}))$$

where $s < t$ and $\Delta = \{\theta_i\}$ is an arbitrary division of $[0,1]$ as above.

Let $dQ = dM + dL$, $dM \in d\mathcal{M}$ and $dL \in d\mathcal{L}$. Then

$$e\,dQ = dL, \quad (I-e)dQ = dM \quad \text{and} \quad e^2 = e.$$

dM and dL are called the <u>martingale part</u> and the <u>bounded variation part</u> of dQ.

It is easy to see

(1) αdQ: $d\mathcal{Q} \to d\mathcal{Q}$, $d\mathcal{M} \to d\mathcal{M}$ and $d\mathcal{L} \to d\mathcal{L}$.

(2) $dQ_1 \cdot dQ_2$: $d\mathcal{Q} \times d\mathcal{Q} \to d\mathcal{L}$, $d\mathcal{Q} \times d\mathcal{L} \to \{0\}$.

(3) $dQ_1 \cdot dQ_2 = (I-e)dQ_1 \cdot (I-e)dQ_2$.

(4) $e \cdot (\alpha \cdot dQ) = \alpha \cdot (e \cdot dQ)$

THEOREM 1. <u>If</u> $f(x_1, x_2, \ldots, x_n) \in C^2(R^n)$ <u>and</u> $Q_i \in \mathcal{Q}_i$, $i = 1,2,\ldots,n$ <u>then</u> $Y = f(Q) = f(Q_1, Q_2, \ldots, Q_n) \in Q$ <u>and</u>

(a) $dY = \sum_i f_i(Q)dQ_i + \frac{1}{2} \sum_{i,j} f_{ij}(Q)dQ_i dQ_j$

<u>where</u> $f_i = f_{x_i}$ <u>and</u> $f_{ij} = f_{x_i x_j}$. <u>If</u> $dQ_i = dM_i + dL_i$, $i = 1,2,\ldots,n$, <u>then</u>

(b) $dY = \sum_i f_i(Q)dM_i + (\sum_i f_i(Q)dL_i + \frac{1}{2} \sum_{i,j} f_{ij}(Q)dM_i \cdot dM_j)$

<u>where the first and the second terms on the right hand side are the martingale part and the bounded variation part of</u> dY.

The formula (a) can be proved by the Taylor expansion as in the usual Brownian motion case; see K. Itô [3] and H.P. McKean, Jr. [5]. The formula (b) follows immediately from (a) by (1), (2) and (3). The formula (a) is more natural and easy to remember. See [4] and [5] for the martingale case where discontinuous processes are also discussed.

THEOREM 2. <u>The stochastic differentials are invariant under stochastic time change by continuously increasing stopping times.</u>

This is a trivial consequence of the invariance of martingales by time change.

THEOREM 3. <u>If</u> $e \cdot dQ_i = 0$, $i = 1,2,\ldots,n$ <u>and if</u> $dQ_i \cdot dQ_j = \delta_{ij} \cdot dt$, <u>then</u> dQ_i, $i = 1,2,\ldots,n$ <u>are Brownian differentials independent of each other</u>.

This is a mere rephrasing of a result by H. Kunita and S. Watanabe [4]. In case $n = 1$, this is equivalent to a theorem of P. Lévy which reads as follows: If $X(t)$ and $X(t)^2 - t$ are both martingales, then $X(t)$ is a Brownian motion.

3. EXAMPLES

<u>1. Application to Diffusion Processes</u>. The Kolmogorov formulation of classical diffusions is essentially as follows

$$edX = adt, \quad e(dX)^2 = bdt$$

where $a(t)$ and $b(t)$ are functions of t and $X(t)$ only. If a and b depend on the whole past history, then X will be a process with continuous paths which is not Markov. We will consider this general case. We assume that $b > 0$. Define $B \in Q$ by

$$dB = b^{-\frac{1}{2}}(dX - adt), \quad B(0) = 0$$

Using the results in Section 2, we have

$$edB = 0 \quad \text{and} \quad e(dB)^2 = dt.$$

Therefore dB is a Brownian differential and we will get a stochastic differential equation:

$$dX = adt + b^{\frac{1}{2}} dB, \quad X(0) = x.$$

$B(t)$ is measurable with respect to $\{X(s), s \leq t\}$ but we are not sure that $X(t)$ is measurable with respect to $\{B(s), s \leq t\}$. This is the case if $a(t) = \alpha(t, X(t))$ and $b(t) = \beta(t, X(t))$ and these functions $\alpha(t,x)$ and $\beta(t,x)$ are sufficiently smooth, as is well known. If $f(x) \in C^2(\mathbb{R})$, then we have

$$e(df(X)) = (f'(X)a + \tfrac{1}{2}f''(X)b)dt,$$

which gives the generator of the process.

2. The Semi-group of Classical Diffusion. Consider a stochastic differential equation

$$dX(t) = a(X(t))dt + \sigma(X(t))dB(t), \quad X(0) = x,$$

where the Lipschitz condition on both a and σ is assumed. Then we have a unique solution which determines a diffusion on $\mathbb{R}$. To be able to apply the Hille-Yosida theorem, we have to prove that the corresponding semi-group carries C_0 into itself, where C_0 is the space of all continuous functions vanishing at ∞. This condition is equivalent to the following one:

$$(5) \qquad P_x(X(t) \in K) \to 0 \quad (x \to \infty)$$

for every t and every compact K fixed. Since $a(x)$ and $\sigma(x)$ are of order $|x|$ as $|x| \to \infty$, we can prove by Theorem 1 that $Y(t) = \log(1 + X(t)^2)$ satisfies

$$dY(t) = \alpha(t)dt + \beta(t)\, dB(t)$$

where $|\alpha(t)|$ and $|\beta(t)|$ are bounded by some constant C. Therefore

$$Y(t) - Y(0) = \int_0^t \alpha(s)ds + \int_0^t \beta(s)d\beta(s)$$

and so

$$E_x(Y(t)-Y(0))^2 \leq 2E_x(\int_0^t \alpha(s)ds)^2 + 2E_x(\int_0^t \beta(s)d\beta(s))^2$$

$$\leq 2C^2(t^2 + t) = C_1$$

namely

$$E_x\left(\log \frac{1+X(t)^2}{1+x^2}\right)^2 \leq C_1$$

$$P_x((1+X(t)^2) \leq (1+x^2)e^{-a}) \leq C_1/a^2$$

Take a such that $C_1/a^2 < \varepsilon$, fix it and then let $x \to \infty$ to obtain (5).

3. Stochastic Differential Equations on a Manifold. We can make the same discussion on a manifold as in $\mathbb{R}$ in terms of the local coordinates. Take an open locally finite covering $\{U_n\}$ of the manifold such that a system of local coordinates is given in each U_n. Take an open V_n for each n such that $\overline{V}_n \subset U_n$ and that $\{V_n\}$ covers the whole manifold. Let $X(t)$ be a diffusion on the manifold starting at a in some V_n, say $V_{n(1)}$. Take the first exit time T_1 from $U_{n(1)}$. Let $V_{n(2)}$ be the first one among all V_n containing $X(T_1)$ and repeat the same procedure to get $n(1), n(2), \ldots$ and $T_1, T_2, \ldots$. Write $X(t)$ in the $U_{n(i)}$-coordinates in $T_{i-1} \leq t \leq T_i$ $(T_0 \equiv 0)$. Now we can make the same argument as in $\mathbb{R}$ to get an n-dimensional Brownian motion, where b is a matrix and $\sigma = b^{\frac{1}{2}}$ is taken in matrix sense so that it is continuous σ in each neighborhood.

REFERENCES

[1] Courrège, Ph., Intégrales stochastiques et martingales de carré intégrable. Séminaire Brelot-Choquet-Deny (théorie du potential) 7e année, 1962-63, exposé 7.

[2] Fisk, D.L., Quasi-martingales. Trans. Amer. Math. Soc., 120 (1965), 369-389.

[3] Itô, K., On a formula concerning stochastic differentials. Nagoya Math. J. 3 (1951), 55-65.

[4] Kunita, H. and Watanabe, S., On square integrable martingales. Nagoya Math. J. 30 (1967) 209-245.

[5] McKean, H.P., Jr., Stochastic integrals, Acad. Press. New York and London, 1969.

[6] Meyer, P.A., Integral stochastiques. I-IV., Lecture Notes in Math. (Springer) 39 Sem. de Prob. (1967), 72-162.

[7] Orey, S., F-processes. Proc. Fifth Berkeley Symp. on Stat. and Prob. 2 (1965) 301-313.

[8] Rao, K.M., Quasi-martingales. Math Scand. 24 (1969) 79-92.

AN APPLICATION OF ITÔ'S FORMULA TO STOCHASTIC CONTROL SYSTEMS

PETR MANDL

Institute of Information Theory and Automation, Prague

Let $\{X_t, t \geq 0\}$ denote the trajectory of a control system governed by Itô's stochastic differential equations

(1) $dX_t^i = b^i(X_t, U_t)\,dt + \sum_{j=1}^{n} \sigma_{ij}(X_t)\,dW_t^j \quad , i = 1, \dots, m,$

where $\{W_t, t \geq 0\}$ is an n-dimensional Brownian motion process

$$\{U_t = U(X_s, s \leq t), t \geq 0\}$$

a non-anticipative control. The control parameter space is denoted by $\mathcal{U}$. The criterion functional (the reward) is assumed to have the form

$$C_T = \int_0^T c(X_t, U_t)\,dt$$

where T is large. Besides measurability postulates, the local boundedness of the coefficients in (1) and of the function $c(x,u)$ is required. If equations (1) contain unknown parameters, the controller has to estimate them from the observed part of the trajectory and to base the control on the estimates. The control should become optimal for $t \to \infty$. The paper deals with the question of how fast the approach to the optimum should be to give the reward with the same asymptotic distribution as under optimal controls.

Introduce

$$a^{ij}(x) = \frac{1}{2}\sum_k \sigma_{ik}(x)\sigma_{jk}(x)$$

We assume that there exist constants $\hat{\theta}, \hat{\varphi}$ and twice continuously differentiable functions $v_1(x), v_2(x)$ on R^m for which

(2) $$\max_{u \in \mathcal{U}} \left\{ \sum_{ij} a^{ij}(x) \frac{\partial^2}{\partial x^i \partial x^j} v_1 + \sum_i b^i(x,u) \frac{\partial}{\partial x^i} v_1 + c(x,u) - \hat{\theta} \right\} = 0 ,$$

(3) $$\min_{u \in \mathcal{U}(x)} \left\{ \sum_{ij} a^{ij}(x) \frac{\partial^2}{\partial x^i \partial x^j} v_2 + \sum_i b^i(x,u) \frac{\partial}{\partial x^i} v_2 + 2(c(x,u) - \hat{\theta}) v_1(x) - \hat{\varphi} \right\} = 0,$$

where $\mathcal{U}(x) = \{u : \varphi_1(x,u) = 0\}$ and

$$\varphi_1(x,u) = \psi_1(x,u) + c(x,u) - \hat{\theta}, \quad \varphi_2(x,u) = \psi_2(x,u) + 2(c(x,u) - \hat{\theta}) v_1(x) - \hat{\varphi}$$

are the expressions in braces in (2) and in (3), respectively. Bellman type equations like (2),(3) have been studied by several authors. The quantities involved have the following interpretation: $\hat{\theta}$ optimal mean reward, $\hat{\varphi}$ optimal mean variance, $v_1(x)$ excess reward, $v_2(x)$ excess variance. We shall abbreviate

$$c(X_t, U_t) = c(t), \quad \varphi(X_t, U_t) = \varphi(t)$$

etc. E will be used to denote the mathematical expectation.

THEOREM 1. Let $z \in (-\infty, \infty), T > 0$. Introduce

$$\chi_t = \exp\{iz(C_t - \hat{\Theta}t)\}.$$

Assume

Then $\int_0^T E\sigma_{ij}^2(X_t)dt < \infty, \quad i,j = 1, \dots, n.$

(4) $E\chi_T = \exp\{-\frac{1}{2}\hat{\varrho}Tz^2\} + \frac{1}{2}\hat{\varrho}z^2 \int_0^T \exp\{-\frac{1}{2}\hat{\varrho}(T-t)z^2\}\varkappa(t,z)dt + \varkappa(T,z)$

holds with

(5)
$$\begin{aligned} \varkappa(t,z) = iz\, E[\int_0^t \chi_s \varphi_1(s)ds + v_1(X_0) - \chi_t v_1(X_t)] - \\ -\frac{1}{2}z^2 E[\int_0^t \chi_s \varphi_2(s)ds - \int_0^t \chi_s dv_2(X_s)], \end{aligned}$$

provided that the expectations on the right-hand side in (5) are finite for $t \in [0,T]$ and $\varkappa(t,z)$ is integrable.

<u>Proof.</u> Partial integration and the use of Itô's formula for stochastic differentials lead to the relations

$$\begin{aligned} E\chi_T &= 1 + iz E\int_0^T \chi_t (c(t) - \Theta)dt \\ 0 &= iz E[\int_0^T \chi_t \psi_1(t)dt - \int_0^T \chi_t dv_1(X_t)] \\ 0 &= -\frac{z^2}{2} E[\int_0^T \chi_t \psi_2(t)dt - \int_0^T \chi_t dv_2(X_t)] \\ -iz E\int_0^T \chi_t dv_1(X_t) &= iz E[v_1(X_0) - \chi_T v_1(X_T)] - \\ &\quad - z^2 E\int_0^T \chi_t (c(t) - \Theta) v_1(X_t)dt \end{aligned}$$

Adding these relations one gets

(6) $$E\chi_T = 1 - \frac{1}{2}\hat{\varrho}z^2 \int_0^T E\chi_t dt + \varkappa(T,z).$$

To obtain (4) one has to solve (6) as first order differential equation for $\int_0^T E\chi_t dt$ and to insert the result into (6) again. □

Comparing the coefficients at $z^m, m = 1,2,\dots,$ in Taylor's development of (6) we get relations for <u>moments</u>

(7) $$EC_T = \hat{\Theta}T + E(v_1(X_0) - v_1(X_T)) + E\int_0^T \varphi(t)dt,$$

$$\frac{1}{m!}E(C_T-\hat{\theta}T)^m = \frac{1}{2}E\int_0^T \frac{(C_t-\hat{\theta}t)^{m-2}}{(m-2)!}\,[\hat{\varphi}dt+\varphi_2(t)dt-dv_2(X_t)] +$$

(8)

$$+E\int_0^T \frac{(C_t-\hat{\theta}t)^{m-1}}{(m-1)!}\,\varphi_1(t)dt - E\frac{(C_t-\hat{\theta}t)^{m-1}}{(m-1)!}\,v_1(X_t), \quad m=2,3,\ldots,$$

under the condition that the expectations involved are finite.

THEOREM 2. Assume $\hat{\varphi}>0$. Let $\{U_t, t\geqq 0\}$ be such that

(9) $\int_0^T E\sigma_{ij}^2(X_t)\,dt < \infty, \quad i,j=1,\ldots,n, \quad T>0,$

(10) $E|v_i(X_t)| <$ const $<\infty$, $i=1,2$, $E|c(t)-\hat{\theta}|\,|v_2(X_t)| <$ const $<\infty$, $t\geqq 0$.

If

(11) $$\lim_{T\to\infty}\frac{1}{\sqrt{T}}\int_0^T E\varphi_1(t)dt = 0 = \lim_{T\to\infty}\frac{1}{T}\int_0^T E|\varphi_2(t)|\,dt,$$

then C_T has asymptotically normal distribution $N(\hat{\theta}T,\hat{\varphi}T)$ for $T\to\infty$.

To demonstrate Theorem 2 one verifies from formula (4) that

$$E\exp\{ix(C_T-\hat{\theta}T)/\sqrt{T}\} \longrightarrow \exp\{-\tfrac{1}{2}\hat{\varphi}x^2\}$$

under the above hypotheses. □

Let $U_t=\hat{u}(X_t)$, $t\geqq 0$, where

(12) $\varphi_1(x,\hat{u}(x)) = 0 = \varphi_2(x,\hat{u}(x)), \quad x\in R^n.$

If (9), (10) hold, then C_T is asymptotically $N(\hat{\theta}T,\hat{\varphi}T)$ for $T\to\infty$. An analysis of (7), (8) implies that $\hat{\theta}$ is the maximal value possible and $\hat{\varphi}$ is the minimal value attainable for controls satisfying

$$U_t\in U(X_t),\ Ev_1(X_t)^2 < \text{const} < \infty, \qquad E|v_2(X_t)| < \text{const} < \infty, \quad t\geqq 0.$$

Hence, $\hat{u}(x)$ is an <u>optimal stationary control policy</u>.

Assume now that an <u>unknown parameter</u> $\alpha\in A$ is present in (1), where A is an open set in R^q. The trajectory and the reward fulfil

$$dX_t^i = b^i(X_t,U_t,\alpha)dt + \sum_j \sigma_{ij}(X_t,\alpha)dW_t^j, \quad i=1,\ldots,n,$$
$$C_T = \int_0^T c(X_t,U_t,\alpha)dt.$$

We denote by α_0 the true value of the unknown parameter. The concepts introduced in the preceding will refer to $\alpha=\alpha_0$. Let $\hat{\mu}(x,\alpha)$ be an optimal stationary policy for $\alpha\in A$ and let

(13) $\varphi_1(x,\hat{u}(x,\alpha_0)) = 0 = \varphi_2(x,\hat{u}(x,\alpha_0)), \quad x \in R^n.$

The controller estimates the unknown parameter. Write α_t^* for his estimate of α_0 based on the trajectory up to time t. If α^0 were known to him, the controller would use $\{\hat{u}(X_t,\alpha_0), t \geqq 0\}$ or another optimal control. Instead he employs the adaptive control

(14) $\{U_t = \hat{u}(X_t,\alpha_t^*), \quad t \geqq 0\}.$

The next theorem contains conditions, under which this control yields the same asymptotic distribution of the reward as $\{\hat{u}(X_t,\alpha_0), t \geqq 0\}$.

THEOREM 3. Let $\hat{\varphi} > 0$. Let, for $x \in R^n$, $\varphi_1(x,\hat{u}(x,\alpha))$ be continuously differentiable with respect to α and let $(\partial/\partial\alpha^j)\varphi_1(x,\hat{u}(x,\alpha))$, $j = 1,\dots,q$ and $\varphi_2(x,\hat{u}(x,\alpha))$ be bounded and continuous at α_0 uniformly in $x \in R^m$. Let the control (14) be such that the trajectory is well defined by (1) together with the initial data and that (9),(10) hold. If

(15) $E\|\alpha_t^* - \alpha_0\|^2 \leqq \dfrac{const}{1+t}, \quad t \geqq 0,$

then C_T has asymptotically normal distribution $N(\hat{\theta}T, \hat{\varphi}T)$ for $T \to \infty$

The <u>proof</u> consists in a verification of (11). One has to use Taylor's development of $\varphi_1(x,\hat{u}(x,\alpha))$ at α_0 and the fact that φ_1 vanishes together with the first derivatives in the point α_0 in virtue of (2),(13). □

The rate required by (15) is the typical convergence rate of the estimates of unknown parameters:

<u>Example 1.</u> Consider the one-dimensional trajectory satisfying

$$dX_t = (b_1(X_t,U_t) + \alpha_0 b_2(X_t))dt + \sigma(X_t)dW_t,$$

where α_0 is unknown,

$$0 < k_1 < \sigma(x) < k_2 < \infty, \quad k_1 < b_2(x) < k_2.$$

An estimate of α_0 is

$$\alpha_T^* = [\int_0^T \sigma(X_t)^{-1} dX_t - \int_0^T b_1(X_t,U_t)\sigma(X_t)^{-1} dt] / \int_0^T b_2(X_t)\sigma(X_t)^{-1} dt.$$

Then

$$E(\alpha_T^* - \alpha_0)^2 = E[W_T / \int_0^T b_2(X_t)\sigma(X_t)^{-1} dt]^2 \leq k_1^{-2} k_2^2 / T . \square$$

The methods are applicable to systems with trajectories restricted to <u>bounded domains</u>, if the behaviour of the system on the boundary is sufficiently regular. The simplest case is the jump into the interior:

<u>Example 2.</u> Let the range of the trajectory $\{X_t, t \geqq 0\}$ fulfilling

$$dX_t = (U_t + \alpha_0)\, dt + \sqrt{D}\, dW_t$$

be the interval $[-1,1]$. After having reached its boundaries the trajectory starts afresh from 0 . Let the criterion be

$C_T = k \int_0^T U_t^2 dt + N \times$ number of jumps into 0 up to time T .

$(D>0, k>0, N>0)$. C_T is to be minimized.

To determine $\hat{u}(x,\alpha)$ we solve

$$\min_u \left\{ \frac{D}{2} \frac{d^2}{dx^2} v_1^\alpha + (u+\alpha) \frac{d}{dx} v_1^\alpha + k u^2 - \theta_\alpha \right\} = 0$$

with boundary conditions $v_1^\alpha(-1) = N + v_1^\alpha(0) = v_1^\alpha(1)$.

We get

$$\hat{u}(x,\alpha) = DK_1 \operatorname{tg}(K_2 - K_1 x) - \alpha ,$$

where

$$K_1 = \arccos\left[e^{-\frac{N}{2Dk}} \cosh\left(\frac{\alpha}{D}\right)\right], \quad K_2 = \operatorname{arctg}\left[e^{-\frac{N}{2Dk}} \sinh\left(\frac{\alpha}{D}\right) / \sin K_1\right],$$

provided $K_2 \pm K_1 \in (-\pi/2, \pi/2)$. For the purpose of adaptive control, α_0 can be estimated as in Example 1. $\square$

Estimates of errors introduced by <u>discrepancies in the drift</u> are also obtainable. Let the actual drift of the system be different from that presumed. Thus, when the controller uses the policy $\hat{u}(x)$ satisfying (12), the system is governed by

$$dX_t^i = (b^i(X_t, \hat{u}(X_t)) + B^i(t))\, dt + \sum_j \sigma_{ij}(X_t)\, dW_t^j , \quad i = 1,\dots,m.$$

It is required that $\{B(t), t \geqq 0\}$ is a random process independent of the future increments of $\{W_t, t \geqq 0\}$. Formulas (4),(7),(8) hold with $\varphi_i(t)$, $i = 1,2$, replaced by

$$\sum_{j} \left(\frac{\partial}{\partial x^j} v_i(X_t) \right) B^j(t), \qquad i = 1,2.$$

The fact that Itô's formula is valid for non-anticipative controls was used by W.H. Fleming in [1]. Detailed proofs of Theorems 1 - 3 can be obtained by adapting the corresponding proofs in author's paper [2] which deals with finite state situation.

Bibliography

[1] Fleming W.H.: Duality and a Priori Estimates in Markovian Optimization Problems, J. of Math. Analysis and Appl. 16, 254-279 (1966).

[2] Mandl P.: On the Adaptive Control of Finite State Markov Processes. Submitted for publication in Z. für Wahrscheinlichkeitstheorie u. verw. G.

STABILITY AND ANGULAR BEHAVIOR OF SOLUTIONS OF STOCHASTIC DIFFERENTIAL EQUATIONS

AVNER FRIEDMAN

Northwestern University

Introduction. The stability of solutions of stochastic differential systems is considered in the book of H. J. Kushner [7], where some examples are also studied. Stability for linear equations was studied by Kasminski [5]. The spiraling of solutions of 2-dimensional systems was treated, in a special case, by Kozin and Prodromou [6]. Finally, for one stochastic equation, detailed results are given in the book of Gikhman and Skorokhod [4].

In this lecture I shall report on some general theorems regarding stability and (for two dimensional systems) spiraling of solutions, recently obtained by Friedman and Pinsky [3].

1. NON-ATTAINABILITY OF THE BOUNDARY

Consider a system of ℓ stochastic differential equations

$$dx_i = \sum_{r=1}^{n} \sigma_{ir}(x)dw^r + b_i(x)dt \qquad (1 \leq i \leq \ell), \tag{1.1}$$

where $w^1(t),\ldots,w^n(t)$ are independent Brownian motions. Assume:

(A) $\sigma_{ir}(x)$, $b_i(x)$ are uniformly Lipschitz continuous on compact subsets, and

$$|\sigma_{ir}(x)| + |b_i(x)| \leq K(1 + |x|) \qquad \text{(K constant).}$$

Let $G_1,\ldots,G_k$ be mutually disjoint sets in R^ℓ; for $1 \leq j \leq k_0$, G_j consists of one point z_j, and, for $k_0 + 1 \leq j \leq k$, G_j is a bounded closed domain with C^3 boundary ∂G_j. Set $\partial G_j = \{z_j\}$ if $1 \leq j \leq k_0$. Let

$$\rho_j(x) = \text{dist}(x, G_j) \quad \text{for } x \notin \text{int } G_j,$$

$$\hat{G}_{j,\varepsilon} = \{x;\ x \notin \text{int } G_j,\ \rho_j(x) \leq \varepsilon\}, \quad \hat{G}_\varepsilon = \bigcup_{j=1}^{k} \hat{G}_{j,\varepsilon},$$

$$(a_{ij}) = \sigma\sigma^*, \quad \sigma = (\sigma_{ir}), \quad \sigma^* = \text{transpose of } \sigma,$$

This work was partially supported by National Science Foundation Grant GP 28484.

$b = (b_1,\ldots,b_\ell)$, $\nu = (\nu_1,\ldots,\nu_\ell)$ = outward normal to ∂G_j $(k_0+1 \le j \le k)$,

$$G = \bigcup_{j=1}^{k} G_j, \quad \tilde{G} = R^\ell - G, \quad \partial\tilde{G} = \bigcup_{j=1}^{k} \partial G_j,$$

$$Lu \equiv \sum_{i,j=1}^{\ell} a_{ij}(x)\frac{\partial^2 u}{\partial x_i \partial x_j} + \sum_{i=1}^{\ell} b_i(x)\frac{\partial u}{\partial x_i} - \frac{\partial u}{\partial t}.$$

We shall need the following assumption:

(B) If $1 \le h \le k_0$ then $b(z_h) = 0$, $\sigma(z_h) = 0$. If $k_0 + 1 \le h \le k$, then

$$\sum_{i,j=1}^{\ell} a_{ij}\nu_i\nu_j = 0 \quad \text{on } \partial G_h, \tag{1.2}$$

$$(b,\nu) + \frac{1}{2}\sum_{i,j=1}^{\ell} a_{ij}\frac{\partial^2 \rho_h}{\partial x_i \partial x_j} \ge 0 \quad \text{on } \partial G_h. \tag{1.3}$$

Note that (1.2) means that $\sigma\nu = 0$. We also remark that (1.3) is satisfied whenever $(b,\nu) = 0$ and G_h is convex.

__Theorem 1__. _Let_ (A), (B) _hold, and let_ $x(t)$ _be any solution of_ (1.1) _with_ $x(0) \in \tilde{G}$. _Then_

$$P\{\exists t > 0;\ x(t) \in \partial\tilde{G}\} = 0. \tag{1.4}$$

Sketch of proof. Let $R(x)$ be a C^2 function in $\tilde{G}$ such that $R(x) = \rho_h(x)$ if $x \in \hat{G}_{h,\varepsilon_0}$ (ε_0 is such that $\hat{G}_{i,\varepsilon_0} \cap \hat{G}_{j,\varepsilon_0} = \phi$ if $i \ne j$), $R(x) = |x|$ if $|x|$ is sufficiently large, and $R(x) > 0$ elsewhere. Let $V(x) = [R(x)]^{-\lambda}$, $\lambda > 0$. Using (1.2), (1.3) one can show that $LV \le \mu V$ in $\tilde{G}$, where μ is a positive constant. Let $0 < \delta < \varepsilon_0$. By Ito's formula,

$$E[e^{-\mu t^*}V(x(t^*))] - EV(x(0)) = E\int_0^{t^*} L[e^{-\mu t}V(x(t))]dt \le 0$$

where $t^* = \tau_\delta \wedge T$; τ_δ is the first time $x(t)$ intersects $\hat{G}_\delta$, and T is any positive number. Now take $T \uparrow \infty$, and then $\delta \searrow 0$.

Remark. The assertion (1.4) is generally false if any one of the conditions in (B) is dropped.

2. STABILITY

Set

$$\mathcal{A} = \sum_{i,j=1}^{\ell} a_{ij}(x)\frac{\partial R}{\partial x_i}\frac{\partial R}{\partial x_j}, \quad \mathcal{B} = \sum_{i=1}^{\ell} b_i(x)\frac{\partial R}{\partial x_i} + \frac{1}{2}\sum_{i,j=1}^{\ell} a_{ij}(x)\frac{\partial^2 R}{\partial x_i \partial x_j},$$

$$Q = \frac{1}{R}\left(\mathcal{B} - \frac{\mathcal{A}}{2R}\right).$$

We assume, for some $\eta > 0$ sufficiently small,

$$Q(x) \le \theta_0 < 0 \quad \text{if } x \in \hat{G}_\eta \qquad (\theta_0 \text{ constant}), \tag{2.1}$$

$$Q(x) \le \theta_\infty < 0 \quad \text{if } |x| > \frac{1}{\eta} \qquad (\theta_\infty \text{ constant}). \tag{2.2}$$

Note that (2.1) and (1.3) imply that

$$(b,\nu) + \frac{1}{2} \sum_{i,j=1}^{\ell} a_{ij} \frac{\partial^2 \rho_h}{\partial x_i \partial x_j} = 0 \quad \text{on } \partial G_h \qquad (k_0 + 1 \le h \le k).$$

We also assume:

(C) For all x with $\eta \le R(x) \le 1/\eta$,

$$\sum_{i,j=1}^{\ell} a_{ij}(x) \frac{\partial R}{\partial x_i} \frac{\partial R}{\partial x_j} > 0. \tag{2.3}$$

(D) The functions

$$\frac{\partial a_{ij}}{\partial x_j}, \quad \frac{\partial^2 a_{ij}}{\partial x_i \partial x_j}, \quad \frac{\partial b_i}{\partial x_i}$$

are uniformly Hölder continuous on compact subsets, and

$$\sum_{i,j=1}^{\ell} \left|\frac{\partial a_{ij}}{\partial x_i}\right| \le C, \qquad \sum_{i,j=1}^{\ell} \frac{\partial^2 a_{ij}}{\partial x_i \partial x_j} - \sum_{i=1}^{\ell} \frac{\partial b_i}{\partial x_i} \le C \qquad (C \text{ constant}).$$

<u>Theorem 2</u>. <u>Let</u> (A), (B), (2.1), (2.2) <u>and</u> (C), (D) <u>hold</u>, <u>and</u> <u>let</u> x(t) <u>be</u> <u>any</u> <u>solution</u> <u>of</u> (1.1) <u>with</u> $x(0) \in \tilde{G}$. <u>Then</u>

$$P\{\lim_{t\to\infty} \operatorname{dist}(x(t), \partial\tilde{G}) = 0\} = 1. \tag{2.4}$$

<u>Sketch of proof</u>. The assumptions of the theorem enable us to construct a continuously differentiable function $\Phi(r)$ $(r > 0)$ with piecewise continuous $\Phi''(r)$ (having discontinuities only at $r = \eta$, $r = 1/\eta$) such that

$$\Phi'(r) > 0, \ r\Phi'(r) \text{ is bounded} \qquad (0 < r < \infty),$$

$$\Phi(r) \to -\infty \quad \text{if } r \to 0,$$

$$L\Phi(R(x)) < -\nu \quad \text{if } R(x) \ne \eta,\ R(x) \ne \frac{1}{\eta}, \tag{2.5}$$

where ν is a positive constant. Let $\Phi^m(r)$ be C^2 functions which converge to $\Phi(r)$ together with the first derivative for all $r > 0$, and with the second derivatives for all $r \ne \eta$, $r \ne 1/\eta$. Define $\Phi_\delta(r) = \Phi(r)$ if $r \ge \delta$, and

$$\Phi_\delta(r) = \log \delta + \frac{r-\delta}{\delta} - \frac{(r-\delta)^2}{2\delta^2} \quad \text{if } 0 < r < \delta.$$

Define $\Phi_\delta^m(r)$ similarly. Let σ^ε be such that $\sigma^\varepsilon(\sigma^\varepsilon)^* = (\varepsilon\delta_{ij} + a_{ij})$, and denote by $x^\varepsilon(t)$ the solution of (1.1) with σ replaced by σ^ε and with $x^\varepsilon(0) = x(0)$. Let $R_\delta(x)$ be a C^2 function in R^ℓ, coinciding with $R(x)$ when $R(x) > \delta$. By Ito's formula,

$$(2.6) \qquad \Phi_\delta^m(R_\delta(x^\varepsilon(t))) - \Phi_\delta^m(R_\delta(x(0)))$$

$$= \sum_{i,j} \int_0^t (\Phi_\delta^m)'(R_\delta(x^\varepsilon(s))) \frac{\partial R_\delta(x^\varepsilon(s))}{\partial x_i} \sigma_{ij}^\varepsilon(x^\varepsilon(s))dw^j$$

$$+ \int_0^t L_\varepsilon \Phi_\delta^m(R_\delta(x^\varepsilon(s)))ds$$

where $L_\varepsilon u = Lu + (\varepsilon/2) \Sigma\, \partial^2 u/\partial x_i^2$. Taking $m \to \infty$ and using the fact (proved by Aronson and Besala [1]) that L_ε has a fundamental solution, we obtain the relation (2.6) for Φ_δ. Using (2.5) and taking $\varepsilon \searrow 0$ over an appropriate sequence (and δ small, depending on ω), and employing Theorem 1, we obtain

$$(2.7) \quad \Phi(R(x(t))) - \Phi(R(x(0))) \leq \sum_{i,j=1}^{\ell} \int_0^t \Phi'(R(x(s))) \frac{\partial R(x(s))}{\partial x_i} \sigma_{ij}(x(s))dw^j - \nu t.$$

Now use the fact that the stochastic integral is $o(t)$.

3. ANGULAR BEHAVIOR IN CASE $\ell = 2$

Consider first the case where $G = \{0\}$. Introducing polar coordinates $x = r\cos\phi$, $y = r\sin\phi$, (1.1) becomes formally

$$(3.1) \quad \begin{aligned} dr &= \sum_{s=1}^{\ell} \tilde{\sigma}_s(r,\phi)dw^s + \tilde{b}(r,\phi)dt, \\ d\phi &= \sum_{s=1}^{\ell} \tilde{\tilde{\sigma}}_s(r,\phi)dw^s + \tilde{\tilde{b}}(r,\phi)dt, \end{aligned}$$

with suitable coefficients. In particular,

$$\tilde{\tilde{\sigma}}_s(r,\phi) = -\frac{\sin\phi}{r}\sigma_{1s} + \frac{\cos\phi}{r}\sigma_{2s},$$

$$\tilde{\tilde{b}}(r,\phi) = -\frac{\sin\phi}{r} b_1 + \frac{\cos\phi}{r} b_2 - \frac{1}{r^2}\langle a(x)\lambda, \lambda^\perp\rangle$$

where

$$\lambda = (\cos\phi, \sin\phi), \quad \lambda^\perp = (-\sin\phi, \cos\phi).$$

Assume:

(E) $\sigma_{is}(x)$ and $b_i(x)$ are differentiable at $x = 0$.

The second equation in (3.1) then has the form

$$(3.2)\qquad d\phi = \Big[\sum_{s=1}^{n} \tilde{\tilde{\sigma}}_s(\phi)dw^s + \tilde{\tilde{b}}(\phi)dt\Big] + \Big[\sum_{s=1}^{n} N_s dw^s + N_0\, dt\Big]$$

where $N_s = o(r)$ $(0 \le s \le n)$ when $r \to 0$.

Now let $y(t) = (r(t),\phi(t))$ be the diffusion process defined by the solution of the stochastic differential equations (3.1) with $r(0) > 0$. The solution never leaves the half-plane $(0,\infty)\times(-\infty,\infty)$. Define $x(t) = (x_1(t),x_2(t))$ where $x_1(t) = r(t)\cos\phi(t)$, $x_2(t) = r(t)\sin\phi(t)$. One can show that $x(t)$ is a solution of (1.1). Thus, $\phi(t)$ is the algebraic angle of the solution of (1.1). It is natural to compare $\phi(t)$ with the solution of the single equation

$$(3.3)\qquad d\phi = \sigma(\phi)dw + b(\phi)dt$$

where

$$(3.4)\qquad \sigma(\phi) = \sqrt{\sum_{s=1}^{n} (\tilde{\tilde{\sigma}}_s(\phi))^2}, \quad b(\phi) = \tilde{\tilde{b}}(\phi).$$

Theorem 3. *Assume that* (A)-(E), (2.1), (2.2) *hold with* $G = \{0\}$, *and that* $\sigma(z) > 0$ *for all* z. *If*

$$(3.5)\qquad \Lambda = 2\int_0^{2\pi} \frac{b(z)}{\sigma^2(z)}\,dz \ne 0,$$

then the algebraic angle $\phi(t)$ *satisfies*

$$(3.6)\qquad P\Big\{\lim_{t\to\infty} \frac{\phi(t)}{t} = c\Big\} = 1$$

where c *is a constant;* $c > 0$ (< 0) *if* $\Lambda > 0$ (< 0).

Sketch of proof. The assumptions of the theorem enable us to construct a C^2 function $f(\phi)$ $(-\infty < \phi < \infty)$ such that f', f'' are bounded, $f' \ge$ const. > 0, $f(\phi)/\phi \to \gamma$ as $\phi \to \infty$ $(\gamma > 0$ if $\Lambda > 0)$, and

$$(3.7)\qquad \tfrac{1}{2}\sigma^2(\phi)f''(\phi) + b(\phi)f'(\phi) = 1.$$

By Ito's formula

$$(3.8)\qquad f(\phi(t)) = f(\phi(0)) + \sum_{s=1}^{n}\int_0^t \tilde{\tilde{\sigma}}_s(r,\phi)f'(\phi)dw^s + \int_0^t \Big[\frac{1}{2}\sum_s(\tilde{\tilde{\sigma}}_s(r,\phi))^2 f''(\phi) + \tilde{\tilde{b}}(r,\phi)f'(\phi)\Big]d\tau.$$

Given $\varepsilon > 0$, let r_0 be such that

$$(3.9)\qquad \Big|\sum_s(\tilde{\tilde{\sigma}}_s(r,\phi))^2 - \sigma^2(\phi)\Big| < \varepsilon, \quad |\tilde{\tilde{b}}(r,\phi) - b(\phi)| < \varepsilon \qquad \text{if } 0 \le r \le r_0,$$

and let $T_\varepsilon = \sup\{t > 0;\ r(t) > r_0\}$. By Theorem 2, $T_\varepsilon < \infty$ a.s. Taking $t > T_\varepsilon$ and using (3.7),(3.9), we find from (3.8),

$$\overline{\lim_{t\to\infty}} \frac{f(\phi(t))}{t} \leq 1 + \varepsilon K \qquad \text{(K constant).}$$

Similarly for $\underline{\lim}$. This readily gives (3.9).

Consider next the case where $G = \{0\}$ and $\sigma(z)$ may vanish at some points. We assume:

(E') The condition (E) holds and

$$\sum_{s=1}^{n} [\tilde{\tilde{\sigma}}_s(r,\phi)]^2 = \sum_{s=1}^{n} [\tilde{\tilde{\sigma}}_s(\phi)]^2[1 + \eta(r,\phi)]$$

where $\eta(r,\phi) \to 0$ if $r \to 0$, uniformly with respect to ϕ.

Theorem 4. *Assume that* (A)-(D), (2.1), (2.2) *and* (E') *hold with* $G = \{0\}$, *that* $\sigma(z) \not\equiv 0$ *and that* $b(z) > 0$ *whenever* $\sigma(z) = 0$. *Then* (3.6) *holds where* c *is a positive constant.*

Sketch of proof. Denote by $\{x_k;\ k = \pm1,\pm2,\ldots\}$ the zeros of $\sigma(z)$. One can prove the existence of a continuously differentiable function $f(x)$, satisfying (3.7) for all $x \neq x_k$, such that $f'(x)$ is positive and periodic, $\sigma^2(x)f''(x) \to 0$ if $x \to x_k$, and $f(x)/x \to \gamma$ (γ positive constant) as $x \to \infty$. Since $f''(x)$ is not continuous at the points x_k, we cannot apply Ito's formula to $f(\phi(t))$. We use "regularization": replace $\tilde{\tilde{\sigma}}_s$ by $\sqrt{\tilde{\tilde{\sigma}}_s^2 + \varepsilon}$ and construct the corresponding f, which we shall denote by f_ε. Then we apply Ito's formula to $f_\varepsilon(\phi^\varepsilon(t))$ and take $\varepsilon \to 0$.

Consider now the case where $G = G_1 = \{x; |x| \leq 1\}$ and $\ell = 2$. We again use polar coordinates and obtain the equation in (3.1). Define $\sigma(\phi)$, $b(\phi)$ by

$$\sigma(\phi) = \sqrt{\sum_{s=1}^{n} (\tilde{\tilde{\sigma}}_s(1,\phi))^2}, \quad b(\phi) = \tilde{\tilde{b}}(1,\phi).$$

Then the assertions of Theorems 3, 4 extend with obvious changes.

Consider next the case where $G = G_1$ consists of a closed bounded domain. If ∂G is given by $x_1 = f(s)$, $x_2 = g(s)$ $(0 \leq s \leq L)$ where $\dot{f}^2 + \dot{g}^2 = 1$, then we can define a local diffeomorphism $x \to y$, mapping

$$x_1 = f(s) + \rho\dot{g}(s), \quad x_2 = g(s) - \rho\dot{f}(s), \qquad (0 \leq \rho \leq \rho_0)$$

into

$$y_1 = (1+\rho)\cos\frac{2\pi s}{L}, \quad y_2 = (1+\rho)\sin\frac{2\pi s}{L}.$$

By a theorem of Schoenflies [8], this diffeomorphism can be extended into diffeomorphism of the whole exterior of ∂G onto the exterior of the circle $|y| = 1$. In the y-space, the new $\sigma(\phi)$, $b(\phi)$ are given by

$$\sigma(\phi) = \sum_{s=1}^{n} [\dot{f}\sigma_{1s} + \dot{g}\sigma_{2s}]^2$$

$$b(\phi) = (\dot{f}b_1 + \dot{g}b_2) - (\dot{g}, -\dot{f})\begin{pmatrix} a_{11} & a_{12} \\ a_{21} & a_{22} \end{pmatrix}\begin{pmatrix} \dot{f} \\ \dot{g} \end{pmatrix}$$

where σ_{is}, a_{ij}, b_i are evaluated on ∂G and $\phi = 2\pi s/L$. We can now immediately state analogs of Theorems 3, 4.

Finally, by combining the cases where G is one point and where G is a domain, we obtain the angular behavior of x(t) in the general case.

Remark 1. Theorems 1- 4 can be extended to the case where some of the domains G_h $(k_0 + 1 \leq h \leq k)$ have piecewise C^3 boundary; see [3].

Remark 2. In the case where $\sigma_{ir}(x)$, $b_i(x)$ are linear homogeneous in x, so that $G = G_1 = \{0\}$, further results have been obtained by Friedman and Pinsky [2], using more probabilistic methods. In this case, when $\ell = 2$, $\phi(t)$ is a diffusion process. Under the assumptions of Theorems 3, 4 it was proved in [2] that the time T it takes $\phi(t)$ to change by 2π has finite expectation. The strong law of large numbers can then be applied.

References

[1] D. G. Aronson and P. Besala, Parabolic equations with unbounded coefficients, J. Diff. Eqs., 3 (1967), 1-14.

[2] A. Friedman and M. A. Pinsky, Asymptotic behavior of solutions of linear stochastic differential systems, Trans. Amer. Math. Soc., to appear.

[3] A. Friedman and M. A. Pinsky, Asymptotic stability and spiraling properties of solutions of stochastic equations, to appear.

[4] I. I. Gikhman and A. V. Skorokhod, Stochastic Differential Equations, "Naukova Dumka," Kiev, 1968.

[5] R. Z. Kasminski, Necessary and sufficient conditions for the asymptotic stability of linear stochastic systems, Theory Probability Appl., 12 (1967), 144-147.

[6] F. Kozin and S. Prodromou, Necessary and sufficient conditions for almost sure sample stability of linear Ito equations, SIAM J. Appl. Math., 21 (1971), 413-424.

[7] H. J. Kushner, Stochastic Stability and Control, Academic Press, New York, 1967.

[8] M. Morse, A reduction to the Schoenflies extension problem, Bull. Amer. Math. Soc., 66 (1960), 113-115.

BOUNDEDNESS PROPERTIES FOR STOCHASTIC SYSTEMS

TOADER MOROZAN

Institute of Mathematics Bucharest

The object of this paper is to present some results on boundedness of solutions of **stochastic systems**.Definitions of boundedness in probability of stochastic dynamical systems, that we shall give, correspond to uniform boundedness or dissipativity in deterministic case.Theorems will be given in terms of Liapunov functions. The general results will be applied to obtain effective conditions for particular systems.

Let $\{\Omega, \mathcal{K}, P\}$ be a probability field. Denote by R^{ℓ} the real ℓ-dimensional space, X the space of ℓ-dimensional random vectors, Ex, $E[x|\mathcal{F}]$, $\mathcal{F}\subset\mathcal{K}$ the mean of x and the conditional mean, respectively.

$S(\alpha)=\{x, x\in X, P\{|x(\omega)|\leq\alpha\}=1\}$, $S_0=\bigcup_{\alpha>0} S(\alpha)$

F will be the interval $[0,\infty)$ or the set N of the positive integers. Let $\mathcal{A}$ be a subset of X.

Definition 1. _A stochastic dynamical system with respect to_ $\mathcal{A}$ is _a set_ $\mathcal{X}$ _of maps_ $x: I(x)\cap F \to X$, _where_ I(x) _are intervals_ $[t_0,\infty)$, $t_0\in F$, _such that:_

(o) _If_ $x\in\mathcal{X}$ _then_ $|x(t)|$ _is a separable process._

(i) _for each_ $t_0\in F$, $\tilde{x}\in\mathcal{A}$ _there exists a_ $x\in\mathcal{X}$ _such that_ $I(x)=[t_0,\infty)$ _and_ $x(t_0)=\tilde{x}$, a.e. _We shall denote this_ x _by_ $x(\cdot, t_0, \tilde{x})$

(ii) _If_ $x_1\in\mathcal{X}$, $x_2\in\mathcal{X}$, $I(x_1)=I(x_2)=[t_0,\infty)$ _and_ $x_1(t_0)=x_2(t_0)$ a.e. _then_ $x_1(t)=x_2(t)$ a.e. for $t\in I(x_1)\cap F$.

If F=N then the condition (o) is automatically satisfied. A stochastic dynamical system will be denoted by $(\mathcal{X}, F, \mathcal{A})$

The elements of a stochastic dynamical system will be called solutions of the system. We remark that the solutions of the Itô equation define a stochastic dynamical system with respect to R^ℓ, and the solutions of the discrete system

$$x(t+1,\omega)=f(t,x(t,\omega),\omega), \quad t \in N, \ \omega \in \Omega \tag{1}$$

where $f: N \times R^\ell \times \Omega \to R^\ell$ is a function continuous in $x \in R^\ell$ and measurable with respect to ω, define a stochastic dynamical system with respect to every $\mathcal{A} \subset X$

<u>Definition 2. The solution of a stochastic dynamical system are</u> :

(1) <u>equi-bounded in probability if</u> $\lim_{n\to\infty} \sup_{\substack{t \in [t_0,\infty)\cap F \\ \tilde{x} \in S(\alpha)\cap\mathcal{A}}} P\{|x(t,t_0,\tilde{x})|>n\}=0$

for all $t_0 \in F$, $\alpha>0$

(2) <u>strongly bounded in probability if</u>

$$\lim_{n\to\infty} \sup_{\tilde{x} \in S(\alpha)\cap\mathcal{A}} P\Big\{ \sup_{t \in [t_0,\infty)\cap F} |x(t,t_0,\tilde{x})| > n\Big\} = 0$$

for all $t_0 \in F$, $\alpha>0$

(3) <u>ultimately bounded in probability if</u>

$$\lim_{n\to\infty} \sup_{\substack{t_0 \in F \\ \alpha>0}} \overline{\lim_{t\to\infty}} \sup_{\tilde{x} \in S(\alpha)\cap\mathcal{A}} P\{|x(t,t_0,\tilde{x})|>n\}=0$$

(4) <u>equiultimately bounded in probability if there exists</u> $M>0$ <u>such that</u> $\lim_{t\to\infty} \sup_{\tilde{x}\in S(\alpha)\cap\mathcal{A}} P\{|x(t,t_0,\tilde{x})|>M\}=0$ for all $t_0 \in F$, $\alpha>0$

(1) and (2) become uniform boundedness and (3) and (4) dissipativity for deterministic systems .

It is easy to prove that if $\lim_{n\to\infty} \sup_{\tilde{x}\in S(\alpha)} P\{|f(t,\tilde{x}(\omega),\omega)|>n\}=0$ for all $t \in N$, $\alpha>0$ and if the solutions of the system (1) are ultimately bounded in probability with respect to $\mathcal{A}$ then they are equi-bounded in probability with respect to $\mathcal{A}$

Let $V: N \times R^{\ell} \to [0,\infty)$ be a continuous function with respect to $x \in R^{\ell}$ and $\gamma(t)$ a sequence of positive numbers with $\sum_{t\in N} \gamma(t) < \infty$

Theorem 1. *If*

$$(1) \quad \lim_{r\to\infty} \inf_{\substack{t\in N \\ |x|\geq r}} V(t,x) = \infty$$

$$(2) \quad E\left[V(t+1,x(t+1,t_0,\tilde{x})) \mid x(j,t_0,\tilde{x}), t_0 \leq j \leq t\right] \leq$$

$$\leq V(t,x(t,t_0,\tilde{x})) + \gamma(t)$$

a.e. *for all* $t \in N$, $t_0 \in N$, $t \geq t_0$, $\tilde{x} \in S_0 \cap \mathcal{A}$

then the solutions of $(\mathcal{X}, N, \mathcal{A})$ *are strongly bounded in probability*

Proof.

Let $\alpha > 0$ and $\tilde{x} \in S(\alpha) \cap \mathcal{A}$

From condition (2) it follows that

$$EV(t,x(t,t_0,\tilde{x})) \leq EV(t_0,\tilde{x}) + \sum_{t\geq 0} \gamma(t)$$

and the sequence $z(t) = V(t,x(t,t_0,\tilde{x})) + \sum_{j\geq t} \gamma(j)$, $t \geq t_0$, is a supermartingale.

From condition (1) it follows that there exists $R_0 > 0$ and a continuous and increasing function $a(r)$ defined for $r \geq R_0$ such that $\lim_{r\to\infty} a(r) = \infty$ and $0 < a(|x|) \leq V(t,x)$ for $t \geq t_0$, $|x| \geq R_0$

For $r \geq R_0$, we have

$$P\left\{\sup_{t\geq t_0} |x(t,t_0,\tilde{x})| > r\right\} = P\left\{\sup_{t\geq t_0} a(|x(t,t_0,\tilde{x})|) > a(r)\right\} \leq$$

$$\leq P\left\{\sup_{t\geq t_0} V(t,x(t,t_0,\tilde{x})) > a(r)\right\} \leq P\left\{\sup_{t\geq t_0} z(t) > a(r)\right\}$$

Since $z(t)$ is a supermatingale, we have [1]

$$a(r)\, P\left\{\max_{t_0\leq j\leq t} z(j) > a(r)\right\} \leq Ez(t_0) + E|z(t)| \leq 2K(t_0,\alpha) + 4\gamma$$

where $K(t_0,\alpha) = \sup_{|x|\leq\alpha} V(t_0,x)$, $\gamma = \sum_{t\in N} \gamma(t)$

Hence

$$P\left\{ \sup_{t\geq t_o} z(t) > a(r) \right\} \leq \frac{2K(t_o,\alpha)+4\gamma}{a(r)}$$

and the theorem is proved.

Theorem 2. If

(1) $\lim_{|x|\to\infty} c(x) = \infty$

(2) $EV(t+1,x(t+1,t_o,\tilde{x})) \leq EV(t,x(t,t_o,\tilde{x}))+\gamma(t)-Ec(x(t,t_o,\tilde{x}))$

for all $t\in N$, $t_o\in N$, $t\geq t_o$, $\tilde{x}\in S_o\cap\mathcal{A}$

where $c:R^{\ell}\to[0,\infty)$ is a continuous function

then the solutions of $(\mathcal{X},N,\mathcal{A})$ are strongly bounded in probability

Proof.

Let $t_o\in N$, $\alpha>0$, $\tilde{x}\in S(\alpha)\cap\mathcal{A}$

From condition (2) it follows that

$$\sum_{t=t_o}^{m} Ec(x(t,t_o,\tilde{x})) \leq EV(t_o,\tilde{x})-EV(m+1,x(m+1,t_o,\tilde{x}))+\sum_{t\geq 0}\gamma(t) \leq$$
$$\leq K(t_o,\alpha)+\gamma$$

where

$K(t_o,\alpha) = \sup_{|x|\leq\alpha} V(t_o,x)$, $\gamma = \sum_{t\in N}\gamma(t)$

Since $c(x)\geq 0$, for $x\in R^{\ell}$, we have

$$\sum_{t=t_o}^{m} b(r)P\{|x(t,t_o,\tilde{x})|>r\} \leq \sum_{t=t_o}^{m} Ec(x(t,t_o,\tilde{x})) \leq K(t_o,\alpha)+\gamma$$

where $b(r)=\inf_{|x|\geq r} c(x)$

Let $A_r=\left\{\sup_{t\geq t_o}|x(t,t_o,\tilde{x})|>r\right\}$, $A_{t,r}=\left\{|x(t,t_o,\tilde{x})|>r\right\}$ and $B_{m,r}=$
$=\bigcup_{t=t_o}^{m} A_{t,r}$

Since $P(A_r)=\lim_{m\to\infty} P(B_{m,r})$, $\lim_{r\to\infty} b(r)=\infty$ and

$$b(r)P(B_{m,r})\leq K(t_o,\alpha)+\gamma \text{ , for all } m\geq t_o$$

Theorem 2 is proved.

Theorem 3. If

(1) $\lim_{r\to\infty} \inf_{\substack{t\in N \\ |x|\geq r}} V(t,x) = \infty$

(2) $EV(t+1,x(t+1,t_0,\tilde{x}))\leq \beta EV(t,x(t,t_0,\tilde{x}))+\gamma$ for all $t\in N$, $t_0\in N$, $t\geq t_0$, $\tilde{x}\in S_0\cap\mathcal{A}$

where $\beta\in(0,1)$, $\gamma\in[0,\infty)$

then the solutions of $(\mathcal{X},N,\mathcal{A})$ are utlimately bounded in probability.

Proof.

Let $t_0\in N$, $\alpha>0$, $\tilde{x}\in S(\alpha)\cap\mathcal{A}$

From condition (2) it follows that

$$EV(t,x(t,t_0,\tilde{x}))\leq \beta^{t-t_0} EV(t_0,\tilde{x})+\gamma(1+\beta+\dots+\beta^{t-t_0-1})\leq$$

$$\leq \beta^{t-t_0} K(t_0,\alpha)+\frac{\gamma}{1-\beta} \quad \text{for all } t\geq t_0, t\in N.$$

Hence

$$a(r)P\{|x(t,t_0,\tilde{x})|>r\}\leq \beta^{t-t_0} K(t_0,\alpha)+\frac{\gamma}{1-\beta}$$

where $a(r)=\inf_{\substack{t\geq 0 \\ |x|\geq r}} V(t,x)$

From this inequality it follows that the solutions of $(\mathcal{X},N,\mathcal{A})$ are ultimately bounded in probability.

Theorem 4. If

(1) There exists $R_0>0$, $\delta>0$ such that $V(t,x)\geq\delta$ for $t\in N$ and $|x|\geq R_0$

(2) $EV(t+1,x(t+1,t_0,\tilde{x}))\leq\beta EV(t,x(t,t_0,\tilde{x}))$ for all $t\in N$, $t_0\in N$, $t\geq t_0$, $\tilde{x}\in S_0\cap\mathcal{A}$ where $\beta\in(0,1)$

then the solutions of $(\mathcal{X},N,\mathcal{A})$ are equiultimately bounded in probability

Proof.

Let $t_0\in N$, $\alpha>0$, $\tilde{x}\in S(\alpha)\cap\mathcal{A}$

We have

$$\delta P\{|x(t,t_0,\tilde{x})| > R_0\} \leq EV(t,x(t,t_0,\tilde{x})) \leq \beta^{t-t_0} K(t_0,\alpha), \text{ for all } t\in N,\ t \geq t_0$$

and thus, Theorem 4 is proved.

Consider the Itô system

$$dx(t)=g(t,x(t))dt+B(t,x(t)dw(t) \qquad (2)$$

where $g: [0,\infty)\times R^\ell \to R^\ell$ is a continuous function, $B(t,x)$ is a ℓxp-matrix whose elements $b_{im}(t,x)$ are continuous functions and $w(t)$ is the p dimensional process of Brownian motion.

Let $W: [0,\infty)\times R^\ell \to [0,\infty)$ be a function of class C^2, and

$$\mathcal{U} W(t,x)= \frac{\partial W(t,x)}{\partial t}+\left(\frac{\partial W(t,x)}{\partial x}, g(t,x)\right) + \frac{1}{2}\sum_{i,j,m} \frac{\partial^2 W(t,x)}{\partial x^i \partial x^j} b_{im}(t,x) b_{jm}(t,x)$$

Theorem 5. If

(1) $\lim_{r\to\infty} \inf_{\substack{t>0 \\ |x|\geq r}} W(t, x)= \infty$

(2) $\mathcal{U}W(t,x) \leq \beta(t)$ for all $t \geq 0$, $x \in R^\ell$

where $\beta:[0,\infty)\to[0,\infty)$ is a continuous function with $\int^\infty \beta(t)dt < \infty$ then the solutions of the system (2) are strongly bounded in probability ($\mathcal{A} = R^\ell$)

Proof.

Let $t_0 \geq 0$, $\alpha>0$, $\tilde{x} \in S(\alpha) \cap R^\ell$

Let $V(t,x)=W(t,x)+ \int_t^\infty \beta(s)ds$, $a(r)=\inf_{\substack{t>0\\|x|\geq r}} V(t,x)$, $K(t_0,\alpha)=\sup_{|x|\leq\alpha} V(t_0,x)$

It is obvious that

$$\lim_{r\to\infty} a(r)=\infty \quad \text{and} \quad \mathcal{U}V(t,x) \leq 0$$

Let $\tau_r(\omega)=\inf\{t, t > t_0, |x(t,t_0,\tilde{x})(\omega)| > r\}$ and $\tau_r(t,\omega)=\min\{t,\tau_r(\omega)\}$

By Itô's formula, we have

$$EV(\tau_r(t),x(\tau_r(t),t_0,\tilde{x}))-V(t_0,\tilde{x})=E\int_{t_0}^{\tau_r(t)}\mathcal{U}V(s,x(s,t_0,\tilde{x}))ds \leq 0$$

Hence

$$a(r)\, P\{\tau_r(\omega)< t\}\leq \int_{\{\tau_r(\omega)<t\}} V(\tau_r(\omega),x(\tau_r(\omega),t_0,\tilde{x})(\omega))P(d\omega)\leq K(t_0,\alpha)$$

for all $t \geq t_0$

But

$$P\{\sup_{t>t_0} |x(t,t_0,\tilde{x})| > r\} = P\{\tau_n(\omega) < \infty\} = \lim_{n\to\infty} P\{\tau_n(\omega) < n\}$$

Thus

$$a(r)\, P\{\tau_n(\omega) < \infty\} \leq K(t_0,\alpha)$$

and Theorem 5 is proved.

Consider the discrete system

$$x(t+1) = A(t)x(t) + G(t,x)\,\eta(t,\omega), \quad t \in N \tag{3}$$

where

$A(t)$ is a $\ell \times \ell$- matrix, $G(t,x)$ is a $\ell \times p$- matrix, whose elements are continuous functions, and $\eta(t,\omega)$ is a p-dimensional random vector.

Suppose that the trivial solution of the deterministic system

$$y(t+1) = A(t)y(t), \quad t \in N$$

is exponentially stable, i.e. there exist $\beta > 0$, $\gamma \in (0,1)$ such that

$$|y(t,t_0,\tilde{y})| \leq \beta\, \gamma^{t-t_0}\, |\tilde{y}| \text{ for all } t \in N, t_0 \in N,\ t \geq t_0,\ \tilde{y} \in R^\ell$$

<u>Theorem 6</u>. If $\sup_{\substack{t \in N \\ x \in R^\ell}} |G(t,x)| < \infty$, $\sup_{t \in N} E\,|\eta(t)| < \infty$

<u>then the solutions of the system (3) are ultimately bounded in probability with respect to X.</u>

<u>Proof</u>.

Let $\tilde{x} \in S_0$, $t_0 \in N$.

Let $x(t)$ be the solution of the system (3), with $x(t_0)(\omega) = \tilde{x}(\omega)$ and $y(t,t_0,\tilde{y})$ the solution of the system $y(t+1) = A(t)y(t)$, with $y(t_0,t_0,\tilde{y}) = \tilde{y}$.

Let $V(t,x) = \sup_{s \in N}(|y(t+s,t,x)|\gamma^{-s})$, $t \in N$, $x \in R^\ell$

It is easy to see that

$$|x| \leq V(t,x) \leq \beta |x| \quad \text{for } t \in N,\ x \in R^\ell$$

$V(t+1,y(t+1,t_o,\tilde{y}))-V(t,y(t,t_o,\tilde{y})) \leq -(1-\gamma)V(t,y(t,t_o,\tilde{y}))$, for $t \geq t_o, t \in N$, $\tilde{y} \in R^\ell$; $|V(t,x_1)-V(t,x_2)| \leq \beta|x_1-x_2|$, $x_1,x_2 \in R^\ell$

We may write

$$V(t+1,x(t+1))-V(t,x(t))=V(t+1,y(t+1,y(t+1,t,x(t)))-V(t,x(t))+$$
$$+V(t+1,x(t+1))-Vt+1,y(t+1,t,y(t)))$$

Thus

$$V(t+1,y(t+1))-V(t,y(t)) \leq -(1-\gamma)V(t,x(t))+ \beta|y(t+1,x(t))-$$
$$-x(t+1)| \leq -(1-\gamma)V(t,x(t))+\beta|G(t,x(t))|\,|\eta(t)| \leq -(1-\gamma)V(t,x(t))+$$
$$+\beta c_1|\eta(t)|$$

where $c_1=\sup\limits_{\substack{t\geq 0\\ x\in R^\ell}} |G(t,x)|$

Hence

$$EV(t+1,x(t+1)) \leq \gamma\, EV(t,x(t))+ \beta c_1 c_2$$

$$c_2 = \sup_{t\geq 0} E\,|\eta(t)|$$

From Theorem 3 it follows that the solution of the system (3) are ultimately bounded in probability with respect to X.

Theorem 7. If

(1) $\sup\limits_{t\in N} |G(t,x)| \leq M\,|x|^{1/2}$

(2) $1 - \gamma - \frac{\beta}{2} M^2 > 0$

(3) $\sum\limits_{t\in N} E\,|\eta(t)|^2 < \infty$

then the solutions of the system (3) are strongly bounded in probability with respect to X .

Proof.

Let $t_o \in N$, $\tilde{x} \in S_o$. Consider $x(t)$ and $V(t,x)$ as in the proof of Theorem 6.

We have

$$V(t+1, x(t+1)) \leq \gamma V(t,x(t))+\beta|G(t,x(t))|\,|\eta(t)| \leq \gamma V(t,x(t)) +$$

$$+ \frac{\beta}{2}\left(|G(t,x(t))|^2 + |\eta(t)|^2\right) \leq V(t,x(t)) + (\gamma-1)|x(t)| + \frac{\beta}{2}M^2|x(t)| + \frac{\beta}{2}|\eta(t)|^2$$

Hence

$$EV(t+1,x(t+1)) \leq EV(t,x(t)) - \left(1-\gamma-\frac{\beta}{2}M^2\right)E|x(t)| + \frac{\beta}{2}E|\eta(t)|^2$$

Using Theorem 2, the theorem is proved.

Consider the discrete system

$$x(t+1)=Ax(t)+a(t,\omega)b\,f(\sigma(t))+Bg(t,\omega), \quad \sigma(t)=c'x(t), \quad t\in N, \qquad (4)$$

where A is a $l\times l$-matrix, B is a $l\times p$-matrix, b and c are l-dimensional vectors, c' represents the transposed of the vector c, $a(t,\omega)$, $t\in N$ is a sequence of random variables, $g(t,\omega)$, $t\in N$ is a sequence of p-dimensional random vectors and $f\in \mathcal{F}_h$ where $\mathcal{F}_h$ is the set of continuous functions $f:R^1 \to R^1$ with $0\leq \sigma f(\sigma)\leq h\sigma^2$ for all $\sigma\in R^1$

Suppose that $a(0,\omega), g(0,\omega), a(1,\omega), \ldots, a(j,\omega), g(j,\omega), \ldots$ are independent and $Eg(t,\omega)=0$, $Ea(t,\omega)=Ea(0,\omega)=\lambda$, $Ea^2(t,\omega)=$ $=Ea^2(0,\omega)=\mu^2$ for all $t\in N$

Let $\mathcal{A}_0$ be the set of $\tilde{x}\in X$ with the property that $\tilde{x}(\omega)$ is independent of $\{a(t,\omega), g(t,\omega), t=0,1,2,\ldots\}$

<u>Theorem</u> 8. <u>If</u>

(1) rank $(b \quad Ab \quad \ldots \quad A^{l-1}b) = l$

(2) $\lambda\neq 0$ <u>and</u> $\frac{1}{h} + \frac{\mu^2}{\lambda}\operatorname{Re} c'(A-e^{it}I)^{-1}b \geq 0$ <u>for all</u> $t\in[-\pi,\pi]$

with $\det(A-e^{it}I)\neq 0$, ($i=\sqrt{-1}$, I is the identity matrix)

(3) <u>there exists</u> $0<\delta\leq h\frac{\mu^2}{\lambda}$ <u>such that the characteristic values</u> z_i <u>of the matrix</u> $A+\delta bc'$ <u>satisfy</u> $|z_j|<1$, $j=1,2,\ldots,l$.

(4) $\sum_{t\in N} E|g(t)|^2<\infty$

<u>then the solutions of the system (4) are stronlgy bounded in probability with respect to $\mathcal{A}_0$ for all $f\in\mathcal{F}_h$</u>

<u>Proof</u>.

Consider the following Popov system

$$y(t+1)=Ay(t)+bu(t) \qquad (5)$$

$$\eta(t)=\sum_{j=0}^{t} \frac{1}{h}u^2(j)-\frac{\mu^2}{\lambda}u(j)c'y(j)\ ,\quad t\in N \qquad (6)$$

where $u(t) \in R^1$.

We shall prove that Popov system (5), (6), is minimal stable, i.e. for all $\tilde{y} \in R^\ell$ there exist $\tilde{u}(t)$, $\tilde{y}(t)$, $t \in N$ such that: $\tilde{y}(t+1)=A\tilde{y}(t)+b\tilde{u}(t)$, $\tilde{y}(o)=\tilde{y}$, $\lim_{t\to\infty} \tilde{y}(t)=o$ and $\eta(t) \leq o$ for all $t \in N$

Indeed, let $\tilde{y} \in R^\ell$. Let $\tilde{y}(t)$ be the solution of the system $y(t+1)=(A+\delta bc')y(t)$ with $y(o)=\tilde{y}$. From condition (3) it follows that $\lim_{t\to\infty} \tilde{y}(t)=o$

Let $\tilde{u}(t)=\delta c'\tilde{y}(t)$. It is obvious that $\tilde{y}(t+1)=A\tilde{y}(t)+b\tilde{u}(t)$, $t \in N$

For $(\tilde{y}(t), \tilde{u}(t))$ we have

$$\eta(t)=\sum_{j=0}^{t}\frac{1}{h}\tilde{u}^2(j)-\frac{\mu^2}{\lambda}\tilde{u}(j)c'y(j)=\sum_{j=0}^{t}\delta\,(c'\tilde{y}(j))^2\left[\frac{\delta}{h}-\frac{\mu^2}{\lambda}\right]\leq 0$$

Thus, from conditions (1) and (2), by Theorem 1 in [3, pp.207] it follows that there exist a positive definite matrix H, a vector q and a real number γ_0 such that

$$y'(t+1)Hy(t+1)-y'(o)Hy(o)=\eta(t)-\sum_{j=0}^{t}(q'y(j)+\gamma_0 u(j))^2$$

for all $t \geq o$, $t \in N$

Thus, we have (for t=o)

$$(y'_0A'+u_0b')H(Ay_0+bu_0)-y'_0Hy_0=\frac{1}{h}u_0^2-\frac{\mu^2}{\lambda}u_0c'y_0-(q'y_0+\gamma_0 u_0)^2 \text{ for}$$

all $u_0 \in R^1$, $y_0 \in R^\ell$

Hence, we obtain that

$$-A'HA+H=qq',\quad A'Hb+\frac{\mu^2}{2\lambda}c=-\gamma_0 q\ ,\quad b'Hb-\frac{1}{h}=-\gamma_0^2$$

Let $t_0 \in N$, $\tilde{x} \in \mathcal{A}_c$, $f \in \mathcal{F}_h$, $x(t,\omega)$ the solution of the system (4) with $x(t_0,\omega)=\tilde{x}(\omega)$.

Since $\{a(t,\omega), g(t,\omega)\}$ is independent of $\{x(j,\omega), t_0 \le j \le t\}$ and $\sigma f(\sigma) - \frac{1}{h} f^2(\sigma) \geq 0$ for all σ

We have

$$E[x'(t+1,\omega)Hx(t+1,\omega) \mid x(j,\omega), t_0 \le j \le t] - x'(t,\omega)Hx(t,\omega) =$$
$$= x'(t,\omega)[A'HA-H]x(t,\omega) + 2\lambda x'(t,\omega)A'Hb f(\sigma(t,\omega)) +$$
$$+ \mu^2 f^2(\sigma(t,\omega))b'Hb + E[g'(t,\omega)B'HBg(t,\omega) \mid x(j,\omega), t_0 \le j \le t] \le$$
$$\le x'(t,\omega)[A'HA-H]x(t,\omega) + 2\lambda x'(t,\omega)f(\sigma(t,\omega))[A'Hb + \frac{\mu^2}{2\lambda}c] +$$
$$+ \mu^2 f^2(\sigma(t,\omega))[b'Hb - \frac{1}{h}] + \gamma(t)$$

where $\gamma(t) = |B'HB| \, E|g(t,\omega)|^2$

Hence

$$E[x'(t+1,\omega)Hx(t+1,\omega) \mid x(j,\omega), t_0 \le j \le t] \le x'(t,\omega)Hx(t,\omega) -$$
$$-(q'x(t,\omega) + \lambda\gamma_0 f(\sigma(t,\omega)))^2 - \gamma_0^2(\mu^2-\lambda^2)f^2(\sigma(t,\omega)) + \gamma(t) \le$$
$$\le x'(t,\omega)Hx(t,\omega) + \gamma(t)$$

Using Theorem 1, the theorem is proved

Theorem 9. If

(1) $\lambda = 0$

(2) The characteristic values z_j of the matrix A satisfy

$$|z_j| < 1, \; j=1,2,\dots,\ell \quad \text{and} \quad \frac{1}{h^2} - \frac{\mu^2}{2\pi}\int_{-\pi}^{\pi} |c'(A-e^{it}I)^{-1}b|^2 dt > 0$$

(3) $\sup_{t\in N} E|g(t,\omega)|^2 < \infty$

then the solutions of the system (4) are ultimately bounded in probability with respect to $\mathcal{A}_0$ for all $f \in \mathcal{F}_h$

Proof.

By $\mathcal{L}(G)$ will be denoted the matrix which verifies

$$\mathcal{L}(G) - A'\mathcal{L}(G)A = G$$

It is easy to prove that $b'\mathcal{L}(cc')b = \frac{1}{2\pi}\int_{-\pi}^{\pi} |c'(A-e^{it}I)^{-1}b|^2\,dt$

Let $\delta = b'\mathcal{L}(cc')b$, $\gamma_0 = b'\mathcal{L}(I)b$, $\delta_0 > 0$ such that $\frac{2}{h} -$

$- \mu^2\delta_0\gamma_0 > 0$ and $(\frac{2}{h} - \mu^2\delta_0\gamma_0)^2 - 4\mu^2\delta > 0$

Let $\lambda_0 > 0$ such that $\lambda_0^2 - \lambda_0(\frac{2}{h} - \mu^2\delta_0\gamma_0) + \mu^2\delta = 0$

Let $H = \frac{\mathcal{L}(cc')}{\lambda_0} + \delta_0\mathcal{L}(I)$

Obviously that H is a positive definite matrix .

It is easy to see that $\frac{2}{h} - \mu^2 b'Hb = \lambda_0$, $A'HA-H = -qq' - \delta_0 I$

where $q = \frac{c}{\sqrt{\lambda_0}}$

Let $x(t,\omega)$, as in the proof of the preceding theorem.

We have

$$E\left[x'(t+1,\omega)Hx(t+1,\omega) \mid x(j,\omega),\ t_0 \le j \le t\right] - x'(t,\omega)Hx(t,\omega) \le$$

$$\le x'(t,\omega)\left[A'HA-H\right]x(t,\omega) + \mu^2 b'Hb\, f^2(\sigma(t,\omega)) + \gamma(t) =$$

$$= x'(t,\omega)\left[A'HA-H\right]x(t,\omega) + f^2(\sigma(t,\omega))\left[\mu^2 b'Hb - \frac{2}{h}\right] +$$

$$+ 2c'x(t,\omega)f(\sigma(t,\omega)) + \gamma(t) - 2\left[\sigma(t,\omega)f(\sigma(t,\omega)) - \frac{1}{h}f^2(\sigma(t,\omega))\right]$$

where $\gamma(t) = |B'HB|\ E\,|g(t,\omega)|^2$

Hence

$$E\left[x'(t+1,\omega)Hx(t+1,\omega) \mid x(j,\omega),\ t_0 \le j \le t\right] - x'(t,\omega)Hx(t,\omega) \le$$

$$\le -(q'x(t,\omega))^2 - \lambda_0 f^2(\sigma(t,\omega)) + 2\sqrt{\lambda_0}\, q'x(t,\omega)f(\sigma(t,\omega)) - \delta_0|x(t,\omega)|^2 +$$

$$+ \gamma(t) \le -\delta_0|x(t,\omega)|^2 + \gamma \le -\delta_1 x'(t,\omega)Hx(t,\omega) + \gamma$$

where $\gamma = \sup_{t\in N}\gamma(t)$, $\delta_1 = \frac{\delta_0}{|H|}$

Using Theorem 3, we conclude that Theorem 9 is proved.

We remark that it can be proved that the trivial solution of the system (4), with $g(t,\omega) \equiv 0$ and $\lambda = 0$ is exponentially stable in mean square for all $f \in \mathcal{F}_h$ if and only if condition (2) of Theorem 9 holds.

Consider the following Itô system

$$dx = Ax + b\,f(\sigma)\,dw(t), \qquad \sigma = c'x \tag{7}$$

where A is a $\ell\times\ell$-matrix, $b,c \in R^{\ell}$, $f\in \mathcal{F}_h$ and w(t) is the process of Brownian motion

Theorem 1o. The trivial solution of the system (7) is exponentially stable in mean square for all $f \in \mathcal{F}_h$ if and only if the matrix A is stable and $\frac{1}{h^2} - \frac{1}{2\pi}\int_{-\infty}^{\infty} |c'(A-i\lambda I)^{-1}b|^2\,d\lambda > 0$, $(i=\sqrt{-1})$

Proof.

By $\mathcal{H}(G)$ will be denoted the matrix which verifies

$$-A'\mathcal{H}(G) - \mathcal{H}(G)A = G.$$

We have $b'\mathcal{H}(cc')b = \int_0^{\infty} |c'e^{At}b|^2\,dt = \frac{1}{2\pi}\int_{-\infty}^{\infty} |c'(A-i\lambda I)^{-1}b|^2\,d\lambda$

Thus, by Theorem in [2], it follows that if the trivial solution of (7), for $f(\sigma)=h\sigma$, is exponentially stable in mean square then the matrix A is stable and $\frac{1}{h^2} - \frac{1}{2\pi}\int_{-\infty}^{\infty} |c'(A-i\lambda I)^{-1}b|^2\,d\lambda > 0$

Suppose, now, that the matrix A is stable and

$$\frac{1}{h^2} - \frac{1}{2\pi}\int_{-\infty}^{\infty} |c'(A-i\lambda I)^{-1}b|^2\,d\lambda > 0$$

Let $\delta = b'\mathcal{H}(cc')b$, $\gamma_0 = b'\mathcal{H}(I)b$, $\delta_0 > 0$ such that $\frac{2}{h} - \delta_0\gamma_0 > 0$ and $\left(\frac{2}{h} - \delta_0\gamma_0\right)^2 - 4\delta > 0$

Let $\lambda_0 > 0$ such that $\lambda_0^2 - \lambda_0\left(\frac{2}{h} - \delta_0\gamma_0\right) + \delta = 0$

Let $H = \frac{\mathcal{H}(cc')}{\lambda_0} + \delta_0\mathcal{H}(I)$ and $W(x) = x'Hx$.

It is easy to see that $\frac{2}{h} - b'Hb = \lambda_0$, $A'H + HA = -qq' - \delta_0 I$,

where $q = \frac{c}{\sqrt{\lambda_0}}$ and $\mathcal{U}W(x) = x'[A'H+HA]x + b'Hb\,f^2(\sigma)$, $\sigma = c'x$

Hence

$$\mathcal{U}W(x) = x'[A'H+HA]x + f^2(\sigma)\left[b'Hb - \frac{2}{h}\right] + 2c'xf(\sigma) - 2\left[\sigma f(\sigma) - \right.$$

$$\left. - \frac{1}{h} f^2(\sigma)\right] \leq -(q'x - \sqrt{\lambda_0}\,f(\sigma))^2 - \delta_0|x|^2$$

From this relation, using the Itô formula, we conclude that

$$\frac{dV(t)}{dt} \leq -\delta_c E\,|x(t,t_0,\tilde{x})|^2 \leq -\delta_1 V(t)$$

where $V(t)=EW(x(t,t_0,\tilde{x}))$, $x(t,t_0,\tilde{x})$ is the solution of (7) with $x(t_0,t_0,\tilde{x})=\tilde{x}$ and thus, we conclude that Theorem lo is proved.

From Theorem 5 and the proof of Theorem lo it follows that if A is stable and $\frac{1}{k^2} - \frac{1}{2\pi}\int_{-\infty}^{\infty} |c'(A-i\lambda I)^{-1}b|^2\, d\lambda > 0$, then the solutions of the system (7) are strongly bounded in probability and equiultimately bounded in probability (with respect to R^ℓ).

REFERENCES

1. Doob, J.L, Stochastic processes, Wiley, New-York,(1953)

2. Levit, M.V. and Yacubovich, V.A., Applied Math.and Mechanics. 1, 142-147 (1972)

3. Popov, V.M. Hyperstability of automatic systems(in Russian), (1970)

SYSTEM IDENTIFICATION

K.J. ÅSTRÖM

Division of Automatic Control, Lund Institute of Technology, Lund, Sweden

1. INTRODUCTION

Many problems of analysis are of the type, given an equation find the solution or characterize its properties. This paper deals with the inverse problem or the identification problem, i.e. given a solution find the equation. Problems of this type do frequently arise in control engineering and in many other fields, e.g. economics and biology.

Much of the control theory that has been developed over the past twenty years, starts with a description of the system to be controlled in terms of a difference or a differential equation. The theory then gives various ways to compute control strategies that are optimal or have other desirable properties. When attempting to apply the theory the first problem that is encountered is to obtain the appropriate mathematical model of the system. Such a model can be obtained from basic physical laws expressing conservation of mass, energy and momentum as state equations. In many cases it turns out, however, that the desired models can not be obtained in this way. For many industrial processes the process dynamics can not be expressed with sufficient accuracy from first principles alone. Even if the process dynamics can be described reasonably well, a characterization of the disturbances is often missing. Many examples are found in paper and pulp and cement industries.

When the desired models can not be determined from à priori knowledge alone, experimentation becomes a necessity. In the experiments an input signal is applied to the process and the process outputs are observed. A description of the process dynamics and the disturbances can then sometimes be obtained through appropriate processing of the experimental data. The situation when the models are derived by experimentation alone is sometimes called the black-box approach. In practice the models are frequently obtained through a combination of experimentation and modelling based on physical laws.

The problem we are faced with is thus a statistic problem although in a fairly complex setting, because we are trying to estimate characteristics of dynamical systems. A significant difference in comparison with the classical statistical problems is also that in most cases it is not reasonable to assume independent observations.

Before the advent of modern control theory most synthesis techniques were limited to linear single-input single-output systems characterized by a transfer function. One of the reasons for the great success of classical control theory was that it was accompanied by a very efficient experimental technique to determine transfer functions experimentally. The determination was done simply by introducing a sinusoidal variation in the input and measuring amplitude and phase relations between the input and the output in steady state. This technique gave the transfer function and the results could be used directly for design.

The advent of modern control theory with its emphasis on state space models has created the need for other identification schemes which results in state space models. The desire to apply modern control theory to practical problems has been a strong driving force for the recent development of identification methods. It is also interesting to note that an almost parallel development has taken place in mathematical economy. See e.g. Fischer [12] and Deistler [9].

The purpose of this paper is not to present a comprehensive review of the fields which are available elsewhere but to discuss a few selected problems that are currently of interest.

The paper is organized as follows. Some preliminary material is given in section 2. Nonuniqueness is a characteristic feature of the inverse problem. It is fairly obvious that there might be many equations which have the same solution. Some aspects on the lack of uniqueness is discussed in section 3, which leads to the notion of identifiability. One source of nonuniqueness is in the selection of classes of models. Model structures for linear multivariable systems are discussed in section 4. The first four parts of the paper deal only with off-line problems i.e. the analysis of the data after the experiment has been performed. Since the motivation for a control engineer to solve an identification problem is to design a control strategy, the relation between identification and control is very important. This problem is discussed in section 5 in the extreme case where it is attempted to identify and control at the same time. This analysis gives aspects on the selection of criteria for the identification problem and it also poses problems of stability of nonlinear stochastic difference equations.

2. PROBLEM FORMULATION

The identification problem can be formulated as to determine a model within a class which fits the experimental data as well as possible. The problem is thus characterized by three elements

- a class of models
- a set of input signals to be used in the experiment
- a criterion

Notice that the selection of the three basic elements is an important part of the problem statement. Once the three elements have been specified we are clearly faced with an estimation problem. A few comments on the basic elements of the problem are given below.

The Class of Models

The models to be considered can be defined in many different ways. Since the ultimate purpose of the identification is to design control strategies it is reasonable to choose models for which control theory is developed. The selection is also guided by physical considerations. In practice it is often an iterative procedure as was mentioned in the introduction.

Some examples of models are given below.

Example 2.1 (Linear stochastic difference equations)

Let a model be characterized by the linear stochastic differential equation

$$x(t+1) = Ax(t) + Bu(t) + v(t)$$

$$y(t) = Cx(t) + e(t) \qquad (2.1)$$

where $\{v(t)\}$ and $\{e(t)\}$ are sequences of independent equally distributed random variables with covariances R_1 and R_2 respectively. The model parameters are the elements of the constant matrices A, B, C, R_1 and R_2, and the class of models is the class of all systems (2.1) where the matrices have given sizes. Notice that the model (2.1) is the canonical structure used in linear stochastic control theory. [3]

An alternative representation of the system is the innovations representation

$$\hat{x}(t+1) = A\hat{x}(t) + Bu(t) + K\varepsilon(t)$$

$$y(t) = C\hat{x}(t) + \varepsilon(t) \tag{2.2}$$

where $\hat{x}(t)$ is the conditional mean of the state $x(t)$ of (2.1) given observations of the output up to time t.

The models (2.1) and (2.2) are equivalent in the sense that they have the same input-output relation and that the stochastic properties of the outputs are the same.

Example 2.2

Let a model be characterized by the difference equation

$$y(t) + a_1 y(t-1) + \ldots + a_n y(t-n) = b_1 u(t-k-1) + \ldots + b_n u(t-k-n)$$

$$+\lambda\left[e(t) + c_1 e(t-1) + \ldots + c_n e(t-n)\right] \tag{2.3}$$

where the model parameters are n, k, λ, and $(a_i, b_i, c_i, i = 1,\ldots,n)$. Let the class of models be all models with $1 \leq n \leq n_o$ and $o \leq k \leq k_o$. Notice that (2.2) is a canonical structure for a single-input single-output system.

Example 2.3

Let the input u and the output y be p- and r-vectors respectively and let a model be characterized by the n-th order vector difference equation [23]

$$y(t) + A_1 y(t-1) + \ldots + A_n y(t-n) = B_1 u(t-1) + \ldots + B_n u(t-n) \tag{2.4}$$

where A_i is an $r \times r$ matrix and B_i an $r \times p$ matrix. Let the class of models be all models where $n \leq n_o$. The model (2.4) is frequently used when modelling macroeconomic systems from experimental data. See e.g. [9]

Example 2.5

Let a model be characterized by the nonlinear stochastic differential equation

$$dx = f(x,u,\alpha)dt + g(x,u,\alpha)dv \qquad (2.5)$$

where $\{v(t)\}$ is a Wiener process and f and g are known functions. Let the class of models be all models with α belonging to a given set.

The Criterion

Since the purpose of the identification is to design a control strategy, it would be natural to evaluate the performance of an identification method with respect to this purpose. To do so, it would be necessary to design a control strategy based on the model obtained in the identification and to evaluate the performance of the closed loop system obtained when the control strategy is applied to the real process. The problem then arises if it is possible to choose criteria for the identification which will ensure good behaviour of the closed loop system. This problem is not solved in general and other criteria are therefore chosen for the identification. Typical examples are minimization of least squares deviation between the outputs of the system and the model. By making suitable assumptions concerning the statistical properties of the disturbances, it is frequently possible to give a statistical interpretation to the different error criteria. It is also possible to use full machinery of statistical parameter estimation techniques, e.g. techniques like the Maximum Likelihood method.

Maximum Likelihood Estimates

It can be shown that the likelihood function for the system in equation (2.2) can be written as

$$-\ln L = \frac{1}{2} \sum_{t=1}^{N} \boldsymbol{\varepsilon}^T(t)\ R^{-1}\varepsilon(t) + \frac{N}{2} \ln \det R + \text{const.} \qquad (2.6)$$

where N is the number of input-output pairs and R the covariance of $\varepsilon(t)$. The identification problem then reduces to the problem of minimizing the function (2.6) with respect to the unknown parameters. The matrix R is frequently not known a priori. This means that it is necessary to minimize (2.6) with respect to R also. Notice that this can be done analytically. We have according to [10]

$$\min_R \left[\frac{1}{2} \sum_{t=1}^{N} \varepsilon^T(t) R^{-1} \varepsilon(t) + \frac{N}{2} \ln \det R \right]$$

$$= \frac{rN}{2} + \frac{N}{2} \ln \det \frac{1}{N} \sum_{t=1}^{N} \varepsilon(t) \varepsilon^T(t) \qquad (2.7)$$

If it is assumed that the input-output data was actually generated by a process having the correct structure it can be shown that

- the estimates converge to the true parameters with probability one as the record length N increases (consistency)
- for large N there are no other estimation procedures that give estimates with smaller variances (asymptotic efficiency)
- the estimate is asymptotically normal with a covariance matrix which is the inverse of the matrix of second derivatives of the log likelihood function (asymptotic normality)

The precise statement of these results for the estimation of the parameters of the model in Example 2.2 are given in [2] . The multivariable case is treated in [8] , [15] and [24] . There are, however, some difficulties with the multivariable case as will be discussed in the next section.

Many other statistical problems can also be posed. For example the determination of the order of a model can be approached as a statistical problem. Let V_n denote the lossfunction obtained for a model with n parameters. The function V_n decreases with increasing n. The problem is to decide if the decrease in V is significant or not. For systems with one output the test quantity

$$F_{n_1,n_2} = \frac{V_{n_1} - V_{n_2}}{V_{n_2}} \cdot \frac{N - n_2}{n_2 - n_1} \qquad n_2 > n_1 \qquad (2.8)$$

where N is the number of observations of the output, can be shown to be asymptotically F-distributed. The quantities F_{n_1,n_2} can thus be used to test if the loss function is reduced significantly when the number of parameters in the model is increased from n_1 to n_2.

Notice that it can be tested if the residuals are gaussian and uncorrelated. It is, however, virtually impossible to assert that the data was actually generated by a model (2.2) with specific parameters. In practice this is of course never true since the model (2.2) is only an approximation of a complex process. The results obtained by using the statistical theories thus must be handled with great care when applied to real data. Notice that when the methods are tried on simulated data it is always possible to assert that the assumptions required by the statistical theory are fulfilled!

Other Interpretations

The minimization of (2.7) can be given a physical interpretation even in the case when no statistical assumptions are made. The equation (2.2) can be rewritten as

$$\hat{x}(t+1) = A\hat{x}(t) + Bu(t) + K\left[y(t) - C\hat{x}(t)\right]$$

$$\varepsilon(t) = y(t) - C\hat{x}(t)$$

The quantity $C\hat{x}(t)$ can be interpreted as a one step prediction of y(t) based on y(t-1), y(t-2),... The quantity $\varepsilon(t)$ can thus be interpreted as the difference between the value of the actual measured process output at time t-1 and the prediction. The minimization in (2.7) thus means that the parameters of the model are changed in such a way that the error in predicting the output one step ahead is as small as possible according to the criterion (2.7). In the case of singel output systems the criterion (2.7) is simply the mean squares prediction error.

A Word of Caution

When the criterion for the identification is chosen ad hoc there is of course no guarantee that the control law computed from the model will give a good result. This is well known to practitioners. An example is given in Fig. 2.1.

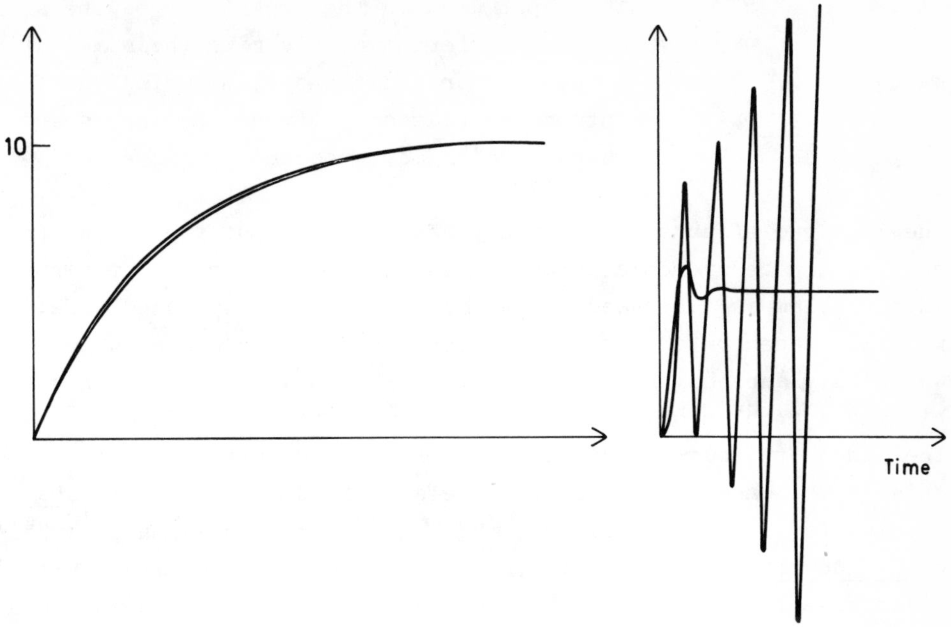

Fig. 2.1

To the left are shown the step responses of two linear systems S_1 and S_2. The curve on the right shows the step responses of the systems obtained when S_1 and S_2 are regulated with a proportional regulator with gain 2.5. The properties of the closed loop system are drastically different although the step responses are almost identical.

3. IDENTIFIABILITY

The identification problem formulated in section 2 may have many different solutions. Sometimes, however, this is not serious since it may happen that both solutions could be acceptable models of the system. Difficulties in the computation of the parameter estimate may however arise. In any case it is useful to know if there is a unique solution or not.

There are several mechanisms which can lead to nonuniqueness for example

- The class of models may be such that there are several models corresponding to different parameter values which have the same input-output relation

- The input used in the experiment may be such that several different models give the same output
- A feedback around the system during the test will introduce relations between the inputs and the outputs which may produce nonuniqueness

The last two effects are entirely due to the input signal and the difficulties can be overcome through a suitable choice of input signals. This problem is discussed in [4] and will not be pursued here. Instead we will concentrate on the first mentioned source of nonuniqueness which only depends on the class of models chosen. The analysis is basically a problem of analysing if the mapping from parameter space to the space of input-output relations for the models is one to one. If this is the case, the class of models is said to be <u>identifiable</u>. If mapping is locally one to one, the class of models is said to be <u>locally identifiable</u>. It is very difficult to obtain general criteria for identifiability. The only cases where this has been done is where the mapping can be obtained analytically. There are, however, some results concerning local identifiability. One criterion can be derived using the theory of (local) observability for nonlinear systems. Consider the class of models

$$\frac{dx}{dt} = f(x, u, \alpha)$$

$$y = g(x, u, \alpha) \tag{3.1}$$

where $\alpha \in R^l$ is a parameter. Augmenting the state vector x by the parameter vector α we get

$$\frac{d}{dt}\begin{bmatrix} x \\ \alpha \end{bmatrix} = \begin{bmatrix} f(x, u, \alpha) \\ 0 \end{bmatrix}$$

$$y = g(x, u, \alpha) \tag{3.2}$$

The system (3.1) is then locally identifiable at α^0 if the system (3.2) is locally observable at (x^0, α^0). A criterion for observability can be obtained by linearizing around a reference trajectory as is done in [17]. See also [16] and [22].

A sufficient criterion is that the matrix W defined by

$$\frac{dW}{dt} = A^T W + WA + C^T C \tag{3.3}$$

is positive definite, where

$$A = \begin{bmatrix} f_x & f_\alpha \\ o & o \end{bmatrix}_{x^o,\ \alpha^o} \tag{3.4}$$

and

$$C = (g_x \quad g_\alpha)_{x^o,\ \alpha^o} \tag{3.5}$$

where f_x denotes the matrix of partial derivatives. All expressions are evaluated at the reference trajectory.

As an illustration we will consider an example of a system which is not identifiable.

Example 3.1

Consider the linear system

$$\frac{dx}{dt} = \begin{bmatrix} o & \alpha_1 \\ -\alpha_2 & o \end{bmatrix} x + \begin{bmatrix} 1 \\ o \end{bmatrix} u$$

$$y = (1 \ \ o)\ x \tag{3.6}$$

The transfer function is

$$G(s) = \frac{s}{s^2 + \alpha_1 \alpha_2} \tag{3.7}$$

By inspection we thus find that all parameters such that $\alpha_1 \alpha_2 =$ const. will give models with the same input output relation. The class of models (3.6) is thus clearly not identifiable.

Notice however that the class of systems defined by

$$\frac{dx}{dt} = \begin{bmatrix} o & \alpha_1 \\ -\alpha_2 & o \end{bmatrix} x + \begin{bmatrix} o \\ 1 \end{bmatrix} u$$

$$y = (1 \quad o)\ x$$

whose transfer function is

$$G(s) = \frac{\alpha_1}{s^2+\alpha_1\alpha_2}$$

is identifiable. Further examples are found in [7].

4. MODEL STRUCTURES

The selection of model structures for linear multivariable systems will now be discussed.

State Equations

Consider the system

$$\frac{dx}{dt} = Ax + Bu$$

$$y = Cx \tag{4.1}$$

where x is an n-vector, the input u is a p-vector and the output y is an r-vector. Since the input-output relation is determined by the completely observable and completely controllable subsystem only, it is clear that the parameters of (4.1) can not be determined unless (4.1) is completely controllable and completely observable in Kalmans sense. If all the elements of the matrices A, B and C are unknown the representation (4.1) contains n(n+p+r) parameters. Since input-output relations of (4.1) is invariant under linear transformations of the state it follows that the input output relation can be characterized by at most n(p+r) parameters. It is also possible to transform the equations to a form which only contains this number of parameters as is illustrated below.

Diagonal forms

If the matrix A has distinct eigenvalues, the system can be transformed to

$$\frac{dx}{dt} = \begin{bmatrix} \lambda_1 & 0 & \dots & 0 \\ 0 & \lambda_2 & \dots & 0 \\ \cdot & & & \cdot \\ \cdot & & & \cdot \\ \cdot & & & \cdot \\ 0 & 0 & \dots & \lambda_n \end{bmatrix} x + \begin{bmatrix} \beta_{11} & \beta_{12} & \dots & \beta_{1p} \\ \beta_{21} & \beta_{22} & \dots & \beta_{2p} \\ \cdot & & & \cdot \\ \cdot & & & \cdot \\ \cdot & & & \cdot \\ \beta_{n1} & \beta_{n2} & \dots & \beta_{np} \end{bmatrix} u$$

$$y = \begin{bmatrix} \gamma_{11} & \gamma_{12} & \dots & \gamma_{1n} \\ \gamma_{21} & \gamma_{22} & \dots & \gamma_{2n} \\ \cdot & & & \cdot \\ \cdot & & & \cdot \\ \cdot & & & \cdot \\ \gamma_{r1} & \gamma_{r2} & & \gamma_{rn} \end{bmatrix} x \qquad (4.2)$$

This representation contains n+np+nr parameters, n of these are redundant since all state variables can be scaled without affecting the input-output relations. The input-output relation can thus be characterized by n(p+r) parameters.

Since the system is completely controllable and observable there is at least one non-zero element in each row of the B matrix and of each column of the C matrix. If the system is such that every state is controllable from one input only and observable from one output only, we find that the system can actually be characterized by 2n parameters only. In this example we thus find that the number of parameters required to characterize the system depends strongly on the manner in which the state variables are coupled to the inputs and the outputs.

Multiple eigenvalues

When the matrix A has multiple eigenvalues the problem of finding minimal parameter representations is more complex. If the matrix A is nonsingular, the system can be transformed to the following form

$$\frac{dz}{dt} = \begin{bmatrix} a_{11} & a_{12}\dots a_{1n} \\ a_{21} & a_{22}\dots a_{2n} \\ \cdot & \cdot \quad\quad \cdot \\ \cdot & \cdot \quad\quad \cdot \\ \cdot & \cdot \quad\quad \cdot \\ a_{r1} & a_{r2}\dots a_{rn} \\ & E \end{bmatrix} z \quad + \begin{bmatrix} b_{11} & b_{12}\dots b_{1p} \\ b_{21} & b_{22}\dots b_{2p} \\ \cdot & \cdot \quad\quad \cdot \\ \cdot & \cdot \quad\quad \cdot \\ \cdot & \cdot \quad\quad \cdot \\ \cdot & \cdot \quad\quad \cdot \\ b_{n1} & b_{n2}\dots b_{np} \end{bmatrix} u$$

$$y = \begin{bmatrix} 1 & 0\dots 0\dots 0 \\ 0 & 1\dots 0\dots 0 \\ \cdot & \cdot \quad \cdot \quad \cdot \\ \cdot & \cdot \quad \cdot \quad \cdot \\ \cdot & \cdot \quad \cdot \quad \cdot \\ 0 & 0\dots 1\dots 0 \end{bmatrix} z \tag{4.3}$$

where E is a matrix with one nonzero element in each row. (This element can be chosen as 1). The representation contains $n(p+r)$ parameters and it is thus a minimum parameter representation. There are, however, two difficulties with this representation. The matrix E reflects the way in which the state variables are coupled to the output. In general there might be

$$N_1 = \binom{n-1}{r-1}$$

different matrices E. Hence if there is no à priori knowledge about the manner in which the outputs are coupled to the state there are N_1 different models of the type (4.3). If $r=1$ then $N_1=1$ and there are no difficulties since there is only one alternative. For moderate values of r and n the number N_1 is, however, fairly large.

There are many canonical structures similar to the one given above. They will, however, all suffer from the same difficulty i.e. unless the internal couplings are known there are many different models. It is also of interest to note that if the internal couplings are known then the number of parameters can be reduced similarly to what was done in the case of diagonal systems. See e.g. [1] and [18].

Summing up, we thus find that the number of parameters required to characterize the input-output relation of a linear system like (4.1) will vary between 2n and n(r+p) depending on the internal couplings of the system. It is possible to obtain canonical representations containing the minimum number of parameters if the internal couplings are known. It is also possible to obtain representations which contain the maximum number of parameters n(p+r). These representations will, however, also depend on the internal couplings. In the general case there are $\binom{n-1}{r-1}$ or $\binom{n-1}{p-1}$ alternatives. In a practical case it is often prohibitive to investigate all possible cases and it is therefore necessary to have à priori knowledge concerning the couplings, to use a method which does not require the investigation of all possibilities or to use other methods which are not based on canonical representations. Mayne has in [19] indicated a recursive procedure which uses the second alternative. The exploitation of realization theory may also circumvent the whole problem.

Vector Difference Equations

For single-output system it is possible to derive an n:th order differential equation (or a difference equation in the sampled data case) which directly characterizes the input-output relation. Compare with Example 2.2. A straight forward generalization to the multivariable case leads to a vector difference equation of the type considered in Example 2.3.

The following example shows that the parametrization problem unfortunately is not solved by vector difference equation either.

Example 4.1

The vector difference equations

$$y(t) + \begin{bmatrix} -3 & 0 \\ 2 & -3 \end{bmatrix} y(t-1) + \begin{bmatrix} 2 & 0 \\ -2 & 0 \end{bmatrix} y(t-2) =$$

$$\begin{bmatrix} 1 & 1 \\ 0 & 2 \end{bmatrix} u(t-1) + \begin{bmatrix} -2 & -1 \\ 2 & 1 \end{bmatrix} u(t-2) \tag{4.4}$$

and

$$y(t) + \begin{bmatrix} -1 & 2 \\ 0 & -5 \end{bmatrix} y(t-1) + \begin{bmatrix} 0 & -6 \\ 0 & 6 \end{bmatrix} y(t-2) =$$

$$\begin{bmatrix} 1 & 1 \\ 0 & 2 \end{bmatrix} u(t-1) + \begin{bmatrix} 0 & 5 \\ 0 & -5 \end{bmatrix} u(t-2) \qquad (4.5)$$

have the same input output relation which is characterized by the pulse transfer function

$$H(z) = \begin{bmatrix} \frac{1}{z-1} & \frac{1}{z-2} \\ 0 & \frac{2z-5}{(z-1)(z-3)} \end{bmatrix} \qquad (4.6)$$

A state space representation of the system is given by

$$x(t+1) = \begin{bmatrix} 1 & 0 & 0 \\ 0 & 2 & 0 \\ 0 & 0 & 3 \end{bmatrix} x(t) + \begin{bmatrix} 1 & 0 \\ 0 & 1 \\ 0 & 1 \end{bmatrix} u(t)$$

$$y(t) = \begin{bmatrix} 1 & 1 & 0 \\ 0 & 1 & 1 \end{bmatrix} x(t) \qquad (4.7)$$

If it is attempted to determine a model of the system by identifying difference equations like (2.4) by modeling outputs one at a time it is clear that difficulties may arise.

5. IDENTIFICATION AND CONTROL

It was mentioned already in the introduction that the main reason for doing identification is to design a control strategy. Several questions then arises. For example, can a suitable criterion for identification be derived from the criterion for control. It may also be questioned if it is reasonable first to make an experiment, then an off-line identification in order to arrive at a model which is finally used to compute the control algorithm. From a practical point of view it would of course be much more reasonable to have algorithms which are generating the desired control algorithm directly on-line. In this section we will discuss such algorithms in a simple case. Consider a single-input single output system described by

$$y(t)+a_1y(t-1)+\ldots+a_ny(t-n) = b_1u(t-1)+\ldots+b_nu(t-n)+e(t) \qquad (5.1)$$

where $\{e(t)\}$ is a sequence of independent uncorrelated disturbances and where the parameters a_i and b_i are assumed to be constant but unknown. Let the criterion be to find a control strategy such that

$$Ey^2(t) \tag{5.2}$$

is minimal at each step and let the admissible control strategies be such that the control signal at time t is a function of the inputs observed up to time t and of the past control signals.

Introduce

$$x = \text{col}\left[a_1,\ a_2,\ldots,a_n,\ b_1,\ b_2,\ldots,b_n\right] \tag{5.3}$$

and

$$\theta(t) = \left[-y(t-1),\ -y(t-2),\ldots-y(t-n),\ u(t-1),\ u(t-2),\ldots,\ u(t-n)\right] \tag{5.4}$$

Let $\hat{x}(t)$ denote the conditional mean of x given observations up to time t and P(t) the covariance of the conditional distribution. It is then shown in [5, Theorem 4] that the optimal control strategy is given by

$$u(t) = \frac{-\sum_{i\neq n+1}' \left[\hat{b}_1\hat{x}_i + p_{n+1,i}\right] \theta_i(t+1)}{\hat{b}_1^2 + p_{n+1,n+1}} \tag{5.5}$$

Notice that if the covariance matrix P goes to zero we get

$$u(t) = -\frac{1}{\hat{b}_1}\left[\hat{a}_1 y(t)+\ldots+\hat{a}_n y(t-n+1)-\hat{b}_2\ u(t-1)-\ldots-\hat{b}_n u(t-n+1)\right] \tag{5.6}$$

which is the optimal control law for the case of known parameters with the true parameters substituted by conditional means. This control law is often called the certainty equivalence control.

It is easy to compute the conditional mean and the conditional covariance in the particular case. They are given by the recursive equations

$$\hat{x}(t+1) = \hat{x}(t) + K(t)\left[y(t) - \theta(t)\ \hat{x}(t)\right]$$

$$K(t) = P(t)\theta^T(t)\ \left[\theta(t)\ P(t)\theta^T(t) + \lambda^2\right]^{-1}$$

$$P(t+1) = \left[I-K(t)\theta(t)\right]P(t) \tag{5.7}$$

See [5, Theorem 5] . Notice that $\hat{x}$ can be interpreted as the least squares estimate of the parameters of the model (5.1), and P as the covariance matrix of the estimate.

For the particular case it is thus seen that the quadratic criterion for the control leads to a control law (5.5), (5.7) which can be considered as an identifier (5.7) followed by a modified certainty equivalence control (5.5). Notice, however, that the control law does not exhibit dual control properties in Feldbaums sense. Dual control will be obtained if the criterion is changed to

$$E \sum_{t=1}^{N} y^2(t) \tag{5.8}$$

If this criterion is used the computational requirements will be almost unsurmountable. Also notice that the results depend heavily on the fact that the residuals $\{e(t)\}$ of the model (5.1) are uncorrelated. The problem complexity is increased significantly if correlated residuals are introduced e.g. by using the model of Example 2.2. It can, however, be shown that the regulator (5.6), (5.7) will have interesting properties when applied to a process described by the equation (2.3)

It was shown in [6] that if the algorithm converges, it will in fact converge to the minimum variance control strategy for (2.3). The problem of convergence is basically a question of stability of a nonlinear stochastic difference equation which brings us back to the main theme of this symposium.

6. NOTES AND REFERENCES

Many references to various aspects of the identification problem are found in the recent survey articles [4], [20] and in the books [14], [21] . There have been two IFAC symposia in Prague in 1967 and 1970. The preprints of the papers presented at these symposia are also a good reference source. Applications of system identification techniques to industrial processes are found in [11], [13], [14], and [19].

1 Ackermann, I.: Die minimale Ein-Ausgangs-Beschreibung von Mehrgrössensystemen und ihre Bestimmung aus Ein-Ausgangs-Messungen. Regelungstechnik und Prozessdatenverarbeitung, 5(1971) 203-206.

2 Åström, K.J., Bohlin, T.: Numerical Identification of Linear Dynamic Systems from Normal Operating Records. Paper IFAC Symposium on Theory of Self-Adaptive Systems, Teddington, England. In Theory of Self-Adaptive Control Systems (Ed P.H. Hammond), Plenum Press, New York, 1966.

3 Åström, K.J.: Introduction to Stochastic Control Theory. Academic Press 1970.

4 Åström, K.J., Eykhoff, P.: System Identification - A Survey. Automatica, 7(1971) 123-162.

5 Åström, K.J. and Wittenmark, B.: Problems of Identification and Control. JMAA 34(1971) 90-113.

6 Åström, K.J. and Wittenmark, B.: On the Control of Constant but Unknown Systems. Paper 37.5. Preprints of the 5th World Congress of IFAC, Paris June 1972.

7 Bellman, R., Åström, K.J.: On Structural Identifiability. Mathematical Biosciences 7(1970), 329-339.

8 Caines, P.E.: The Parameter Estimation of State Variable Models of Multivariable Linear Systems. Control Systems Centre Report No 146, The University of Manchester. Institute of Science and Technology, April 1971.

9 Deistler, M., Ober-Hofer, W.: Macro Economic Systems. Survey Paper 5th World Congress of IFAC Paris June 1972.

10 Eaton, J.: Identification for Control Purposes. IEEE Winter Meeting, New York 1967.

11 Eklund, K.: Linear Drum Boiler - Turbine Models. Report 7117, Division of Automatic Control, Lund Institute of Technology, Nov. 1971.

12 Fischer, F.M.: The Identification Problem in Econometrics, Mc-Graw Hill, New York (1966).

13 Gustavsson, I.: Comparison of Different Methods for Identification of Industrial Processes. Automatica 8, (1972) 127-142.

14 Isermann, R.: Experimentelle Analyse der Dynamik von Regelsystemen - Identification Hochschultaschenbücher, Bibliographisches Institut Mannheim (1971).

15 Kashyap, R.L.: Maximum Likelihood Identification of Stochastic Linear Systems. IEEE Trans. Autom. Control AC-15 (1970) 25 pp.

16 Kostyukovski, Yu.M.-L.: Observability of Nonlinear Controlled Systems. Avtomatika i Telemekhanika 9 (1968) 29-42.

17 Lee, E.B., Marcus, L.: Foundations of Optimal Control Theory. Wiley, New York 1967.

18 Luenberger, D.G.: Canonical Forms for Linear Multivariable Systems. IEEE Trans., AC-12 (1967) 290-293.

19 Mayne, D.Q.: The Identification of Industrial Processes. Preprints JACC (1971).

20 Nieman, R.E., Fischer, D.G., Seborg, D.E.: A Review of Process Identification and Parameter Estimation Techniques. Int. J. Control 13n2 (1971) 209.

21 Sage, A.P., Melsa, J.L.: System Identification. Academic Press, New York (1971).

22 Schoenwandt, U.: On Observability of Nonlinear Systems. Paper 8.2 Preprints 2nd IFAC Symposium on Identification and Process Parameter Estimation, Prague (1970).

23 Valis, J.: Identification of Multivariable Linear Systems of Unknown Structure by the Prior Knowledge Fitting Method, Report 7005, Sept., (1970), LTH. Also available in Proc. 2nd IFAC Symposium on Identification on Process Parameter Estimation, Prague Paper.

24 Woo, K.T.: Maximum Likelihood Identification of Noisy Systems. 2nd IFAC Symposium on Identification and Process Parameter Estimation, Prague, paper 3.1 (1970).

PARAMETRIZATION AND IDENTIFICATION OF LINEAR MULTIVARIABLE SYSTEMS

D.Q. MAYNE

Department of Computing and Control
Imperial College of Science and Technology

1. INTRODUCTION

Let $\mathcal{R}$ = (A,B,C) denote the following representation of a deterministic linear discrete-time dynamical system:

$$x(k+1) = Ax(k) + Bu(k) \quad (1)$$

$$y(k) = Cx(k) \quad (2)$$

where $x(k) \in R^n$, $u(k) \in R^m$, $y(k) \in R^r$, and $(A,B,C) \in W \triangleq \{(F,G,H) : F \in R^{nxn}, G \in R^{nxm}, H \in R^{rxn}\}$. n is the state dimension of $\mathcal{R}$. (A,B,C) has a (unique) transfer function Z(z):

$$Z(z) = C(zI-A)^{-1}B \quad (3)$$

where Z is rational (has rational entries) and proper ($Z(\infty) = 0$). Z has the formal expansion:

$$Z(z) = \sum_{i=1}^{\infty} Y_i z^{-i} \quad (4)$$

where

$$Y_i = CA^{i-1}B \in R^{rxm} \quad (5)$$

i = 1,2 ... are the Markov parameters, or weighting pattern:

$$y(k) = A^k x(0) + \sum_{i=1}^{k-1} Y_i u(k-i) \quad (6)$$

The realization problem is: given Z, rational and proper (given $\{Y_i, i=1,2...\}$) determine a (A,B,C) of minimal state dimension n, satisfying (3) (satisfying (5)). The non-uniqueness of (A,B,C) (similar representations have the same transfer function - (A_1,B_1,C_1), $(A_2,B_2,C_2) \in W$ are similar if there exists a non singular $T \in R^{nxn}$ such that $A_2 = TA_1T^{-1}$, $B_2 = TB_1$, $C_2 = CT^{-1}$) results in a degree of arbitrariness in realization algorithms, and the resultant realization, which cannot be removed by naive normalization. This paper shows that each (rational, proper) Z has an output structure, characterized by the set of integers $(n_1,...n_r,m)$, where $\sum_{i=1}^{r} n_i = n$, the degree of Z (state dimension of minimal realization). If $(n_1,...n_r,m)$ is known (rather than (n,m,r)) a canonical form can be specified such that one, and only one, minimal realization of Z has this form. This extends the useful results of Caines (1970). All arbitrariness in the realization algorithm is removed.

Uniqueness is also necessary (and, with the extra condition

on $\{u_k\}$ of 'persistent excitation' (Åström, 1965), sufficient (Caines, 1970)) for consistent estimation of the parameters $\theta \triangleq (A,B,C,D,K,\Sigma) \in W_s = \{\theta$: A is stable, zeros of $|I + C(zI-A)^{-1}K|$ lie strictly inside unit circle, Σ is symmetric and positive definite$\}$ of the following stochastic linear discrete-time system $\mathcal{R}_s$:

$$x(k+1) = Ax(k) + Bu(k) + Ke(k) \tag{7}$$

$$y(k) = Cx(k) + Du(k) + e(k) \tag{8}$$

$x(k) \in R^n$, $u(k) \in R^m$, $y(k)$, $e(k) \in R^r$, and $\{e(k), k=1,2..\}$ is a sequence of independent normal random variables of zero mean and variance Σ.

2. PRELIMINARIES

Let $(A,B,C) \in W$

$$\Gamma_q \triangleq [C^T,(CA)^T \ldots (CA^{q-1})^T]^T, \qquad \Delta_q \triangleq [B,AB,\ldots A^{q-1}B] \tag{9}$$

Γ_n, Δ_n are, respectively, the observability and controllability matrices. (A,C) is an observable pair ((A,B) is a controllable pair) if rank Γ_n (rank Δ_n) is n. (A,B,C) is a minimal realization if and only if (A,B), (A,C) are, respectively, controllable and observable pairs, which is equivalent to $\Gamma_n \Delta_n$ having rank n.

Let c_i denote the ith row of C.

$$M_i \triangleq \begin{bmatrix} c_i \\ c_i A \\ \vdots \\ c_i A^{n-1} \end{bmatrix}, i = 1 \ldots r. \tag{10}$$

$$\text{Let rank } [M_1^T, M_2^T \ldots M_i^T] = n_1 + n_2 + \ldots n_i, \quad i = 1 \ldots r \tag{11}$$

$$I \triangleq \{i : i \in \{1 \ldots r\}, \quad n_i \neq 0\} = \{p_1, p_2 \ldots p_s\} \tag{12}$$

$$\bar{M}_i \triangleq \begin{bmatrix} c_i \\ \vdots \\ c_i A^{n_i - 1} \end{bmatrix} \quad i \in I \tag{13}$$

n_i, $i \in I$, is least integer such that $c_i A^{n_i}$ (and $c_i A^j$, $\forall j > n_i$) is linearly dependent on the rows of $\{\bar{M}_j ; j \in I_i\}$, $I_i \triangleq \{p_1, p_2 \ldots i\}$.

(A,C) observable implies that:

$$\sum_{i=1}^{n} n_i = \sum_{i \in I} n_i = n \tag{14}$$

and hence $M \in R^{nxn}$ has rank n:

$$M \triangleq [\bar{M}_{p_1}^T \ldots \bar{M}_{p_s}^T]^T = P\Gamma_n \tag{15}$$

for some $P \in R^{nxnr}$. The Hankel matrix $S_q \in R^{qrxqm}$ of Z is:

$$S_q = \begin{bmatrix} Y_1, & Y_2 & \cdots & Y_q \\ Y_2, & Y_3 & \cdots & Y_{q+1} \\ \vdots & \vdots & & \vdots \\ Y_q, & Y_{q+1} & \cdots & Y_{2q-1} \end{bmatrix} \tag{16}$$

If (A,B,C) is a realization of Z, $S_n = \Gamma_n \Delta_n$.
The matrices N_i, $\bar{N}_i$ consist of the same rows of S_n as M_i, $\bar{M}_i$ do of Γ_n. Let s_i be i^{th} row of S_n.

$$N_i \triangleq \begin{bmatrix} s_i \\ s_{i+r} \\ \vdots \\ s_{i+(n-1)r} \end{bmatrix}, i = 1 \ldots r; \quad \bar{N}_i = \begin{bmatrix} s_i \\ s_{i+r} \\ \vdots \\ s_{i+(n_i-1)r} \end{bmatrix}, \; i \in I \tag{17}$$

The following result is well known (see e.g. Mayne (1968)).

<u>Proposition 1</u> Let $(A,B,C) \in W$ be a minimal realization with rank $(M_1^T, .. M_i^T) = n_1 + \ldots + n_i$. Then ($f_i$ is i^{th} row of F, etc.):

(i) (A,B,C) is equivalent to (F,G,H) where:

$$(\sigma_i \triangleq n_{p_1} + n_{p_2} + \ldots n_{p_i}, \; i = 1 \ldots s; \; \sigma_o \triangleq 0) \tag{18}$$

$$f_i = e_{i+1}, \quad i \in \{1 \ldots n\} \setminus \{\sigma_1, \sigma_2 \ldots \sigma_s\} \tag{19}$$

$$f_{\sigma_i} = (\alpha_{i1}, \alpha_{i2} \ldots \alpha_{ii}, 0), \quad i = 1 \ldots s \tag{20}$$

$$h_{p_i} = e_{1+\sigma_{i-1}}, \quad i = 1 \ldots s \tag{21}$$

$$h_i = (\beta_{i1}, \beta_{i2}, \ldots \beta_{i[i]}, 0), \quad i \in \{1 \ldots r\} \setminus I \tag{22}$$

where α_{ij}, $\beta_{ij} \in R^{n_{p_j}}$, $e_i \in R^n$ is the ith unit vector.

$$[i] \triangleq \text{largest integer} \in I \text{ and less than } i \tag{23}$$

α_{ij}, β_{ij} are obtained by solving:

$$c_i A^{n_i} = \sum_{j \in I_i} \alpha_{ij} \bar{M}_j, \quad i \in I \tag{24}$$

$$c_i = \sum_{j \in \bar{I}_i} \beta_{ij} \bar{M}_j, \quad i \in \{1 \ldots r\} \setminus I \tag{25}$$

$$I_i \triangleq \{p_1, p_2 \ldots i\}, \quad i \in I; \quad \bar{I}_i \triangleq \{p_1, p_2 \ldots [i]\} \tag{26}$$

(ii) If Γ_n^*, Δ_n^* are the observability and controllability matrices of (F,G,H),

and $P_1 \Gamma_n = \begin{bmatrix} M \\ R \end{bmatrix}$, then $P_1 \Gamma_n^* \triangleq \begin{bmatrix} M^* \\ R^* \end{bmatrix} = \begin{bmatrix} I_n \\ R^* \end{bmatrix}$ (27)

where P_1 is a row permutation matrix.

<u>Proof</u> Construct similar system with T = M.

3. OUTPUT STRUCTURE

Let Z be rational, proper, have degree n, and have Hankel Matrix S_q. Z is said to have output structure $(n_1, n_2 \ldots n_r; m)$ if rank $[N_1^T, N_2^T, \ldots N_i^T] = n_1 + n_2 + \ldots n_i$, $i = 1 \ldots r$. Clearly $\sum_{i=1}^{r} n_i = n$.

Proposition 2 Let $(A,B,C) \in W$ be any minimal realization of Z with Hankel matrix S_q. Then:

$$\alpha S_{n+1} = 0 \quad \text{if and only if } \alpha \Gamma_{n+1} = 0$$

where $\alpha \in R^{(n+1)r}$.

Proof Since (A,B,C) is a minimal realization of Z

$$S_{n+1} = \Gamma_{n+1} \Delta_{n+1}$$

and Γ_{n+1}, Δ_{n+1} have maximal rank n. Hence:

$$\alpha S_{n+1} = (\alpha \Gamma_{n+1}) \Delta_{n+1} = 0$$

if and only if: $\alpha \Gamma_{n+1} = 0$.

Proposition 3 Let Z (rational, proper) have output structure $(n_1 \ldots n_r; m)$. Then any minimal realization of Z is similar to (F,G,H) of form defined in Proposition 1 (Eqs (19)-(22)).

Proof Let (A,B,C) be any minimal realization of Z. A direct consequence of Proposition 2 is that rank $(M_1^T \ldots M_i^T)$ = rank $(N_1^T \ldots N_i^T)$, $i = 1 \ldots r$. The result follows from Proposition 1.

Comment Since any minimal realization is similar to (F,G,H), any two minimal realizations are similar.

Proposition 4 The following algorithm yields a minimal realization of Z (rational, proper).

(i) Determine n_i, the least integer such that $s_{i+n_i r}$ is linearly dependent on the rows of $(\bar{N}_j,\ j \in I_i)$, $i = 1 \ldots r$. $n_i = 0$ if s_i is linearly dependent on rows of $(\bar{N}_j, j \in \bar{I}_i)$.

(ii) Solve the following equations for α_{ij}, β_{ij}

$$s_{i+n_i r} = \sum_{j \in I_i} \alpha_{ij} \bar{N}_j \,, \quad i \in I \tag{28}$$

$$s_i = \sum_{j \in \bar{I}_i} \beta_{ij} \bar{N}_j \,, \quad i \in \{1 \ldots r\} \setminus I \tag{29}$$

(iii) Construct F,H according to (19)-(22)

$$G = \text{first n columns of } N \triangleq [\bar{N}_{p_1}^T \ldots \bar{N}_{p_s}^T]^T$$

Proof (i) follows from Proposition 3.

(ii) follows from Proposition 2 since relation (28) (which is of the form $\alpha S_{n+1} = 0$) is true if and only if (24) is true.

Similarly for β_{ij}.

(iii) From Proposition 1 (ii):

$$P_1 S_n = \begin{bmatrix} N \\ X \end{bmatrix} = P_1 \Gamma^*_n \Delta^*_n$$

$$= \begin{bmatrix} I_n \\ R^* \end{bmatrix} [G, FG \ldots F^{n-1}G]$$

i.e. $N = [G, FG, \ldots F^{n-1}G]$

<u>Proposition 5</u> Let Z (rational, proper) have output structure $(n_1, n_2 \ldots n_r; m)$. Then there exists one and only one minimal realization (F,G,H) of Z where (F,G) has the form defined in (19)-(22).

<u>Proof</u> Existence is shown in Proposition 4. Suppose there exist two realizations (A,B,C), (F,G,H) with the form of (19)-(22). These realizations are similar. Hence

$$\Gamma_n = \Gamma^*_n T$$

where T defines the similarity transformation. From Proposition 1 (ii), since (A,B,C), (F,G,H) both have form of (19)-(22):

$$P_1 \Gamma_n = \begin{bmatrix} I_n \\ N \end{bmatrix}, \qquad P_1 \Gamma^*_n = \begin{bmatrix} I_n \\ N^* \end{bmatrix}$$

for P_1 non singular. Hence:

$$P_1 \Gamma_n = P_1 \Gamma^*_n T \rightarrow T = I$$

i.e. (F,G,H) is unique.

4. IDENTIFICATION OF (i) $\mathcal{R}$: (ii) $\mathcal{R}_s$ FROM INPUT-OUTPUT DATA

(i) Without loss of generality the case $x(0) = 0$ is considered ($G \rightarrow [G, x(0)]$, $u(k) \rightarrow (u(k), u^*(k))$, $u^*(k) = \delta_{o,k}$ transforms general problem into zero $x(0)$ problem, (Ho (1966))). Also:

$$[y_1, y_2 \ldots y_q] = [Y_1, Y_2 \ldots Y_q] U_q, \quad U_q \triangleq \begin{bmatrix} u_o & u_1 & \ldots\ldots & u_{q-1} \\ 0 & u_o & \ldots\ldots & u_{q-2} \\ 0 & \ldots\ldots\ldots & & u_o \end{bmatrix}$$

If $U_q \in R^{qm \times q}$ satisfies $U_q Q = [I_{jm}, 0]$(*) (Ho (1966)) for some Q, j, then for given (A,B,C) and $\{u_k\}$, Y_k satisfies (5) k=1..j if and only if y_k satisfies (1),(2) (with $x(0) = 0$), k=1...q. Let $\{u_k\}$ satisfy (*), j = 2n. Then:

(a) Suppose $n_1 \ldots n_{i-1}$ have been determined. Let $\mathcal{R}^i = (A^i, B^i, C^i)$, parameters θ^i, denote the corresponding canonical form for Z^i, the transfer function relating $\{u(k)\}$ to $\{y^i(k)\}$. $y^i \in R^i$ consists of first i elements of $y \in R^r$. n_i is least integer ≥ 0 such that min $V^i(\theta^i) \triangleq \frac{1}{2} \sum_{k=1}^{q} ||(y(k) - Cx(k))^i||^2$ with respect to θ^i, subject to (1), $x(0) = 0$, is zero.

(b) Parameters θ of (A,B,C), in canonical form $(n_1 \ldots n_r;m)$ are given by unique minimum with respect to θ of $V^r(\theta)$.

(ii) The stochastic case (identification of $\mathcal{R}_s$) is the natural analogue of (1). Condition (*) is replaced by the 'persistent excitation' condition (Åström 1965). V^i is now the negative of the log likelihood function for $\mathcal{R}_s^i$, the modified stochastic system having only the first i outputs $(y_1(k) \ldots y_i(k)) \triangleq y^i(k)$. Assuming that $n_1 \ldots n_{i-1}$ have been ascertained, n_i is increased until the change in the minimum of V^i with respect to θ^i (the parameters of $\mathcal{R}_s^i$) is statistically insignificant (other tests may be more sensitive). $\hat{\theta}$, the maximum likelihood estimate of the parameters of $\mathcal{R}_s$, is obtained by minimising V^r, when $n_1 \ldots n_r$ have been ascertained. Note that both in the deterministic and stochastic case minimising V^i yields a representation relating $\{u(k)\}$ to $\{y^i(k)\}$.

The independent work of Bucy (1968) and Bucy and Ackerman (1971), in which the algorithm of Proposition 4 is first described, is closely related. Caines has obtained a unique model for the cyclic case and used it effectively for identification (Caines 1970, 1971). Other canonical forms exist (e.g. the dual); the advantage of the above is the linking of the structure numbers $(n_1 \ldots n_r)$ with models having specified number of outputs, permitting sequential determination of these numbers*.

5. REFERENCES

Ackerman, J.E., Bucy, R.S., Information and Control, 19, 224-231, (1971).

Åström, K.J. et al., IBM Nordic Laboratory Report, Tp 18.150,(1965).

Bucy, R.S., IEEE Transactions, AC 13, 567-569, (1968).

Caines, P.E., Ph.D. Thesis, Imperial College, (1970).

Caines, P.E., International Journal Control, 13, 529-547, (1971).

Caines, P.E., in IEE Conference Proceedings, Series No.78, (1971).

Ho, B.L., Ph.D. Thesis, Stanford University, (1966).

Ho, B.L., Kalman, R.E., Regelungstechnik, 14, 545-548, (1966).

Mayne, D.Q., Proceedings I.E.E., 115, 1363-1368, (1968).

---

* Z may be defined as having output structure $(n_1, n_2 \ldots n_r;m)$ if Z^i (matrix of first i rows of Z) has (MacMillan) degree $n_1+n_2+\ldots n_i$, $i = 1 \ldots r$.

OPTIMIZATION OF SENSORS' LOCATION IN A DISTRIBUTED FILTERING PROBLEM

ALAIN BENSOUSSAN

University of Paris IX and IRIA *

INTRODUCTION **

The aim of this paper is to study the following control problem for a Riccati equation in an infinite dimensional space :

Denoting by v the control (in a function space), one considers the following Riccati equation

$$(1)\quad \frac{dP}{dt} + PA^* + AP + PD(t;v)P = D_o$$

$$P(0) = P_o$$

with the following payoff

$$(2)\quad J(v) = N(v) + \operatorname{tr} P(T\ ;\ v)$$

This problem comes in quite naturally from the following filtering problem

* This work has been made during a stay of the author at MIT, Electronics system Laboratory, on a joint research contract supported by the National Science Foundation and IRIA.

** The author is indebted to J. WILLEMS who suggested improvement of some proofs and to M. ATHANS, L. GOULD and S.K. MITTER for a very fruitful exchange of ideas.

$$(3)\quad \begin{cases} \dfrac{dy}{dt} + A(t)y = f(t) + B(t)\,\xi(t) \\ \\ y(0) = y_o + \zeta \end{cases}$$

$$(4)\quad z(t) = C(t\ ;\ v)\ y(t) + \eta\,(t\ ;\ v)$$

where y(t) is the stochastic process to filter and $z(t)$ is the observation process, which depends on the decision variable v.

In the payoff J(v), the term tr P(T ; v) is a measure of the accuracy, and N(v) represents the cost of the control.

In the case of finite dimensional systems, such a problem has been considered by M. ATHANS [1], [2].

The introduction of infinite dimensional spaces in that context is quite interesting, because it includes as a particular case the problem of optimizing the location of sensors in a distributed parameter system, which is the main motivation of this article.

We first start with the particular case when N(v) = 0. We then prove an existence theorem for an optimal control, and we check the assumptions in the example of sensors' location. We then consider the problem of getting necessary conditions of optimality.

In the previously quoted references, M. ATHANS has derived necessary conditions by using the matrix maximum principle. However, his derivations are formal, particularly because v will not in general belong to a vector space (it may be a boolean function).

Using here a different argument (purely algebraic) we prove a necessary condition which differs from ATHANS' one by a second order term.

1. Setting of the problem

1.1. Notations - Assumptions

Let $(\Omega, \mathcal{A}, \mu)$ be a probability space. Let H, V be two Hilbert spaces such that

(1.1.) $V \subset H$, V is dense in H with a continuous injection. We shall identify H with its dual space, and denoting by V' the dual space of V, we get the following inclusion

(1.2.) $V \subset H \subset V'$

each space being dense in the next one with a continuous injection. The time $t \in [0, T]$; We consider a family A(t) of operators such that

$$(1.3.) \begin{cases} A(.) \in L^{\infty}(0, T ; L(V ; V')) \\ \langle A(t)z, z \rangle \geq \alpha \| z \|^2, \quad \forall z \in V. \end{cases}$$

Let now E , F be two Hilbert spaces (F is finite dimensional) and B(t), C(t) two family of operators such that

$$(1.4.) \begin{cases} B(.) \in L^{\infty}(0, T ; L(E ; H)) \\ C(.) \in L^{\infty}(0,T ; L(H ; F)). \end{cases}$$

On E and F, one introduces two Wiener processes

$\xi(t, \omega)$, $\eta(t ; \omega)$ such that

$$(1.5.) \quad E\, \xi(t) = 0, \quad E\, \eta(t) = 0 \quad , \quad \forall t$$

$$(1.6.)\quad E\,(\xi(t), e_1)\,(\xi(s), e_2) = \int_0^{\min(t,s)} (Q(\tau)e_1, e_2)\,d\tau\,,\ \forall e_1, e_2 \in E$$

$$E\,(\eta(t), f_1)\,(\eta(s), f_2) = \int_0^{\min(t,s)} (R(\tau)\,f_1, f_2)\,d\tau\,,\ \forall f_1, f_2 \in F.$$

in (1.6.) $Q(.) \in L^\infty(0, T\,;\,L(E\,;\,E))$, $Q(t) \geq 0$, self adjoint and <u>nuclear</u>. Similarly, $R(.) \in L^\infty(0, T\,;\,L(F\,;\,F))$, $R(t) \geq 0$, self adjoint and <u>nuclear</u>. But we assume that $R(t)$ is <u>invertible</u>. Since $R(t)$ is at the same time nuclear and invertible, it follows that F must be finite dimensional.

We assume that $\xi(t)$ and $\eta(t)$ are not correlated and thus independant one from each other.

Let then $y_o \in H$, $f(.) \in L^2(0, T\,;\,H)$, and ζ a gaussian random variable with values in H, such that

$$(1.7.)\quad E\,\zeta = 0$$

$$(1.8.)\quad E\,(\zeta, h_1)\,(\zeta, h_2) = (P_o\,h_1, h_2)\,,\ \forall h_1, h_2 \in H.$$

In (1.8.) $P_o \in L(H\,;\,H)$, $P_o \geq 0$, self adjoint and <u>nuclear.</u> Furthermore, we assume that ζ is independant from ξ and η.

Let us now introduce a set $\mathcal{U}$, called <u>the set of controls</u>, and for $v \in \mathcal{U}$, a family $C(t\,;\,v)$ and $\eta(t\,;\,v)$ with exactly the same properties as $C(t)$ and $\eta(t)$.

Let $y(t)$ be the solution of the stochastic differential equation

$$(1.9)\quad \begin{cases} dy(t) + A(t)\,y\,dt = f(t)\,dt + B(t)\,d\xi(t) \\ y(0) = y_o + \zeta\,. \end{cases}$$

The solution of (1.9.) is unique and is a stochastic process with values in H (see A. BENSOUSSAN [1]).

The observation process is defined by the following stochastic differential equation

(1.10.) $d\mathfrak{z}(t) = C(t;v)dy(t) + d\eta(t;v)$.

For any fixed v, the conditional expectation of y (t) on $\mathfrak{z}(s)$, $0 \leqslant s \leqslant t$, denoted by $\hat{y}(t)$ is a stochastic process with values in H, solution of the Kalman filter (generalized as in BENSOUSSAN [1]) defined by the following stochastic differential equation

$$(1.11.)\quad \begin{cases} d\hat{y} + A\hat{y}dt + P(t;v)D(t;v)\hat{y}dt = fdt + P(t;v)C^*(t;v)R^{-1}(t;v)d\mathfrak{z}(t;v) \\ \hat{y}(0) = y_o \end{cases}$$

where

$$(1.12.)\quad D(t;v) = C^*(t;v)R^{-1}(t;v)C(t;v)$$

and P(t;v) is solution of the Riccati equation

$$(1.13)\quad \begin{cases} \frac{dP}{dt} + PA^* + AP + PD(t;v)P = BQB^*(t) = D_o(t) \\ P(0) = P_o . \end{cases}$$

If one considers the estimation error $\epsilon(t;v) = \hat{y}(t;v) - y(t;v)$, then for any t, P(t) is the <u>covariance operator</u> of $\epsilon(t;v)$

1.2. The control problem

Our problem is to choose v. We must define a choice criterion. A natural one is to consider a compromise between the cost of control and the effectiveness of the measurement being measured by $E|\epsilon(T;v)|^2 = \text{tr}\, P(T;v)$.

Thus we shall consider functionals of the form

$$J(v) = N(v) + \text{tr}\, P(T;v). \tag{1.14}$$

The problem thus amounts to a control problem on the Riccati equation.

1.3. A particular case

Let us consider the case when $N(v) = 0$. We then get

Theorem 1.1. : If there exists u such that

$$D(u) \geqslant D(v) \quad \forall v \tag{1.15}$$

in the sense of the order relationship of non negative self adjoint operators then u is solution of the control problem.

Proof :

Let us set $D_v = D(v)$ and $P_v = P(v)$ solution of (1.13). According to the decoupling theory (see J.L. LIONS [1]), if one considers the system of equations

$$\begin{cases} \dfrac{d\varphi_v}{dt} + A\varphi_v + D_o \psi_v = 0 \\ \\ \dfrac{-d\psi_v}{dt} + A^* \psi_v - D_v \varphi_v = 0 \\ \\ \varphi_v(0) + P_o \psi_v(0) = 0 \\ \psi_v(T) = h \end{cases} \tag{1.16}$$

then one gets

$$(1.17) \quad \varphi_v(T) = -P_v(T)\,h.$$

Therefore

$$(1.18) \quad ((P_v(T) - P_u(T))\,h, h) = +(h, \varphi_u(T) - \varphi_v(T)).$$

Setting

$$\gamma = \varphi_u - \varphi_v$$

$$\delta = \Psi_u - \Psi_v$$

one gets

$$(1.19) \begin{cases} \dfrac{d\gamma}{dt} + A\gamma + D_o\,\delta = 0 \\ -\dfrac{d\delta}{dt} + A^*\delta - D_u\gamma = (D_u - D_v)\,\varphi_v \\ \gamma(0) + P_o\,\delta(0) = 0 \\ \delta(T) = 0. \end{cases}$$

Thus

$$\int_0^T ((D_u - D_v)\varphi_v, \varphi_v)\,dt = -\int_0^T (D_u\gamma, \varphi_v)\,dt + \int_0^T \left(-\frac{d\delta}{dt} + A^*\delta, \varphi_v\right) dt$$

taking account of (1.16) and integrating by parts, one gets

$$(1.20)\quad \int_0^T ((D_u - D_v)\varphi_v, \varphi_v)dt = -\int_0^T (D_u\gamma, \varphi_v)dt + (\delta(0), \varphi_v(0))$$

$$+ \int_0^T (\delta, - D_o\psi_v)\, dt$$

$$= - \int_0^T (D_u\gamma, \varphi_v)dt + (\delta(0), \varphi_v(0)) + \int_0^T (\psi_v, \frac{d\gamma}{dt} + A\gamma)dt$$

$$= - \int_0^T (D_u\gamma, \varphi_v)dt + (\delta(0), \varphi_v(0)) + (\gamma(T); h)$$

$$- (\gamma(0), \psi_v(0)) + \int_0^T (\gamma, D_v\, \varphi_v)\, dt.$$

But

$$(\gamma(0), \varphi_v(0)) = - (P_o\psi_v(0), \delta(0)) = - (\psi_v(0), P_o\delta(0))$$

$$= (\psi_v(0), \gamma(0))$$

and therefore (1.20) becomes

$$(1.21)\quad \int_0^T ((D_u - D_v)\, \varphi_v, \varphi_v)\, dt = (\gamma(T), h) - \int_0^T (\varphi_v, (D_u - D_v)\gamma)\, dt.$$

Taking account of (1.19) one gets

$$\int_0^T ((D_u - D_v)\varphi_v, \gamma)\, dt = \int_0^T (\frac{-d\delta}{dt} + A^* \delta - D_u\gamma, \gamma)\, dt$$

$$= - \int_0^T (D_u\gamma, \gamma)dt - (P_o\delta(0); \delta(0))$$

$$- \int_0^T (D_o\delta, \delta)\, dt.$$

Therefore we finally get, recalling (1.17)

$$(1.22)\quad ((P_v(T) - P_u(T))h,h) = \int_0^T ((D_u - D_v)\varphi_v, \varphi_v)\,dt$$

$$+ (P_o\,\delta(0),\delta(0)) + \int_0^T (D_o\delta,\delta)\,dt$$

$$+ \int_0^T (D_u\gamma, \gamma)dt.$$

If (1.15) is true, it immediately follows from (1.22) that

$$P_v(T) \geqslant P_u(T)$$

and thus

$$\mathrm{tr}\; P_v(T) \geqslant \mathrm{tr}\; P_u(T).$$

1.4. An example :

It is clear that condition (1.15) will not be true in general. However this may happen sometimes. Let us give an example of this situation.

Let the model (1.9) correspond to the diffusion operator, i.e.,

$$(1.23)\quad +\frac{\partial y}{\partial t} - \Delta y = f(x,t) + \xi(t,x) \quad , x \in \Omega \subset R^n$$

$$y\,|_{\Sigma} = 0, \qquad \Sigma = \,]0,T[\, x\ \partial\Omega$$

$$y(x,0) = y_o(x) + \zeta(x).$$

Suppose now that we can use one of m possible observers defined by

(1.24) $z_j(t) = \alpha_j(t)y(0,t) + \eta_j(t), \quad j = 1 \ldots m,$

where $r_j(t)$ is the covanance of $\eta_j(t)$.

The control variable is thus $v_j(t) = \begin{cases} 1 \text{ if we choose observer } j \\ 0 \text{ if we do not choose ———} \end{cases}$

We can write the observation process as follows

$$(1.25) \quad \begin{cases} z(t) = \sum_{j=1}^{m} v_j(t)\ \alpha_j\ (t)\ y\ (0,t) + \sum_{j=1}^{m} v_j(t)\ \eta_j(t) \\ v_j(t) = 0 \text{ or } 1 \quad , \quad \sum_{j=1}^{m} v_j(t) = 1. \end{cases}$$

Thus we get

$$(1.26) \quad D(t;v)z = \sum_{j=1}^{m} v_j(t) \frac{\alpha_j^2(t)}{r_j(t)} (z(0))^2 \quad \forall z \in L^2(\Omega)$$ (1)

Let us define u(t) in the following way

$$u(t) = \begin{cases} 0 \\ \vdots \\ 0 \\ 1 \quad \text{index } j \text{ such that } \dfrac{\alpha_j^2(t)}{r_j(t)} = \max\limits_k \dfrac{\alpha_k^2(t)}{r_k(t)} . \\ 0 \\ \vdots \\ 0 \end{cases}$$

It is clear that (1.15) is satisfied.

Remark 1.1. : A problem of m observers with a choice of shifting from one to another when time varies has been studied by M. ATHANS [1].

---

(1) (1.26) is formal. To be correct one should change $z(0)$ into $\frac{1}{\text{meas } \mathcal{O}} \int_{\mathcal{O}(0)} z(x)dx$ where $\mathcal{O}(0)$ is a ball of center 0.

2. Existence result

2.1. Assumptions

The problem we consider is the same as in § 1.2 with the following assumptions

(2.1) $v \in \mathcal{U}$ Hilbert space

(2.2) $v \to N(v)$ is weakly lower semi continuous and coercive.

We shall denote by $e_1 \dots e_n \dots\dots$ an orthonormal basis for H. According to (1.16) we shall consider the family of two point boundary value problems

$$(2.3) \quad \begin{cases} \dfrac{d\varphi_j}{dt} + A\,\varphi_j + D_o\,\Psi_j = 0 \\[2mm] \dfrac{-d\Psi_j}{dt} + A^*\,\Psi_j - D(v)\,\varphi_j = 0 \\[2mm] \varphi_j(0) + P_o\Psi_j(0) = 0 \\[2mm] \Psi_j\,(T) = e_j. \end{cases}$$

where $\varphi_j = \varphi_j(t;\, v) \qquad \Psi_j = \Psi_j(t\;;\, v)$.

We then get

$$(2.4) \quad \varphi_j(T) = -\,P(T)e_j$$

and thus

$$(2.5) \quad \mathrm{tr}\, P\,(T;\, v) = -\sum_{j=1}^{\infty} (\varphi_j(T;v),\, e_j).$$

The control problem can be formulated as follows :

$$(2.6)\quad \text{Min } J(v) = N(v) - \sum_{j=1}^{\infty} (\varphi_j(T;v),\ e_j).$$

We shall now make the fundamental assumption

$$(2.7)\quad \begin{cases} \text{for any } \Psi \in L^2(H) \quad v \to D(.;\ v)\ \Psi(.)\ :\ \mathcal{U} \to L^2(H) \\ \text{is continuous when } \mathcal{U} \text{ and } L^2(H) \text{ are provided with the weak} \\ \text{topologies.} \end{cases}$$

2.2. <u>Existence of an optimal Control</u>

<u>Theorem 2.1</u> : <u>With the Notations and Assumptions of sections</u> 1.1. <u>and</u> 2.1, <u>there exists an optimal solution for problem</u> (2.6).

<u>Proof</u> :

Let v^α be a minimizing sequence. From the fact that $N(v^\alpha)$ remains in a bounded set and from assumption (2.2.) it follows that v^α remains in a bounded set of $\mathcal{U}$.

From (2.3) it follows the identity

$$(2.8)\quad -(\varphi_j(T;v),\ e_j) = (P_o\Psi_j(0), \Psi_j(0)) + \int_0^T (D_o\Psi_j(t), \Psi_j(t))dt$$

$$+ \int_0^T (D(t;v)\varphi_j(t), \varphi_j(t))dt.$$

Therefore

(2.9) $P_o\Psi_j^\alpha(0)$ remains in a bounded set of H

(2.10) $D_o\Psi_j^\alpha$ remains in a bounded set of $L^2(H)$.

(2.11) $D(v^\alpha)\varphi_j^\alpha$ remains in a bounded set of $L^2(H)$.

Considering (2.3) and using the energy equation, it follows from (2.9), (2.10) and (2.11) that

(2.12) ψ_j^α and φ_j^α remain in bounded sets of

$$W(0,T) = \{ z \in L^2(0,T\ ;\ V) \mid \frac{dz}{dt} + Az \in L^2(H) \}.$$

Therefore we can extract subsequences, still denoted v^α, φ_j^α, ψ_j^α such that

(2.13) $$\begin{cases} v^\alpha \to u & \text{in } \mathcal{U} \text{ weakly} \\ \psi_j^\alpha \to \psi_j & \text{in } W(0,T) \text{ weakly} \\ \varphi_j^\alpha \to \varphi_j & \text{in } W(0,T) \text{ weakly.} \end{cases}$$

Using a compactness result of LIONS [2] it follows from (2.13) that

(2.14) $\varphi_j^\alpha \to \varphi_j$ in $L^2(H)$ strongly.

Moreover

$$\int_0^T (D_j(t\ ;\ v^\alpha)\ \varphi_j^\alpha(t),\ \psi(t)dt = \int_0^T (\varphi_j^\alpha(t), D_j(t;\ v^\alpha)\psi(t))dt$$

which proves using (2.14) and assumption (2.7) that

(2.15) $D_j(.;v^\alpha)\psi_j^\alpha(.) \underset{\alpha \to \infty}{\to} D_j(.\ ;u)\ \varphi_j(.)$ in $L^2(H)$ weakly

We then can go to the limit in equation (2.3), getting at once

(2.16) $$\varphi_j \equiv \varphi_j\ (.;u)$$

$$\psi_j \equiv \psi_j(.;u).$$

Therefore for any fixed N, one gets

$$- \sum_{j=1}^{N} (e_j, \varphi_j^{\alpha}(T)) \to - \sum_{j=1}^{N} (e_j, \varphi_j(T;u)).$$

Thus setting

$$J^N(v) = N(v) - \sum_{j=1}^{N} (e_j, \varphi_j(T;v))$$

one gets

$$(2.17) \quad J^N(u) \leq \liminf_{\alpha \to \infty} J^N(v^{\alpha}) \leq \liminf_{\alpha \to \infty} J(v^{\alpha}) = \inf J(v)$$

from which it follows going to the limit when $N \to \infty$

$$(2.18) \quad J(u) = \inf J(v).$$

2.3. Location of sensors in a distributed parameter system :

Let us consider the model (1.23). The observation procedure consists in measuring the value of y as a function of time, in m points $b_1 \ldots b_m$. The problem consists in choosing the best location of $b_1 \ldots b_m$.

Thus we have (if $x \in R^n$),

$$(2.19) \quad v = b_1 \ldots b_m \in R^{nxm} = \mathcal{U}.$$

The observation is then

$$(2.20) \quad \mathfrak{z} = \begin{matrix} \mathfrak{z}_1(t) \\ \vdots \\ \mathfrak{z}_j(t) \\ \vdots \\ \mathfrak{z}_m(t) \end{matrix} = \int_{\mathcal{O}(b_j)} y(x,t)dx + \eta_j(t).$$

Let $\Psi_1, \Psi_2 \in L^2(Q)$ and

$$(2.21)\quad \int_0^T (D(t;v)\,\Psi_1(t),\,\Psi_2(t)dt = \int_0^T dt(R^{-1}(t;v)\,C(t;v)\Psi_1(t),C(t;v)\Psi_2(t)).$$

Assuming that the variance of $\eta_j(t)$ does not depend on b_j ; the formula (2.21) becomes $(R(t) = r)$

$$\frac{1}{r}\sum_{j=1}^{m}\int_0^T dt \int_{\mathcal{O}(b_j)} \Psi_1(x,t)dx \int_{\mathcal{O}(b_j)} \Psi_2(x,t)dx.$$

But

$$b \to \int_{\mathcal{O}(b)} \Psi(x,t)dx$$

is continuous from R^n into $L^2(0,T)$ and thus

$v \to \int_0^T (D(t;v)\,\Psi_1(t),\,\Psi_2(t)dt$ is continuous from $(R^n)^m$ into

R , $\forall\ \Psi_1, \Psi_2 \in L^2(Q)$. Thus condition (2.7) is satisfied.

When the cost of sensors satisfies (2.2.) (for instance $v \to N(v)$ is continuous >0 and coercive), Theorem 2.1. proves the existence of a best location for m given sensors.

3. Necessary Conditions

Orientation : We shall consider here the problem of finding necessary conditions for u to be an optimal control. We shall compare our result to that of M. ATHANS [1] , showing in fact that Athans' necessary condition obtained through Pontryagin's maximum principle (adapted to matrices) differ from ours (obtained through a direct algebraic argument) by a second term order.

3.1. Notations : Assumptions :

We go back to the situation of § 1.1., assuming in fact that v belongs to some set $\mathcal{U}$ of controls (no longer necessarily a hilbert space).

Let us denote by $\pi(t)$ the difference (depending on u and v)

(3.1.) $\pi(t) = P(t;v) - P(t;u).$

We notice that

$$\sum_{j=1}^{\infty} (\pi(T)e_j', e_j) = \text{tr } P(T;v) - \text{tr } P(T;u).$$

Therefore one can write

(3.2.) $J(v) - J(u) = N(v) - N(u) + \sum_{j=1}^{\infty} (\pi(T)e_j, e_j).$

Let us write down the equation for π

(3.3.) $$\begin{cases} \frac{d\pi}{dt} + \pi A^* + A\pi = P_u D_u P_u - P_v D_v P_v \\ \pi(0) = 0 \end{cases}$$

which we may also write

(3.4.) $$\begin{cases} \frac{d\pi}{dt} + (A + P_u D_v)\pi + \pi(A^* + D_v P_u) = P_u D_u P_u - P_v D_v P_v \\ \qquad + P_u D_v P_v + P_v D_v P_u \\ \qquad - 2 P_u D_v P_u \\ \pi(0) = 0. \end{cases}$$

Let us also introduce Σ solution of

$$(3.5)\quad \begin{cases} \dfrac{-d\Sigma}{dt} + \Sigma\,(A + P_u D_v) + (A^* + D_v P_u)\,\Sigma = 0 \\ \\ \Sigma\,(T) = I. \end{cases}$$

Let us set

$$(3.6)\quad \Delta = P_u D_u P_u - P_v D_v P_v + P_u D_v P_v + P_v D_v P_u - 2\,P_u D_v P_u$$

3.2. Auxiliary result

Our aim here is to prove the

Lemma 3.1. : We have

$$(3.7)\quad \sum_{j=1}^{\infty} (\pi(T)e_j, e_j) = \sum_{j=1}^{\infty} \int_0^T (\Sigma\,(t)\,\Delta\,(t)e_j, e_j)dt$$

Proof :

Let us first recall that $\pi(T)$ is a nuclear operator (see GELFAND - VILENKIN [1]) and the sum in (3.7) is independant of the chosen basis $e_1 \ldots e_j \ldots$ So is $\int_0^T \Sigma(t)\,\Delta\,(t)dt$, since $\Delta\,(t)$ is a.e. a nuclear operator. By a standard argument one can show that in the second member of (3.7),it is possible to invert the summation and the integration. Thus (3.7) can be written

$$(3.8)\quad \text{tr}\ \pi(T) = \int_0^T \text{tr}\ \Sigma\,(t)\ \Delta\,(t)dt.$$

Let us set

$$(3.9)\quad \tilde{A}(t) = A(t) + P_u(t)D_v(t).$$

Thus (3.4.) and (3.5.) can be rewritten as follows

$$(3.10)\quad \begin{cases} \dfrac{d\pi}{dt} + \tilde{A}\pi + \pi \tilde{A}^* = \Delta \\ \pi(0) = 0 \end{cases}$$

$$(3.11)\quad \begin{cases} \dfrac{-d\Sigma}{dt} + \Sigma \tilde{A} + \tilde{A}^* \Sigma = 0 \\ \Sigma(T) = I. \end{cases}$$

Introducing $\tilde{\Phi}(t,s)$ the Green operator associated with $\tilde{A}$ (see J.L. LIONS [3]), one easily checks (see for instance A. BENSOUSSAN [1]) that

$$(3.12)\quad \pi(t) = \int_0^t \tilde{\Phi}(t,s)\, \Delta(s)\, \tilde{\Phi}^*(t,s)ds$$

$$(3.13)\quad \Sigma(t) = \tilde{\Phi}^*(T,t)\, \tilde{\Phi}(T,t).$$

Therefore

$$(3.14)\quad \pi(T) = \int_0^T \tilde{\Phi}(T,t)\, \Delta(t)\, \tilde{\Phi}^*(T,t)dt$$

and

$$(3.15)\quad \sum_{j=1}^{\infty} (\pi(T)e_j, e_j) = \sum_{j=1}^{\infty} \int_0^T (\tilde{\Phi}(T,t)\, \Delta(t)\, \tilde{\Phi}^*(T,t)\, e_j, e_j)dt.$$

$$= \int_0^T \sum_{j=1}^{\infty} (\tilde{\Phi}(T,t)\, \Delta(t)\, \tilde{\Phi}^*(T,t)\, e_j, e_j)dt$$

$$= \int_0^T \sum_{j=1}^{\infty} (\tilde{\Phi}^*(T,t)\, \tilde{\Phi}(T,t)\, \Delta(t)\, e_j, e_j)dt$$

by a classical property of nuclear operators (see GELFAND - VILENKIN [1]) and thus we get (3.8).

3.3. Necessary conditions of optimality

Using (3.2) and (3.8) we get

$$(3.16)\quad J(v) - J(u) = N(v) - N(u) + \int_0^T \operatorname{tr} \Sigma(t)\, \Delta(t)\, dt.$$

$$= N(v) - N(u) + \int_0^T \operatorname{tr} \Sigma(t)\, P_u(t)\, (D_u(t) - D_v(t))P_u(t)dt$$

$$- \int_0^T \operatorname{tr} \Sigma(t)\, (P_v(t) - P_u(t))D_v(t)(P_v(t) - P_u(t))dt.$$

But

$$\operatorname{tr} \Sigma\, (P_v-P_u)D_v(P_v-P_u) = \operatorname{tr} \Sigma^{1/2}\, (P_v-P_u)D_v(P_v-P_u)\, \Sigma^{1/2}$$

$$> 0.$$

Thus we get

<u>Theorem 3.1.</u> : <u>With the notations and assumptions of § 3.1., a necessary condition of optimality for u is</u>

$$(3.17)\quad N(v) - N(u) + \int_0^T \operatorname{tr} \Sigma_{u,v}(t)\, P_u(t)(D_u(t) - D_v(t))\, P_u(t)\, dt \geq 0$$

$$\forall v \neq u \quad ▨$$

<u>Remark 3.1</u> :

We have written $\Sigma_{u,v}$ instead of Σ for the solution of (3.5) in order to point out the dependance of Σ on u and v. Condition (3.17) differs from the one obtained by ATHANS [1] using Matrix maximum principle by the fact that in Athans' condition one takes $\Sigma_{u,u}$ instead of $\Sigma_{u,v}$

3.4. An other necessary condition

Orientation : We shall see here under what conditions one can change $\Sigma_{u,v}$ into $\Sigma_{u,u}$ in (3.17)

Let us set $\Sigma_u = \Sigma_{uu}$ and

$$R = \Sigma_{uv} - \Sigma_u$$

then R is solution of

$$(3.18)\begin{cases} -\dfrac{dR}{dt} + R(A + P_u D_v) + (A^* + D_u P_u)R = (D_u - D_v)P_u\,\Sigma + \Sigma\,P_u(D_u - D_v) \\ R\,(T) = 0 \end{cases}$$

and

$$(3.19)\quad \int_0^T \operatorname{tr} \Sigma\,P_u(D_u - D_v)P_u\,dt = \int_0^T \operatorname{tr} \Sigma_u P_u\,(D_u - D_v)\,P_u\,dt$$

$$+ \int_0^T \operatorname{tr} R\,P_u\,(D_u - D_v)\,P_u\,dt.$$

We want that (3.17) implies

$$(3.20)\quad N(v) - N(u) + \int_0^T \operatorname{tr} \Sigma_u P_u (D_u - D_v)\,P_u\,dt > 0 \quad (1)$$

For that it is necessary and sufficient that

$$(3.21)\quad \frac{\int_0^T \operatorname{tr} R\,P_u\,(D_u - D_v)\,P_u\,dt}{N(v) - N(u) + \int_0^T \operatorname{tr} \Sigma_u\,P_u(D_u - D_v)P_u\,dt} > -1.$$

(1) Condition given by ATHANS [1].

Let us notice that

$$(3.22)\quad \int_0^T \mathrm{tr}\, \Sigma_u P_u (D_u - D_v) P_u \, dt = \int_0^T \mathrm{tr}\, \Phi_u^{*}(T,t)\Phi_u(T,t)P_u(t)(D_u(t)-D_v(t))P_u(t)dt$$

where $\Phi_u(t_1,t_2)$ denotes the Green operator associated to $A + P_u D_u$.

But (3.22) can also be written as follows

$$\int_0^T \mathrm{tr}\, \Phi_u(T,t)\ P_u(t)\ (D_u(t)-D_v(t))P_u(t)\Phi_u^{*}(T,t)\ dt.$$

$$= \mathrm{tr}\, F_v(T) - \mathrm{tr}\, F_u(T)$$

where $F(t)$ is solution of

$$(3.23)\quad \begin{cases} \dfrac{dF}{dt} + (A + P_u D_u)F + F\,(A^{*} + D_u P_u) = P_u(+D_v)P_u \\ F(0) = 0 \end{cases}$$

Now setting $\Delta = (D_u - D_v)P_u\Sigma + \Sigma P_u(D_u - D_v)$

we get from (3.18)

$$R(t) = \int_0^T \Phi_u^{*}(s,t)\ \Delta(s)\,\Phi_u(s,t)\ ds.$$

Therefore

$$\int_0^T \mathrm{tr}\, R\, P_u(D_u - D_v)P_u\ dt = \mathrm{tr}\int_0^T dt \int_t^T ds\, \Phi_u^{*}(s,t)\ \Delta(s)\ \Phi_u(s,t)P_u(t)$$

$$x(D_u(t)-D_v(t))P_u(t)$$

$$= \int_0^T ds \int_o^s dt\ \mathrm{tr}\ \Delta(s)\ G_u(s,t)P_u(t)(D_u(t)-D_v(t))P_u(t)G_u^{*}(s,t)$$

$$= \int_0^T ds\ \mathrm{tr}\ \Delta(s)\ (F_v(s) - F_u(s))$$

Therefore (3.21) becomes

$$(3.22)\quad \frac{\int_0^T \mathrm{tr}\left[(D_u-D_v)P_u\Sigma + \Sigma P_u(D_u-D_v)\right](F_v(t)-F_u(t))dt}{N(v)-N(u) + \mathrm{tr}\, F_v(T) - \mathrm{tr}\, F_u(T)} > -1.$$

Let us consider v such that

$$(3.23)\quad \sup_t \mathrm{ess}\ \| D_v(t) - D_u(t) \| \leq \lambda$$

then one easily checks that

$$(3.24)\quad \left| \int_0^T \mathrm{tr}\left[(D_u-D_v)P_u\Sigma + \Sigma P_u(D_u-D_v)\right](F_v(t)-F_u(t))dt \right| \leq C\lambda^2 ,$$

where C is constant (depending on u)

Therefore if the following condition is satisfied

$$(3.25)\quad \frac{| N(v) - N(u) + \mathrm{tr}\, F_v(T) - \mathrm{tr}\, F_u(T) |}{\lambda} \geq C_1 \text{ when } \lambda \to 0$$

then for some λ condition (3.22) will be satisfied.

Thus we have proved

<u>Proposition 3.1.</u> : <u>If</u> (3.23) <u>and</u> (3.25) <u>are satisfied</u>, <u>then</u>, <u>for</u> u <u>to be an optimal solution of Problem</u> (1.13),(1.14), <u>it is necessary that for any</u> v <u>such that</u>

$$\sup_t \mathrm{ess}\ \|D_v(t)-D_u(t)\| \leq Ct \quad \text{(depending on u)}$$

<u>the condition</u>

$$(3.26)\quad N(v) - N(u) + \int_0^T \mathrm{tr}\Sigma_u P_u(D_u-D_v)P_u\, dt \geq 0$$

<u>be satisfied</u>. ▨

<u>Remark 3.2</u> : From Proposition 3.1 it appears clearly that (3.26) is an approximation of (3.17) neglecting higher oder terms in D_v-D_u.

References

M. ATHANS [1] Lincoln Laboratory report

[2] IFAC Congress, Paris, June 1972

A. BENSOUSSAN [1] Filtrage optimal des systèmes linéaires, Dunod, Paris, 1971

I.M. GELFAND-N.Y.VILENKIN [1] Generalized functions, Applications of Harmonic Analysis, Academic Press, N.Y.

J.L. LIONS [1] Contrôle optimal des systèmes gouvernés par des équations aux dérivées partielles, Dunod, Paris, 1968

[2] Quelques méthodes de résolution de problèmes aux limites non linéaires Dunod, Paris, 1969

[3] Equations différentielles opérationnelles, Springer Verlag, Berlin, 1961

SOME BANACH-VALUED PROCESSES WITH APPLICATIONS+

T.E. DUNCAN

* Department of Applied Mathematics and Statistics,
State University of New York at Stony Brook,
Stony Brook, New York

1. Introduction

A generalization of the recursive filtering results of Kalman and Bucy [8] will be considered here where the stochastic processes are Banach valued. Some different features appear in contrast to the case where the processes are $\mathbb{R}^n$ valued, in particular a dense Hilbert space plays a fundamental role analogous to the notion of white noise when the processes are $\mathbb{R}^n$ valued. This notion of a Hilbert space for Gaussian measure was abstracted by Segal [11] and Gross [6] obtained conditions for the countable additivity of measures whose cylinder set measures were the canonical normal distribution. The canonical normal distribution has a formal rotational invariance property analogous to the fundamental solution of the heat equation on $\mathbb{R}^n$. A Gaussian measure on a Banach space will be characterized as a measure such that for any finite collection of continuous linear functionals on the Banach space the real valued random variables obtained have a joint Gaussian distribution.

+ Research partially supported by
National Science Foundation
Grant GK-32136

2. Preliminaries

Given a Gaussian cylinder set measure on a Hilbert space. To extend this cylinder set measure to a regular Borel measure on the completion of the Hilbert space with respect to a continuous seminorm, the following notion of measurable seminorm is fundamental. For the canonical normal distribution it would be sufficient to use a simpler notion of Gross [6] but the following definition [1] will indicate some possible generalizations. As some preliminaries to the definition a dual system or duality over the reals is two vector spaces X and Y and a bilinear form $\langle\cdot,\ \cdot\rangle : X \times Y \to \mathbb{R}$ that separates points for both spaces. FD(X) denotes the collection of finite dimensional subspaces of X, M(Y,G) denotes the smallest σ-algebra on Y for which all the elements of $G \subset X$ are measurable and $C(Y,X) = \cup\{M(Y,G):G\varepsilon FD(X)\}$. A cylinder set measure on Y is a non negative finitely additive set function, m, on C(Y,X) with m(Y) = 1 and countably additive on M(Y,G) for each fixed $G\varepsilon FD(X)$. Now the definition of measurable seminorm :

Definition : Given a duality $\langle X,Y\rangle$ and a cylinder set measure m on Y, a seminorn $|.|$ on Y is called m-measurable if for each $\varepsilon > o$ there is a $G\varepsilon FD(Y)$ such that if $F\varepsilon FD(X)$ and $F\perp G$ then

$$m(\{y : | y - F^{\perp}| < \varepsilon \}) > 1 - \varepsilon$$

or equivalently

$$m^*(\{y : | y - G | < \varepsilon \}) > 1 - \varepsilon$$

where m* is the outer measure induced from m by the algebra C(Y,X).

Let $|.|_1$ be a continuous norm on the separable Hilbert space H_1. Consider a Gaussian cylinder set measure μ and the family of canonical normal distributions $(\rho_t)_{t\varepsilon \mathbb{R}_+}$ on H_1 indexed by the variance parameter. Assume that the norm $|.|_1$, is μ-measurable and $\rho.$-measurable. (A seminorm instead of a norm could have been considered with the obvious changes in the subsequent discussion). A regular Borel measure, μ_{X_0}, which is the extension of the cylinder set measure μ, exists on the Banach space $\mathbb{B}$, which is the completion of H_1 with respect to $|.|_1$[1, 6]. Likewise there is a (version of the) $\mathbb{B}_1$-valued Brownian motion with continuous sample functions [7] whose cylinder set measures are the family of canonical normal distributions $(\rho_t)_{t\varepsilon \mathbb{R}_+}$ on H_1. Let $|.|_2$ be a continuous norm on a separable Hilbert space H_2 that is measurable with respect to the family of canonical normal distributions $(\eta_t)_{t\varepsilon \mathbb{R}_+}$ on H_2. Thus there is a $\mathbb{B}_2$-valued Brownian motion with continuous sample functions where $\mathbb{B}_2$ is the completion of H_2 with respect to $|.|_2$. All σ-algebras used in the subsequent discussion will be assumed to be completed with respect to the relevant probability measures.

The model for the stochastic filtering will be characterized in terms of two vector-valued stochastic differential equations. The process $(X_t)_{t \in \mathbb{R}_+}$ will be the state and the process $(Y_t)_{t \in \mathbb{R}_+}$ will be the observations.

$$dX_t = i_1\, a(t) X_t dt + dB_1(t) \tag{1}$$

$$dY_t = i_2\, c(t) X_t dt + dB_2(t) \tag{2}$$

with initial conditions X_o and Y_o where X_o is a Gaussian random variable independent of the Brownian motions B_1 and B_2 and has zero mean and covariance operator $\Gamma_o \in L(H_1, H_1)$ (the family of continuous linear maps from H_1 to H_1) and $Y_o \equiv 0$. The linear maps a and c are assumed to be continuous in the uniform operator topology and

$$\begin{aligned} a(t) &: \mathbb{B}_1 \to H_1 \\ c(t) &: \mathbb{B}_1 \to H_2 \\ i_1 &: H_1 \to \mathbb{B}_1 \\ i_2 &: H_2 \to \mathbb{B}_2 \end{aligned}$$

Lemma The solution of the stochastic differential equation (1) exists and is unique for all intervals $[o,t]$ where $t < \infty$.

Proof : The verification follows by the usual technique of Picard iteration using the fact that there is some $\alpha > o$ such that

$$E\, e^{\alpha |X_o|_1^2} < \infty$$

(Fernique [4]). ∎

Remark : The processes $(X_t)_{t \in \mathbb{R}_+}$ and $(Y_t)_{t \in \mathbb{R}_+}$ can be easily shown to be Banach-valued Gaussian processes, for example, by finite sum approximations to the integrals and then passage to the limit. Since the drift terms in the stochastic differential equations are Hilbert space valued the Hilbert spaces associated with the processes $(X_t)_{t \in \mathbb{R}_+}$ and $(Y_t)_{t \in \mathbb{R}_+}$ are H_1 and H_2 respectively.

In the next result an expression for a Radon-Nikodym derivative is obtained which generalizes a similar result in the finite dimensional case [2]. It explicitly indicates the interplay between the Hilbert and the Banach spaces. The probability space will be considered to be the space of Banach-valued continuous functions.

Proposition : Let μ_Y and μ_{B_2} be the measures on the space of $\mathbb{B}_2$-valued continuous functions with the Borel σ-algebra for the processes $(Y_t)_{t\in[0,T]}$ and $(B_2(t))_{t\in[0,T]}$ where $T < \infty$.

Then μ_Y is mutually absolutely continuous with respect to μ_{B_2} and the Radon-Nikodym derivative is

$$\frac{d\mu_Y}{d\mu_{B_2}} = E_{\mu_X} \exp\left[\int_0^T (c_s X_s, dY_s) - \frac{1}{2}\int_0^T |c_s X_s|_{H_2}^2 \, ds\right]$$

$$= \exp\left[\int_0^T (c_s \hat{X}_s, dY_s) - \frac{1}{2}\int_0^T |c_s \hat{X}_s|_{H_2}^2 \, ds\right] \qquad (3)$$

where E_{μ_X} denotes integration with respect to the measure μ_X for the process $(X_t)_{t\in[0,T]}$

$$\hat{X}_t = E[X_t \mid Y_u \; u < t] = \frac{E_{\mu_X} X_t M_t}{E_{\mu_X} M_t} \qquad (4)$$

and Y in equation (3) has the μ_{B_2} distribution and $M_t = E\left[\frac{d\mu_Y}{d\mu_{B_2}} \mid Y_u, u \le t\right]$ and $|.|_{H_2}$ is the norm for the Hilbert space H_2.

Proof : Let μ_{XY} and μ_{XB_2} be the measures on the space of $\mathbb{B}_1 \times \mathbb{B}_2$ valued continuous functions with the Borel σ-algebra completed with respect to μ_{XB_2} for the processes $(X_t, Y_t)_{t\in[0,T]}$ and $(X_t, B_2(t))_{t\in[0,T]}$ respectively. Using an extension [3] of a result of Girsanov [5] it follows that $\mu_{XY} << \mu_{XB_2}$ (and also $\mu_{XB_2} << \mu_{XY}$) and the Radon-Nikodym derivative is

$$\frac{d\mu_{XY}}{d\mu_{XB_2}} = \exp\left[\int (c_s X_s, dB_2(s)) - \frac{1}{2}\int | c_s X_s |_{H_2}^2 ds\right]$$

where $|.|_{H_2}$ is the norm for the Hilbert space H_2.

Since μ_{XY} and μ_{XB_2} are Gaussian measures the above absolute continuity can be easily verified i.e. the Radon-Nikodym derivative formally transforms B_2 into Y by the usual transformation of measures and since the processes are Gaussian if the transformation works "locally" then there is absolute continuity by the dichotomy result for Gaussian measures [11].

Since the processes X. and $B_2(.)$ are independent

$$\frac{d\mu_Y}{d\mu_{B_2}} = E_{\mu_X}\left[\frac{d\mu_{XY}}{d\mu_{XB_2}}\right]$$

where E_{μ_X} denotes integration with respect to the measure μ_X. Applying the formula for change of variables for an arbitrary continuous local martingale [9]

$$\frac{d\mu_Y}{d\mu_{B_2}} = E_{\mu_X}\left[1 + \int M_s(c_s X_s, dB_2(s))\right]$$

where

$$M_t = \exp\left[\int_0^t (c_s X_s, dB_2(s)) - \frac{1}{2}\int_0^t |c_s X_s|_{H_2}^2 ds\right]$$

Let $f : \mathbb{B}_1 \to \mathbb{B}_3$ be a Borel measurable function and $\mathbb{B}_3$ be a separable Banach space. Assume that $f(X_t)$ is Bochner integrable. Then

$$\hat{f}(X_t) = E[f(X_t)|Y_u, u \leq t]$$

exists and

$$\hat{f}(X_t) = \frac{E_{\mu_X} f(X_t) M_t}{E_{\mu_X} M_t} \qquad \text{a.s } \mu_{XY} \qquad (5)$$

The existence of the conditional expectation follows from Scalora (Theorem 2.1 [10]). To give the characterization (5) for the conditional expectation the absolute continuity that has been obtained will be used. Let $\Lambda \varepsilon B(Y_u\ u < t)$ (the augmented smallest σ-algebra for which the random variables Y_u $u < t$ are measurable).

$$\int_{\mathbb{B}_1}\int_{\Lambda} f(X_t)\,d\mu_{XY} = \int_{\mathbb{B}_1}\int_{\Lambda} E\left[f(X_t) \mid Y_u,\ u < t\right] d\mu_{XY}$$

$$= \int_{\mathbb{B}_1}\int_{\Lambda} f(X_t) M_t\, d\mu_X\, d\mu_{B_2}$$

$$= \int_{\mathbb{B}_1}\int_{\Lambda} E\left[f(X_t) \mid Y_u,\ u \le t\right] M_t\, d\mu_X\, d\mu_{B_2}$$

$$= \int_{\Lambda} E\left[f(X_t) \mid Y_u\ u \le t\right] E_{\mu_X} M_t\, d\mu_{B_2}$$

$$E\left[f(X_t) \mid Y_u,\ u \le t\right] = \frac{E_{\mu_X}\left[f(X_t)M_t\right]}{E_{\mu_X}\left[M_t\right]} \qquad \text{a.s } \mu_Y$$

Assume initially that the process $(c_t X_t)_{t\varepsilon[o,T]}$ is $\mathbb{R}^n$-valued. By the result for finite dimensional Brownian motion [2]

$$E_{\mu_X} \frac{d\mu_{XY}}{d\mu_{XB_2}} = \exp\left[\int_o^T (c_s\hat{X}_s,\ dY_s) - \frac{1}{2}\int_o^T |c_s\hat{X}_s|^2_{H_2}\, ds\right]$$

Conditional expectation commutes with continuous linear transformations for Bochner integrable random variables (Theorem 2.3 [10]).

Consider now the general case. Since for some $\alpha > o$ $Ee^{\alpha|X.|_1^2} < \infty$ it easily follows that for $t < \infty$

$$\int_o^t E|X_s|_1^2\, ds < \infty$$

and also that

$$\int_o^t E|c_s X_s|^2_{H_2}\, ds < \infty$$

By the monotone convergence theorem there is a sequence, (P_m), of finite dimensional projections on H_2 such that

$$\int E\,|(I - P_n)\, c_s X_s|^2_{H_2}\, ds \to o \qquad \text{as } n \to \infty$$

Let $M.^{(n)}$ be the Radon-Nikodym derivative for the process $Y.^{(n)}$ with the drift term P_n c. X. . By Doob's inequality $M.^{(n)} \to M.$ uniformly in t in probability and thus

$$\int |P_n c_s \hat{X}_s^{(n)} - c_s \hat{X}_s|_{H_2}^2 \, ds \to o \qquad \text{a.s.}$$

Combining these facts

$$\exp\left[\int (P_n c_s \hat{X}_s^{(n)}, dB_2(s)) - \frac{1}{2}\int |P_n c_s \hat{X}_s^{(n)}|_{H_2}^2 \, ds\right]$$

$$\to \exp\left[\int (c_s \hat{X}_s, dB_2(s)) - \frac{1}{2}\int |c_s \hat{X}_s|_{H_2}^2 \, ds\right]$$

in probability.
Since $M_t^{(n)} \to M_t$ in $L^1(P)$ for each t

$$E_{\mu_X} M_t^{(n)} \to E_{\mu_X} M_t \quad \text{in } L^1(P) \text{ by}$$

Fubini's theorem. Therefore

$$E_{\mu_X} M_t = \exp\left[\int_o^t (c_s \hat{X}_s, dB_2(s)) - \frac{1}{2}\int_o^t |c_s \hat{X}_s|_{H_2}^2 \, ds\right] \qquad \blacksquare$$

3. Main Result

The following main result gives the conditional mean and the conditional covariance for the filtering problem in terms of differential equations.

Theorem : Consider the processes described by (1) and (2). The conditional mean, $(\hat{X}_t = E[X_t \mid Y_u \; u \le t])$, and the conditional covariance, (P_t), exist as a $\mathbb{B}_1$-valued process and as a self adjoint linear operator on H_1 respectively. Let $\ell, \ell_1, \ell_2 \in H_1$, then the conditional mean and the conditional covariance satisfy the following equations

$$d(\ell, \hat{X}_t) = (\ell, i_1 a_t \hat{X}_t) \, dt + (c_t i_1 P_t \ell, dY_t - c_t \hat{X}_t \, dt) \tag{6}$$

$$\frac{d}{dt}(\ell_1, P_t \ell_2) = (\ell_1, a_t i_1 P_t \ell_2) + (\ell_1, P_t j_1 a_t^* \ell_2) - (\ell_1, P_t j_1 c_t^* c_t i_1 P_t \ell_2) + (\ell_1, I \ell_2) \tag{7}$$

where $i_1 : H_1 \to \mathbb{B}_1$, $j_1 = i_1^*$,

$$\hat{X}_o = o \text{ , and } P_o = \Gamma_o$$

Proof : The existence of the conditional mean and the conditional covariance follow from the norm integrability of X. [4] and (Theorem 2.1 [10]). Since X. exists it is sufficient to take a countable number of linear functionals that separate points in $\mathbb{B}_1$ (by the Hahn-Banach theorem) to identify $\hat{X}$.

Consider

$$(\ell, \hat{X}_t) = \frac{E_{\mu_X}(\ell, X_t)M_t}{E_{\mu_X} M_t}$$

Define

$$\overline{M}_t = \frac{M_t}{E_{\mu_X} M_t}$$

$$(\ell, \hat{X}_t) = E_{\mu_X} (\ell, X_o + \int_o^t i_1 a_s X_s ds + B_1(t)) (1 + \int_o^t \overline{M}_s (c_s X_s - c_s \hat{X}_s, dY_s - c_s \hat{X}_s ds) \qquad (8)$$

Since X_o is independent of Y. and has zero mean it can be neglected in the subsequent calculations. Applying the formula for change of variables for an arbitrary continuous local semimartingale [9] the conditional mean can be expressed as

$$(\ell, \hat{X}_t) = E_{\mu_X}\left[\int_o^t (\ell, i_1 a_s X_s)\overline{M}_s ds + \int_o^t \overline{M}_s (\ell, dB_1(s)) + \int_o^t (\ell, X_s)\overline{M}_s (c_s X_s - c_s \hat{X}_s, dY_s - c_s \hat{X}_s ds) \right]$$

Certain interchanges of integration will now be considered. The following interchanges of integration can be justified by Fubini's theorem.

$$E_{\mu_X} \int_o^t (\ell, i_1 a_s X_s)\overline{M}_s ds = \int_o^t E_{\mu_X} (\ell, i_1 a_s X_s)\overline{M}_s ds$$

$$= \int_o^t (\ell, i_1 a_s \hat{X}_s) ds$$

$$E_{\mu_X} \int_0^t (\ell, X_s)\overline{M}_s\ (c_s X_s - c\hat{X}_s,\ c_s \hat{X}_s)\ ds$$

$$= \int_0^t E_{\mu_X}\ (\ell, X_s)\overline{M}_s\ (c_s X_s - c_s \hat{X}_s,\ c_s \hat{X}_s)\ ds$$

$$= \int_0^t E_{\mu_X}\ (\ell, X_s)\overline{M}_s\ (X_s - \hat{X}_s,\ c_s^{*} c_s \hat{X}_s)\ ds$$

$$= \int_0^t (\ell,\ P(s) j_1 c_s^{*} c_s \hat{X}_s)\ ds$$

where letting $\tilde{X}_s = X_s - \hat{X}_s$ P is defined for $\ell_1,\ \ell_2 \varepsilon H_1$ as

$$(\ell_1,\ P(s)\ell_2) = E_{\mu_X}(\ell_1,\ \tilde{X}_s)(\ell_2, \tilde{X}_s)\overline{M}_s$$

$$= E(\ell_1, \tilde{X}_s)(\ell_2, \tilde{X}_s)$$

The last inequality follows because X. and Y. are Gaussian processes.

Only the stochastic integrals remain to justify interchanges of integration.

Define a sequence of stopping times (T_n) as

$$T_n = \inf\{t : |X_t|_1 > n\}$$

$$= T \text{ if the above set is empty}$$

Let $X_t^{(n)} = X_{t \wedge T_n}$ then it easily follows that

$$\int_0^T E|c_s X_s^{(n)} - c_s X_s|_{H_2}^2\ ds \to 0 \qquad \text{as } n \to \infty$$

since

$$\int_0^T E|c_s X_s|_{H_2}^2\ ds < \infty$$

Let $M_s^{(n)}$ be the Radon-Nikodym derivative for the process with drift term $c.X.^{(n)}$. It is known [2,5] that

$$E|M_t^{(n)}|^2 < \infty$$

Therefore

$$E_{\mu_X} \int_0^t \overline{M}_s^{(n)}\ (\ell, dB_1(s)) = 0 \qquad \text{a.s } \mu_Y$$

$$E_{\mu_X}\int_0^t (\ell,X_s^{(n)})\overline{M}_s^{(n)}\,(c_s\tilde{X}_s^{(n)}\,,\,dY_s)$$

$$=\int_0^t E_{\mu_X}\,(\ell,X_s^{(n)})\overline{M}_s^{(n)}(c_s\tilde{X}_s^{(n)},\,dY_s)$$

$$=\int_0^t E_{\mu_X}\,(\ell,\tilde{X}_s^{(n)})\overline{M}_s^{(n)}\,(c_s\tilde{X}_s^{(n)},\,dY_s)$$

$$=\int_0^t E_{\mu_X}\,(\ell,\tilde{X}_s^{(n)})\overline{M}_s^{(n)}\,\sum_i\,(\gamma_i,c_s\tilde{X}_s^{(n)})(\gamma_i,dY)$$

where (γ_i) is an orthonormal basis in H_2.

Since the stochastic integral is continuous with respect to $L^2(P)$ convergence we have

$$E_{\mu_X}\int_0^t (\ell,\tilde{X}_s^{(n)})\overline{M}_s^{(n)}\,(c_s\tilde{X}_s^{(n)},\,dY_s)$$

$$=\int_0^t \sum_i\,(j_1c_s{}^*\gamma_i,\,P_s^{(n)}\ell)(\gamma_i,dY)$$

$$=\int_0^t (c_s i_1\,P_s^{(n)}\ell,\,dY_s)$$

As $n\to\infty$ by (8) this last integral converges in $L^1(\mu_{XY})$ to

$$\int_0^t (c_s i_1 P_s\,\ell,\,dY_s)$$

Combining these results the equation for the conditional mean is obtained.

Consider now the conditional covariance operator. Let $\ell_1,\ \ell_2,\ \varepsilon H$

$$E(\ell_1,\tilde{X}_t)(\ell_2,\tilde{X}_t) = E\{\left[(\ell_1,X_o + \int_0^t i_1 a_s\tilde{X}_s ds + B_1(t)) - \int_0^t (c_s i_1 P_s\ell_1,dY_s - c_s\tilde{X}_s ds)\right]$$

$$\left[(\ell_2,\,X_o + \int_0^t i_1 a_s\tilde{X}_s ds + B_1(t)) - \int_0^t (c_s i_1 P_s\ell_2,dY_s - c_s\hat{X}_s ds)\right]\}$$

$$= (\ell_1,P_t\ell_2)$$

Applying the formula for change of variables and suitably grouping terms

$$E(\ell_1,\tilde{X}_t)(\ell_2,\tilde{X}_t) = E\Bigg[\int_0^t (\ell_1, i_1 a_s \tilde{X}_s)(\ell_2,\tilde{X}_s)ds$$

$$+ \int_0^t (\ell_2, i_1 a_s \tilde{X}_s)(\ell_1,\tilde{X}_s)ds + \int_0^t (\ell_2,\tilde{X}_s)(\ell_1, dB_1(s))$$

$$+ \int_0^t (\ell_1,\tilde{X}_s)(\ell_2, dB_1(s)) + \int_0^t (I\ \ell_1,\ell_2)\ ds$$

$$- \int_0^t (\ell_2,\tilde{X}_s)(c_s i_1 P_s \ell_1,\ c_s\tilde{X}_s ds + dB_2(s))$$

$$- \int_0^t (\ell_1,\tilde{X}_s)(c_s i_1 P_s \ell_2,\ c_s\tilde{X}_s ds + dB_2(s))$$

$$+ \int_0^t (c_s i_1 P_s \ell_1, c_s i_1 P_s \ell_2)ds + (\ell_1, X_o)(\ell_2, X_o)\Bigg]$$

The stochastic integrals are easily verified to be square integrable and using Fubini's theorem

$$(\ell_1, P_t \ell_2) = \int_0^t (j_1 a_s^* \ell_1, P_s \ell_2)\ ds$$

$$+ \int_0^t (\ell_1, P_s j_1 a_s^* \ell_2)\ ds$$

$$+ \int_0^t (I\ \ell_1,\ell_2)\ ds$$

$$- \int_0^t (c_s i_1 P_s \ell_1, c_s i_1 P_s \ell_2)\ ds$$

$$+ (\ell_1,\ \Gamma_o\ \ell_2)$$

REFERENCES

1. R.M. Dudley, J. Feldman and L. LeCam, On seminorms and probabilities, and abstract Wiener spaces, Ann. of Math., 93 (1971) 390-408.

2. T.E. Duncan, Evaluation of likelihood functions. Information and Control 13 (1968) 62-74.

3. T.E. Duncan, Transforming Frechet-valued Brownian motion by absolute continuity of measures. to appear

4. X. Fernique, Intégrabilité des vecteurs gaussiens, C.R. Acad. Sc. Paris Ser. A 270 (1970) 1698-1699.

5. I.V. Girsanov, On transforming a certain class of stochastic processes by absolutely continuous substitution of measures, Theor. Probability Appl. 5 (1960) 285-301.

6. L. Gross, Abstract Wiener spaces, Proc. Fifth Berkeley Symposium on Math. Stat. and Prob. University of California Press 1 (1965) 31-42.

7. L. Gross, Potential theory on Hilbert space, J. Functional Analysis, 1 (1967) 123-181.

8. R.E. Kalman and R.S. Bucy, New results in linear filtering and prediction theory, J. Basic Eng. ASME Ser. D 83 (1961) 95-108.

9. H. Kunita and S. Watanabe, On square integrable martingales, Nagoya Math. J. 30 (1967)

10. F.S. Scalora, Abstract martingale convergence theorems, Pacific J. Math. 11 (1961) 347-374.

11. I.E. Segal, Distributions in Hilbert space and canonical systems of operators, Trans. Amer. Math. Soc. 88 (1958) 12-41.

STOCHASTIC STABILITY

H.J. KUSHNER

Center for Dynamical Systems
Division of Applied Mathematics
Brown University
Providence, Rhode Island 02912

INTRODUCTION

In this paper, we will discuss and prove some of the basic results in the theory of stochastic stability for systems governed by continuous time Markov processes. Our concern will be mainly with the asymptotic behavior of the paths of the process. The development will be along the lines of [1], [2]. A detailed and introductory discussion of stochastic stability for discrete parameter systems appears in [3]. In fact, [3] contains introductory discussions of asymptotic stability, the invariance theorems, the existence of (and convergence of the measures of the process to) invariant measures, and a number of examples.

Next, we give some definitions, then mention some of the problems with which stochastic stability deals. Then we give a brief introduction to some deterministic results, and discuss some of the probabilistic structures to be used in the sequel. Then some results on asymptotic stability w.p.1. will be discussed and proved, and an invariance theorem proved. Finally we give two examples, one dealing with a non-linear diffusion, and the other with a problem arising in the identification of the parameters of a linear differential equation. The paper will be as self contained as space permits. For the most part, the discussion will concern the case where the transition functions are homogeneous - since statements and proofs are notationally simpler there, but some results for the non-homogeneous case will also be stated.

Stochastic stability is a long way from being a mature subject, even from the theoretical point of view. Let u_t denote a Markov process. Then under some suitable condition on $f(\cdot)$, the process $x_t = (y_t, u_t)$ where y_t

given by $\dot{y} = f(y,u)$ is a Markov process, and we may desire to investigate whether $y_t \to 0$ w.p.l. In many applications u_t is stationary, or at least its paths do not converge. Yet the Liapunov functions must take both u_t and y_t into account. It is not clear what the appropriate theorems are for such cases, nor is it understood how to find (even in relatively simple cases) useful Liapunov functions. It is hoped that a combination of the stability results of part I, and the invariant set results of part II will be helpful here, since the "derivatives" of the Liapunov functions will probably be semidefinite in such cases. A relatively simple example is given in Example 2 in part III. Investigations into further possibilities are continuing.

I. STOCHASTIC STABILITY

1. Markov and Strong Markov Processes

For our purposes a Markov process can be defined in the following way. Let X be a topological space (called the state space) and $\Omega, \mathscr{B}$ and $P_x\{\cdot\}$, a sample space, a σ-algebra on Ω and family of probability measures on $(\Omega,\mathscr{B})$ (for each $x \in X$ there is a measure $P_x\{\cdot\}$). Let x_t, $t \geq 0$ be a family of random variables from $(\Omega,\mathscr{B})$ to X with $\mathscr{B}_t \subset \mathscr{B}$ being the least σ-algebra which measures x_s, $x \leq t$. Let there exist a real valued function $P(\cdot,\cdot,\cdot)$ (called the transition function) on $X \times [0,\infty) \times \mathscr{B}(X)$, where $\mathscr{B}(X)$ is the σ-algebra on X which is induced by the topology on X. Let $P(\cdot,t,\Gamma)$ be measurable on $\mathscr{B}(X)$ for each $t \geq 0$ and $\Gamma \in \mathscr{B}(X)$, and let $P(x,t,\cdot)$ be a probability measure on $\mathscr{B}(X)$. Furthermore, let $P(x,t,\Gamma) = P_x\{x_t \in \Gamma\}$ and

$$(1) \qquad P_x\{x_{t+s} \in \Gamma | \mathscr{B}_s\} = P(x_t,s,\Gamma) \quad \text{w.p.l.}$$

for each $s \geq 0$, $t \geq 0$, $x \in X$, $\Gamma \in \mathscr{B}(X)$. Then we say that $(\Omega,P_x,\mathscr{B}_t,\mathscr{B},x_t)$ (or simply that the process x_t) is a homogeneous Markov process.

Note: By (1) we can write (1) as $P_{x_s}\{x_t \in \Gamma\}$. The subscript x_s denotes that the initial condition is a random variable with the distribution of x_s. The argument $x_t \in \Gamma$ indicates the event whose probability is being written, under the given initial condition.

The definition of a non-homogeneous Markov process is similar - but we must keep track of two time indices - rather than one. Then we define the transition function $P(\cdot,\cdot;\cdot,\cdot)$ by $P_{x,t}\{x_s \in \Gamma\} = P(x,t;t+s,\Gamma)$ (the probability that with initial condition x at time t, the process is in Γ s units of time later).

Eqn. (1) implies that $P(\cdot,\cdot,\cdot)$ satisfies the Chapman-Kolmogorov equation

(2) $$P(x,t+s,\Gamma) = \int P(x,t,dy)P(y,s,\Gamma), \quad s \geq 0,\ t \geq 0.$$

For our purposes, the Markov process is slightly too broad a class of processes, for the following reason. Let x_t be a Markov process whose state space is the real line. Define $\tau(\omega)$ by

$$\tau(\omega) = \inf\{t: x_t = b > 0\}$$

and suppose that $\tau(\omega) < \infty$ w.p.1. Then it is not necessarily true (even if the terms are well defined) that w.p.1.

(3) $$P_x\{x_{\tau+t} \in \Gamma | x_s,\ \ s \leq \tau\} = P_{x_\tau}\{x_t \in \Gamma\}.$$

(Observe that $x_\tau = b$ in the example.) In words - the distribution of the process t units after first hitting b, conditional upon all the path data up to the first hitting time, may depend on how we arrived at b and not simply on the fact that $x_\tau = b$. (For an example see Loeve [4], p. 578). Indeed, (3) should be true for a process arising in a physical application, and we will restrict our attention (without apparent loss of generality) to processes where (3) is true for a large class of random variables τ.

<u>Definition</u>. A non-negative random variable τ (defined on a set $\Omega_\tau \subset \Omega$) is called a <u>Markov time</u> if

$$\{\tau \le t\} \in \mathscr{B}_t;$$

i.e., τ is a Markov time if we can tell whether or not $\tau \le t$ by watching the process x_s up to time t only, for each $t \ge 0$.

Definition. If (3) holds for all Markov times τ, then x_t is said to be a strong Markov process.

Definition. We will consider only strong Markov processes. If x_t is a Markov process for which the function of x given by $E_x g(x_t)$ is continuous for each $t > 0$ and real valued continuous and bounded $g(\cdot)$, then x_t is a Feller process.

A Feller process whose paths are continuous from the right is a strong Markov process. ([5], Theorem 3.10).

Definition. Let B denote the Banach space of real valued bounded measurable functions on X, and B_0 the subset of B for which

$$E_x f(x_t) \to f(x), \quad f \in B_0$$

weakly as $t \to 0$. If the weak limit

$$\frac{E_x f(x_t) - f(x)}{t} \overset{w}{\to} g(x),$$

exists (i.e., there is pointwise convergence, and the left hand side is bounded as $h \to 0$) and is in B_0, we say that $g(\cdot)$ is in the domain $\mathscr{D}(\tilde{A})$ of the weak infinitesimal operator $\tilde{A}$, and write $\tilde{A}f = g$.

Suppose τ is a Markov time and $E_x\tau < \infty$, and x_t is right continuous w.p.1., and $f \in \mathscr{D}(\tilde{A})$, (the continuity conditions can be weakened; see [5], p. 133) then we have the important relation (4), known as Dynkins formula ([5], p. 133),

$$E_x f(x_\tau) - f(x) = E_x \int_0^\tau \tilde{A}f(x_s)ds. \tag{4}$$

The operator $\tilde{A}$ plays a role for Markov processes, similar to the role the differentiation operator plays for differentiable non-random real valued functions. Equation (4) is an analog of the deterministic integral - differential relationship and, as such, will play an important role in the sequel.

<u>The non-homogeneous case</u>. If x_t is non-homogeneous, or if we wish to apply (4) to functions $f(x,t)$ of both state and time, then we can proceed as follows. Define t to be a state of the process (replace X by $X \times [0,\infty)$). Redefine B, B_0 appropriately, let $f \in \mathscr{D}(\tilde{A})$ and $g = \tilde{A}f$ if $f(\cdot,\cdot) \in B$ and

$$\frac{E_{x,t}f(x_s,t+s) - f(x,t)}{s} \rightarrow g(x,t)$$

weakly as $s \rightarrow 0$, and $E_{x,t}g(x_s,t+s) \rightarrow g(x,t)$ weakly as $s \rightarrow 0$. Then under the conditions on τ in (4),

$$E_{x,t}f(x_\tau,t+\tau) - f(x,t) = E_{x,t}\int_0^\tau \tilde{A}f(x_s,t+s)ds. \tag{4'}$$

In (4') we understand that time is measured from the origin t; i.e., the value of x_s in (4') is the value of the state s units of time after the initial time t.

2. A Few Sources of Stochastic Stability Problems

Stochastic stability problems occur in almost all phases of physics, control theory, numerical analysis and economics where dynamical models subject to random disturbances appear, and the process is of interest over a long period of time. Only a few simple problem types will be mentioned here.

Suppose that y_t' is a Markov process which drives the differential equation $\dot{y}_t = f(y_t',y_t,\alpha)$, where α is a parameter. y_t' may represent an external driving term, or random variations in some parameter of the equation. We may be interested in the range of α for which $y_t \rightarrow 0$ w.p.1., or for which $|y_t|$ remains bounded in some statistical sense. For α fixed at α_0, we may be interested in the range of initial conditions y,y' for which (for some $1 > \delta > 0$)

$$P_{y,y'}\{\sup_{\infty > t \geq 0} |y_t| \geq \lambda\} \leq \delta.$$

E.g., y_t may represent a stress in a mechanical structure, and it may be of interest to keep the stress less than $\lambda > 0$. Also, it may be desirable to know whether

$$P_{y,y'}\{\sup_{\infty > t \geq 0} |y_t| \geq \lambda\} \to 0$$

as y or y' or both tend to zero (a type of stability of the origin w.p.1.).

The above stability properties are all properties of the paths of the processes. There are many problems of interest concerning the asymptotic behavior of the moments and of the measures of the process - and even in cases where the process is of interest for only a finite time [1], [3].

A large class of stability problems arise in tracking situations. For example, suppose that we are driving on a road and sample our instruments and errors (e.g., distance from the center of the lane and from other cars) somewhat irregularly (as is usually done), then can we track the center of the lane within a certain error, etc. Tracking problems arise in radar and machine tool systems.

Many types of stochastic convergence can be studied. Here we deal with w.p.1. convergence mainly. Other stability problems deal with (a) convergence w.p.1. to a set, (b) recurrence - the process always returns to a bounded set w.p.1., (c) no finite escape time w.p.1., (d) convergence or boundedness of certain moments of the process, (e) convergence of the distributions of the process to an invariant measure. Types (d-e) are usually more difficult to treat than w.p.1. convergence, but also are of considerable practical importance.

3. A Brief Review of Deterministic Stability

Some results in deterministic stability are briefly reviewed because, in a certain abstract sense, the stochastic results are analogies of the deterministic results.

Let R^r denote Euclidean r-space, $f(\cdot)$ a continuous function from R^r to R^r and suppose that there is a continuous solution to the homogeneous differential equation $\dot{x} = f(x)$. Let $V(\cdot)$ denote a continuous, non-negative, real valued, continuously differentiable function on R^r (whose gradient is denoted by $V_x(\cdot)$) satisfying $V(0) = 0$, $V(x) > 0$, for $|x| \neq 0$. Let the set defined by $Q_\lambda \equiv \{x: V(x) < \lambda\}$ be bounded with the derivative of $V(x_t)$ non-positive along trajectories in Q_λ; namely

$$\dot{V}(x_t) = V_x'(x_t)f(x_t) \equiv -k(x_t) \leq 0 \tag{5}$$

for $x_t \in Q_\lambda$. Let $x_0 = x$ be in Q_λ. The following statements can be made:

$V(x_t)$ is non-increasing. Then $x_t \in Q_\lambda$ for all $t \geq 0$. From

$$V(x) - V(x_t) = \int_0^t k(x_s)ds \geq 0 \tag{6}$$

we have that $\int_0^\infty k(x_s)ds < \infty$. This, and the uniform continuity of $k(x_s)$ on $[0,\infty)$ imply that $k(x_s) \to 0$ as $s \to \infty$, and $x_s \to \{x: k(x) = 0\} \cap Q_\lambda \equiv K_\lambda$.

Furthermore as $x \to 0$, the maximum excursions of $|x_t|$ decrease to zero.

Define an <u>invariant set</u> of points G in R^r as follows. Let $x \in G$. Then there is a function x_t, $t \in (-\infty,\infty)$ which satisfies the equation $\dot{x} = f(x)$ with $x_0 = x$, and furthermore $x_t \in G$ for all $t \in (-\infty,\infty)$. Thus G contains entire trajectories over the doubly infinite time interval $(-\infty,\infty)$.

Let the trajectory x_t be <u>bounded</u>. In particular, let $x_0 = x \in Q_\lambda$, and assume (5). Then the <u>invariance theorem</u> [7] states that the path tends to the largest invariant set contained in K_λ.

The theorem is important since it is often used to show that the x_t tend to a much smaller set than K_λ. It gives a very nice characterization of the sets to which x_t can tend.

Example. Define the differential equation on R^2,

$$\dot{x}_1 = x_2$$

$$\dot{x}_2 = -g(x_1) - ax_2$$

where

$$\int_0^t g(s)ds \to \infty \quad \text{as} \quad t \to \infty, \quad sg(s) > 0 \quad \text{for} \quad s \neq 0$$

$$g(0) = 0, \quad a > 0.$$

Define the Liapunov function

$$V(x) = x_2^2 + 2 \int_0^{x_1} g(s)ds.$$

Then

$$V_x'(x)f(x) = -k(x) = -2ax_2^2.$$

We can conclude that $x_{2t} \to 0$. But what about x_{1t}?

It is natural to expect that $x_{1t} \to 0$ also, and indeed (although the Liapunov function argument does not directly yield it) it can be proved using a limiting argument, using the facts that $V(x_t)$ is non-increasing and $x_{2t} \to 0$. Yet it would be much simpler to merely substitute $x_{2t} \equiv 0$ in the differential equation, and see what trajectories are possible; namely, put the limit of x_{2t} into the equation, directly. The invariance theorem allows us to do this, and to conclude that $x_{1t} \to 0$ also. In examples involving functional differential or more complicated systems, the invariance theorems can save an enormous amount of work.

In the sequel, we will develop stochastic counterparts of all the concepts which we just used. While homogeneity is required for the invariance theorem, there are straightforward non-homogeneous extensions of the Liapunov function theorems.

4. Stopped Processes

The weak infinitesimal operator $\tilde{A}$ and Dynkins formula (4) will be used to replace (5), (6) for the stochastic problem. The domain $\mathcal{D}(\tilde{A})$ was defined to be a subset of a set of bounded functions on X. However, the Liapunov functions $V(\cdot)$ which are most likely to be used, and to which $\tilde{A}$ is to be applied, are usually unbounded (as is usual in the deterministic case). Even if $V(\cdot)$ were bounded, the process may have a stability property only in a bounded or compact set Q in X. I.e., $\tilde{A}f(x)$ may be non-positive only in some neighborhood Q of the origin.

There is no loss of generality in studying the process only while the paths are in such a set Q. For we can often (always, if X is σ-compact) find a sequence of sets $Q_n \uparrow X$, and, if desired, study the behavior of X by studying the "limits" of the behavior of the process up to, say, τ_n, where $\tau_n = \inf\{t: x_t \notin Q_n\}$. Thus, we can bound $V(x)$ for x "sufficiently far" from Q, or we can define a new process by merely stopping x_t on first exit from Q. The latter approach is much more convenient.

Let Q be a set in X. Dynkin ([5], Chapter 4) gives various general conditions under which $\tau = \inf\{x_t \notin Q\}$ is a Markov time. We mention only the following. Let x_t be right continuous w.p.1.

(a) Q is open and has compact closure. (Lemma 4.1)

(b) Q is open, X is a metric space (or metrizable) and X - Q is compact. (Lemma 4.1)

(c) Q is open, and x_t is continuous. (p. 111)

Define the stopped process $\tilde{x}_t = x_{t\cap\tau}$, where $t \cap \tau = \min(t,\tau)$. Let either (a)-(c) above hold, and let x_t be right continuous w.p.1. Then x_t is a strong Markov process ([5], Theorem 10.2). Unless otherwise mentioned, $\tilde{A}_Q$ will be used to denote the weak infinitesimal operator of the process $\tilde{x}_t$. Let x_t be continuous w.p.1. Then to apply Dynkins formula to an unbounded function $V(\cdot)$, we only need check that the restriction of $V(\cdot)$ to Q is in $\mathcal{D}(\tilde{A}_Q)$. If x_t is right continuous w.p.1., we need to check whether the restriction of $V(\cdot)$

to the union over $x = x_0$ in Q of the almost sure range x_s, $s \leq \tau$, is in $\mathscr{D}(\tilde{A}_Q)$. Such verification usually seems to be straightforward in examples.

5. Stochastic Stability and Asymptotic Stability

Unless otherwise mentioned, we will use the following assumptions in this section. After the theorems are proved, extensions to more general cases will be discussed.

(A1) X is Euclidean r-space.

(A2) $V(\cdot)$ is a non-negative real valued and continuous function on R^r.

(A3) Define $Q_\lambda = \{x: V(x) < \lambda\}$ and assume that Q_λ is not empty. Let x_t denote a right continuous homogeneous strong Markov process on the state space X, defined until at least the first time of exit from Q_λ. Write $\tilde{A}_\lambda$ for the weak infinitesimal operator of $\tilde{x}_t$, where $\tau_\lambda = \inf\{t: x_t \notin Q_\lambda\}$, and $\tilde{x}_t = x_{t \cap \tau_\lambda}$.

(A4) $V(\cdot) \in \mathscr{D}(\tilde{A}_\lambda)$ (where the definition of $V(\cdot)$ is assumed restricted to the union over $x_0 = x$ in Q of the almost sure range of $\tilde{x}_t$).

(A5) $\sup_{x \in Q_\lambda} P_x\{\sup_{t \geq s \geq 0} ||x_s - x|| > \varepsilon\} \to 0$ as $t \to 0$ for any $\varepsilon > 0$.

Observe that, if $y \notin Q_\lambda$, but is in the almost sure range of x_s, $s \leq \tau_\lambda$ for some $x_0 = x \in Q_\lambda$, then $\tilde{A}_\lambda V(y) \equiv -k(y) = 0$. We will use this fact implicitly in the following theorems.

Theorem 1. Assume (A1)-(A4). Let $\tilde{A}_\lambda V(x) \leq 0$ (recall that the operation $\tilde{A}_\lambda V(x)$ is defined for the stopped process). Then $V(\tilde{x}_t)$ converges w.p.1., as $t \to \infty$. Hence $V(x_t)$ converges for almost all paths remaining in Q_λ. For $x \in Q_\lambda$,

$$(7) \qquad P_x\{\sup_{\infty > t \geq 0} V(x_t) \geq \lambda\} = P_x\{\sup_{\infty > t \geq 0} V(\tilde{x}_t) \geq \lambda\} \leq V(x)/\lambda.$$

If $V(0) = 0$ and $V(x) \neq 0$ for $x \neq 0$, then as $|x| \to 0$, the probability in (7) goes to zero (a type of stability of the origin).

Proof. Applying (4) gives

$$\text{(8)}\qquad E_x V(\tilde{x}_t) - V(x) = E_x \int_0^t \tilde{A}_\lambda V(\tilde{x}_s)ds = E_x \int_0^{t\cap\tau_\lambda} \tilde{A}_\lambda V(x_s)ds \le 0.$$

Thus, w.p.1.,

$$E_{\tilde{x}_s} V(\tilde{x}_t) \le V(\tilde{x}_s)$$

or, equivalently, since $\tilde{x}_s$ is Markov, ($\mathscr{B}_s$ is the smallest σ-algebra which measures x_r, $r \le s$)

$$E[V(\tilde{x}_{t+s}) \mid \mathscr{B}_s] \le V(\tilde{x}_s), \text{ w.p.1.}$$

Thus $\{V(\tilde{x}_t), \mathscr{B}_t\}$ is a non-negative super-martingale. This gives the convergence of $V(\tilde{x}_t)$. (7) is the super-martingale probability inequality. The rest of the statements are obvious. Q.E.D.

Non-homogeneous case. Suppose that the Liapunov function $V(\cdot,\cdot)$ depends on x and t or that x_t is non-homogeneous. We state the following Theorem 2, without proof.

Theorem 2. Let the real valued continuous functions (on R^r, $R^r \times [0,\infty)$, R^r, resp.) $V_1(\cdot)$, $V(\cdot,\cdot)$, $V_2(\cdot)$ satisfy, for some real $t_0 \ge 0$, and $\lambda > 0$,

$$V_1(x) \le V(x,s) \le V_2(x)$$

for $s \ge t_0$ and $x \in Q'_\lambda \equiv \{x: V_1(x) < \lambda\}$. Let x_t be a right continuous strong Markov process defined until at least the first exit time τ_λ from Q'_λ. Let $\tilde{A}_\lambda$ denote the weak infinitesimal operator of the process $(\tilde{x}_t, t\cap\tau_\lambda)$, which is (x_t, t) stopped on first exit from Q'_λ. Suppose $V(x,t) \in \mathscr{D}(\tilde{A}_\lambda)$ and $\tilde{A}_\lambda V(x,t) \le 0$ for $t \ge t_0$. Then, for $t \ge t_0$,

$$(9)\qquad P_{x,t_0}\{\sup_{\infty > s \geq 0} V_1(x_s) \geq \lambda\} \leq P_{x,t_0}\{\sup_{\infty > s \geq 0} V(x_s, s+t_0) \geq \lambda\}$$

$$\leq V(x,t_0)/\lambda.$$

Also $V(x_t, t+t_0)$ *converges for almost all paths for which* $V(x_t, t+t_0) \leq \lambda$ *for all* $t \geq 0$, *where we use* $x_0 = x$ *for the initial condition at the initial time* t_0; *thus there is convergence with at least probability* $1 - V(x,t_0)/\lambda$.

Let $V_2(x) \to 0$ *as* $|x| \to 0$; *then the right hand side of* (9) *goes to zero as* $|x| \to 0$. *Let also* $V_1(0) = 0$ *and* $V_1(x) > 0$ *for* $|x| \neq 0$; *then for any* $\varepsilon > 0$ *and any neighborhood of the origin* A_1, *there is a neighborhood* $A_2 \subset A_1$ *so that, if* $x \in A_2$, *the probability of* x_t *ever leaving* A_1 *is no greater than* ε.

Theorem 3. (*Asymptotic Stability*). *Assume* (A1)-(A5). *Let* $\tilde{A}_\lambda V(x) \leq -k(x) \leq 0$ *in* Q_λ. *Then* $k(\tilde{x}_t) \to 0$ *in probability and* $V(\tilde{x}_t)$ *converges* w.p.1. *Thus* $k(x_t) \overset{P}{\to} 0$ (*and also* $V(x_t)$ *converges*) *for almost all paths which never leave* Q_λ. (*Equation* (7) *gives a lower bound* $1-V(x)/\lambda$ *to the probability of never leaving* Q_λ.)

Let $k(\cdot)$ *be uniformly continuous in* Q_λ. *Then* $x_t \to [\bigcap_{\varepsilon>0} \{x: k(x) < \varepsilon\}] \cap Q_\lambda \equiv P_\lambda$ *for almost all paths which never leave* Q_λ. *If the hypotheses hold for all* $\lambda < \infty$ *and* $V(x) \to \infty$ *as* $|x| \to \infty$, *then* $x_t \to \bigcap_{\varepsilon>0} \{x: k(x) < \varepsilon\} = K$ w.p.1. *The convergence in the last two sentences is in the topology for the compactified* R^r, *if the* $Q_\lambda \cap \{x: k(x) < \varepsilon\}$ *are unbounded. If they are bounded, replace* $\bigcap_\varepsilon \{k(x) < \varepsilon\}$ *by* $\{x: k(x) = 0\}$.

Proof. The key to the proof is the fact that the total time which the process $\tilde{x}_s$ can spend outside of the set $K_\varepsilon = \{x: k(x) \geq \varepsilon > 0\} \cap Q_\lambda$ is finite w.p.1. for any $\varepsilon > 0$. This follows from the inequality

$$(10)\qquad V(x) \geq -E_x V(\tilde{x}_t) + V(x) = E_x \int_0^t k(\tilde{x}_s)\,ds \geq \varepsilon E_x T'(t,\varepsilon),$$

where $T_x'(t,\varepsilon)$ is the total time that $k(\tilde{x}_s) \geq \varepsilon$ in $[0,t]$. That $k(\tilde{x}_t) \overset{P}{\to} 0$ follows from (10).

We next prove the first statement of the second paragraph of the theorem. Let $T(t,\varepsilon)$ denote the total time that $k(\tilde{x}_s) \geq \varepsilon$ in $[t,\infty)$. Then $T(t,\varepsilon) \to 0$ w.p.1. for any $\varepsilon > 0$ and $x \in Q_\lambda$. The rest of the proof combines this fact with (A5), the uniform stochastic stability assumption, to yield the w.p.1. convergence. Let C_ε denote the set

$$C_\varepsilon = \{x: k(x) < \varepsilon\} \cap Q_\lambda.$$

Assume that $k(x) < 0$ for some $x \in Q_\lambda$, for otherwise the theorem is trivial. Then, by uniform continuity of $k(\cdot)$ in Q_λ, there is some ε_0 so that the distance between $Q_\lambda \cap C_\varepsilon^c$ and $C_{\varepsilon/2}$ (C_ε^c is the complement of C_ε) is positive (say $\geq \delta(\varepsilon)$) for $0 < \varepsilon < \varepsilon_0$, and $Q_\lambda \cap C_\varepsilon^c$ is not empty.

Define the Markov times σ_n, σ_n' (finite on sets Ω_n, Ω_n', resp.) as follows. (If σ_n or σ_n' is not defined at ω, set it equal to ∞ there.) $\sigma_0 = 0$, $\sigma_0' = \inf\{t: \tilde{x}_t \in C_{\varepsilon/2}\}$, $\sigma_1 = \inf\{t: \tilde{x}_t \in Q_\lambda \cap C_\varepsilon^c, t \geq \sigma_0'\}$, $\sigma_n' = \inf\{t: \tilde{x}_t \in C_{\varepsilon/2}, t \geq \sigma_{n-1}\}$, $\sigma_n = \inf\{t: \tilde{x}_t \in Q_\lambda \cap C_\varepsilon^c, t \geq \sigma_n'\}$, etc. There is some $\rho > 0$ so that

$$\sup_{x \in Q_\lambda} P_x\{\sup_{\rho \geq s \geq 0} |x_s - x| \leq \delta(\varepsilon)/2\} \geq 1/2. \tag{11}$$

Define

$$A_n = \{\omega: \tilde{x}_{\sigma_n + s} \in C_{\varepsilon/2}^c \cap Q_\lambda,\ 0 \leq s \leq \rho,\ \sigma_n < \infty\}.$$

If $\omega \in A_n$ infinitely often, then the total time out of $C_{\varepsilon/2} \cup Q_\lambda^c$ is infinite for the corresponding path $\tilde{x}_t(\omega)$. Then $\omega \in A_n$ only finitely often w.p.1. But $\Sigma I_{A_n} \to \infty$ w.p.1. if and only if ($\mathscr{B}_{\sigma_n}$ measures $\tilde{x}_s$, $s \leq \sigma_n$, thus all A_i, $i = 0,\dots,n-1$, are in $\mathscr{B}_{\sigma_n}$)

$$\Sigma P_x\{A_n | \mathscr{B}_{\sigma_n}\} \to \infty \quad \text{w.p.1.}$$

([5], p. 398-399), and, by the strong Markov property ($P_{x_t}\{A\} = 0$ for $t = \infty$)

$$(12)\qquad \Sigma P_x\{A_n|\mathscr{B}_{\sigma_n}\} \geq \Sigma P_{x_{\sigma_n}}\{\sup_{\rho \geq s \geq 0} |x_s - x| > \frac{\delta(\varepsilon)}{2}\} I_{\{\sigma_n < \infty\}}$$

$$\geq \frac{1}{2}\Sigma I_{\{\sigma_n < \infty\}} .$$

Thus $\sigma_n < \infty$ only finitely often w.p.1. The remaining statements of the theorem follow easily from what we have already proved. Q.E.D.

Discussion and Extensions

(1) It is not necessary that $k(\cdot)$ be continuous in Q_λ, nor even that $k(x) = 0$ anywhere. See the hypothesis and proof of [1], Theorem 2, Chapter 2. There are examples arising in control theory where $X = R^r$ has a hole in it, i.e., a target set S_1 is deleted, and $k(x) \geq 1$ for $x \notin S_1$. The set S_2 is absorbing, so there is a discontinuity of $k(\cdot)$ on the boundary of X.

(2) If the hypotheses of Theorem 3 hold for all λ and $V(x) \to \infty$ as $|x| \to \infty$, then $x_t \to \{x: k(x) = 0\}$ w.p.1.

(3) If $\tilde{A}_\lambda V(x) \leq -k(x) \leq 0$ in Q_λ, then $k(\tilde{x}_t) \overset{P}{\to} 0$ for any state space, provided that the Dynkins formula is valid for the $\tilde{x}_t$ process.

(4) If Q_λ is unbounded, (A5) may not hold. Furthermore $\{x: k(x) < \varepsilon\} \cap Q_\lambda$ may be unbounded. Suppose that (A5) holds if Q_λ is replaced by any compact subset S of X. Then, we can obtain the following. For any compact S and $\varepsilon > 0$, there is a random variable $\tau_{S,\varepsilon} < \infty$ w.p.1., so that $x_t \notin S - C_\varepsilon$ for $t \geq \tau_{s,\varepsilon}$. Thus $x_t \to \{x: k(x) = 0\} \cup \{\infty\}$ w.p.1. in the one-point-compactification topology of R^r.

Sometimes subsidiary conditions can be used to eliminate the point $\{\infty\}$. Refer to the next section for the definition of the terms "weakly bounded" and "invariant set". Let the measures of the process $\tilde{x}_t$ be weakly bounded, and let $k(\tilde{x}_t) \overset{P}{\to} 0$. Then x_t tends in probability to the support of the largest invariant set whose support is contained in $[\{x: k(x) = 0\} \cap Q_\lambda] \cup Q_\lambda^c$. Thus, $\tilde{x}_t$ tends in probability to the union of Q_λ^c and a subset of $\{x: k(x) = 0\} \cap Q_\lambda$.

The remarks and results for unbounded Q_λ are motivated by the stability problem for a process of the type $\dot{y} = f(u,y)$, where u_t, and the pair $(u_t,y_t) = x_t$ are Markov processes. The process u_t may serve as a time varying parameter, and not converge in any sense. We may be concerned with the convergence of the component y_t only, but the Liapunov function may depend on both components.

(5) If X is a metric space, the proof still goes through under (A2)-(A5), if we replace R^r in (A2) by a metric space X. It may be difficult to verify (A5) and the uniform continuity of $k(\cdot)$ in this case, and the closure of Q_λ will not usually be bounded. But it sometimes happens that if $x = x_0 \in Q_\lambda$, then the path $\tilde{x}_s$, $\tau_\lambda > s \geq 0$, is contained in a bounded subset of Q_λ w.p.l. Then Q_λ is "effectively" contained in a bounded subset, and if $k(\cdot)$ is uniformly continuous and (A5) holds on this subset, then the proof goes through. See [6] for a specific example. (A5) plays a crucial role in the proof (since we need to guarantee that $\tilde{x}_t$ does not jump (w.p.l.) from $C^c_{\varepsilon/2}$ to C^c_ε and back to $C^c_{\varepsilon/2}$ infinitely often in a total integrated time which is finite), and some form of uniform stochastic continuity condition is probably essential.

II. INVARIANT SET THEOREMS AND APPLICATIONS TO STOCHASTIC DYNAMICAL SYSTEMS

In this Section we will develop a stochastic theory of invariance analogous to the deterministic theory in [7], [8]. The main conclusion is that, under given conditions, the measures of the process x_t tend to an invariant set of measures, and that x_t tends to the closure of the support set of this set of measures in probability as $t \to \infty$.

Note that we are using the terms "invariance" and "invariant" according to their usage in the general theory of dynamical systems. The term has nothing to do with the stochastic notion of invariant measure. In this Section x_t will be a homogeneous strong Markov process. We essentially follow the development in [2], with some changes and corrections.

1. Definitions

Let X, the state space of the process x_t, be a separable metric space. Let ϕ denote the initial measure of the process; i.e., $P\{x_0 \in A\} = \phi(A)$. Let $m(t,\phi,\cdot)$ denote the measure induced on the Borel sets of X by the process at time t, with initial measure ϕ. The semigroup property[+]

$$m(t+s,\phi,\cdot) = m(t,m(s,\phi),\cdot)$$

holds.

Let $\mathcal{M}$ denote the space of probability measures on X. A sequence $\{\psi_n\}$ in $\mathcal{M}$ is said to converge weakly to ψ if $\int f(x)\psi_n(dx) \to \int f(x)\psi(dx)$ for every $f(\cdot)$ in C_X, the space of continuous bounded functions on X. We may abbreviate the convergence relation as $f[\psi_n] \to f[\psi]$. A set $M = \{\psi_\alpha\}$ in $\mathcal{M}$ is weakly bounded if, for each $\varepsilon > 0$, there is a compact set $K_\varepsilon \subset X$ for which $\psi_\alpha(X-K\ ) \le \varepsilon$ for all α. Define an ω-limit set[++] as a set $W(\phi)$ in $\mathcal{M}$ with the property: $\psi \in W(\phi)$ if there is a sequence $t_n \to \infty$ so that $f[m(t_n,\phi)] \to f[\psi]$ (ψ is a weak limit of a sequence of measures taken along the trajectory) for all $f(\cdot) \in C_X$. A set $M \subset \mathcal{M}$ is an invariant set if for each $\psi \in M$, there is a sequence of measures $m'(t,\cdot)$, for $t \in (-\infty,\infty)$ where $m'(0,\cdot) = \psi(\cdot)$, the initial measure, and $m(t,m'(s,\psi),\cdot) = m'(t+s,\cdot)$ for any $t \ge 0$ and $s \in (-\infty,\infty)$. Thus for each $\psi \in M$, there is a trajectory of measures defined for all $t \in (-\infty,\infty)$ and satisfying the law of motion of the process x_t and initial condition ψ. Let ψ be in $\mathcal{M}$. $x \in X$ is in the support set $S(\psi)$ of ψ if $\psi(N) > 0$ for each neighborhood of N of x. Similarly $S(Q) = \bigcup_{\psi \in Q} S(\psi)$ is the support set of a set Q in $\mathcal{M}$. The set $S(\psi)$ is closed, but $S(Q)$ is not necessarily closed. The process x_t is a Feller process if $E_x f(x_t)$ is continuous in x for $t > 0$ and $f(\cdot) \in C_X$.

[+] Occasionally for simplicity ϕ is written for $\phi(\cdot)$ and $m(s,\phi)$ for $m(s,\phi,\cdot)$ or $m(s,\phi(\cdot),\cdot)$.

[++] It is important to keep in mind that the ω-limit set is an ω-limit set of a trajectory of measures.

Next, the main theorem and a useful corollary will be given. Then the conditions of the theorem will be replaced by more easily verifiable conditions.

2. The Invariance Theorem

Theorem 4. Assume (B1)-(B3).

(B1) The trajectory $\{m(t,\phi),\ t \geq 0\}$ is weakly bounded.

(B2) For each $f(\cdot)$ in C_X, $f[m(t,\phi)]$ is continuous in t on any finite t $(t \geq 0)$ interval, uniformly in ϕ, for ϕ in any weakly bounded set.

(B3) $f[m(t,\phi)]$ is weakly continuous in ϕ for each fixed $t \geq 0$. [I.e., as $\phi_n \overset{w}{\rightarrow} \phi$, $f[m(t,\phi_n)] \rightarrow f[m(t,\phi)]$ for each $f(\cdot) \in C_X$ and each $t \geq 0$.]

Then $W(\phi)$ is a non-empty, weakly bounded, weakly compact invariant set and there is a sequence $\tilde{\psi}(t)$ in $W(\phi)$, $t \geq 0$, so that

$$f[m(t,\phi)] - f[\tilde{\psi}(t)] \rightarrow 0$$

for all $f(\cdot) \in C_X$, as $t \rightarrow \infty$.

Proof. According to Theorem 1, Section 1, Chapter 9 of [10], a sufficient condition for a sequence in $\mathcal{M}$ to have a weakly convergent subsequence is that it be weakly bounded. Thus $W(\phi)$ is not empty.

Let $\{\varepsilon_i\}$ denote a real sequence which tends to zero. By (A1), there are compact sets G_i so that $G_{i+1} \supset G_i$ and $m(t,\phi,X-G_i) \leq \varepsilon_i$, all $t \geq 0$. For each G_i, there is a countable family $\mathscr{F}_i'$ of continuous functions, defined on G_i, and dense in C_{G_i}. Each element of $\mathscr{F}_i'$ can be extended to a continuous function on X without increasing its norm (using the normality of the metric space and [9], Theorem 1.5.3). Let $\mathscr{F}_i$ denote the countable family of such extensions and $\mathscr{F} = \bigcup_i \mathscr{F}_i$. Write $G = \bigcup_i G_i$. Observe that, for any $f(\cdot) \in C_X$,

$$\int_G f(x)m(t,\phi,dx) = \int f(x)m(t,\phi,dx). \qquad (*)$$

(*) also holds for $m(t,\phi,\cdot)$ replaced by an element in the weak closure of $\{m(t,\phi,\cdot)\}$.

Let $m(t_n,\phi,\cdot)$ converge weakly to $\psi(\cdot)$ in the ω-limit set $W(\phi)$. Define the function $F_n(\cdot,\cdot)$ by

$$F_n(t,f) = \int f(x)m(t_n+t,\phi,dx).$$

If $t_n - T \geq 0$, then $F_n(t,f) = \int f(x)m(t+T,m(t_n-T,\phi),dx)$. Since $\{m(t_n-T,\phi)\}$ is weakly bounded, (B2) implies that $F_n(t,f)$ is continuous in t on $[-T,T]$ uniformly in n, for each $f(\cdot)$. Thus Ascoli's Theorem implies that there is a uniformly convergent subsequence on $[-T,T]$. By successive applications of the diagonal procedure, we can extract a subsequence (of t_n) for which $F_n(t,f)$ converges to a continuous function of t, $F(t,f)$ for each $f(\cdot) \in \mathscr{F}$, and uniformly on <u>any</u> compact $[-T,T]$ interval. Since, for any $f(\cdot) \in C_X$ and $\varepsilon > 0$, there is an $f_\varepsilon(\cdot)$ in $\mathscr{F}$ for which $|F_n(t,f_\varepsilon) - F_n(t,f)| < \varepsilon$ for all n and $t \geq -t_n$, the asserted convergence is for all $f \in C_X$.

Define the set function $\psi(t,\cdot)$ by

$$\psi(t,A) = \inf_{f \geq I_A} F(t,f)$$

where $f(\cdot) \in C_X$, and I_A is the indicator function of the Borel set A in X.

The argument in [10], pp. 441-444, can be used to prove that, for each $t \in (-\infty,\infty)$, $\psi(t,\cdot)$ is a unique probability measure, $\psi(t,G) = 1$ and

$$F(t,f) = \int f(x)\psi(t,dx)$$

for each $f(\cdot) \in C_X$. Thus $m(t_n+t,\phi,\cdot) \xrightarrow{w} \psi(t,\cdot)$ for each $t \in (-\infty,\infty)$, where $\psi(0,\cdot) = \psi(\cdot)$. The weak closure of $\{m(t,\phi,\cdot)\}$ is also weakly bounded and is supported in G. Thus, by (B3), we can write, for any $t \in (-\infty,\infty)$, $s \geq 0$, and $f(\cdot)$ in C_X,

$$f[m(s,m(t_n+t,\phi))] \to f[m(s,\psi(t))]$$

$$f[m(s,m(t_n+t,\phi))] = f[m(0,m(t_n+t+s,\phi))]$$

$$\to f[m(0,\psi(t+s))] = f[\psi(t+s)],$$

which implies that $\psi(t+s,\cdot) = m(s,\psi(t),\cdot)$ since the continuous functions determine the measures uniquely. Thus $\{\psi(t)\}$ obeys the law of the process and each $\psi(t)$ is in an invariant set.

Let $f[\psi_n(\cdot)]$ converge for each $f(\cdot) \varepsilon C_X$, as $n \to \infty$, where $\psi_n(\cdot) \varepsilon W(\phi)$ (thus $\{\psi_n\}$ are weakly bounded). There is a measure $\psi(\cdot)$ for which $f[\psi_n(\cdot)] \to f[\psi(\cdot)]$ on C_X, and $\psi(G) = 1$. We need to show that $\psi(\cdot) \varepsilon W(\phi)$. For each n, $m(t_i(n),\phi,\cdot) \overset{w}{\to} \psi_n(\cdot)$ as $i \to \infty$, for some real sequence $t_i(n) \to \infty$. Since

$$\lim_n \lim_i \int f(x)m(t_i(n),\phi,dx) = \lim_n \int f(x)\psi_n(dx)$$
$$= \int f(x)\psi(dx),$$

for each $f \varepsilon \mathscr{F}$, we can extract a subsequence $\{t_\alpha\}$ of the double sequence $\{t_i(n)\}$ for which $m(t_\alpha,\phi,\cdot) \overset{w}{\to} \psi(\cdot)$.

Only the last assertion of the theorem remains to be proved. Suppose that there is a sequence $\{t_n\}$ so that for any subsequence $\{t_n'\}$, and some $f(\cdot)$ in $\mathscr{F}$ or C_X,

$$(*) \qquad \limsup_n \inf_{\psi \varepsilon W(\phi)} |f[m(t_n',\phi)] - f[\psi(\cdot)]| > 0.$$

By weak boundedness of $\{m(t_n',\phi,\cdot)\}$, there is a subsequence which converges weakly to some $\psi(\cdot) \varepsilon \mathscr{M}$. This $\psi(\cdot)$ must also be in $W(\phi)$, a contradiction to (*). Q.E.D.

<u>Theorem</u> 5. <u>Assume</u> (B1)-(B3) <u>of Theorem</u> 4. <u>Then</u>

(i) $x_t \overset{P}{\to} \overline{S}(W(\phi)) \equiv C$, <u>the closure of the support set of the invariant</u>

set $W(\phi)$, i.e., $P_\phi\{\inf_{y\varepsilon C} |x_t-y| > \varepsilon\} \to 0$ as $t \to \infty$, for any $\varepsilon > 0$.

(ii) Let $k(\cdot)$ be a real valued, non-negative and continuous function on X and let $k(x_t) \overset{P}{\to} 0$. Let G_n denote compact sets in X for which $m(t,\phi,X-G_n) < \varepsilon_n \to 0$, $G_{n+1} \supset G_n$. Then x_t converges in probability to the largest support set of an invariant set whose support is contained in $\lim_n G_n \cap \{x: k(x) = 0\}$.

Proof. (i) Let $N_\varepsilon(C)$ denote an ε-neighborhood of C. We will show that, for each $\varepsilon > 0$,

$$(*) \qquad \lim_{t \to \infty} P_\phi\{x_t \; \varepsilon \; X-\overline{N}_\varepsilon(C)\} = 0,$$

since (*) implies (i). Suppose (*) is violated. Then there are $t_n \to \infty$ and $\varepsilon_0 > 0$ so that $P_\phi\{x_{t_n} \; \varepsilon \; X-\overline{N}_\varepsilon(C)\} \geq \varepsilon_0 > 0$. There is a function $f(\cdot) \; \varepsilon \; C_X$ satisfying $0 \leq f(x) \leq 1$, $f(x) = 0$ on $\overline{N}_{\varepsilon/2}(C)$, $f(x) = 1$ on $X-N_\varepsilon(C)$. For some subsequence $\{t_n'\}$ of $\{t_n\}$, $m(t_n',\phi,\cdot)$ converges weakly to a $\psi(\cdot)$ in $W(\phi)$ and $f[m(t_n',\phi)] \to f[\psi(\cdot)] \geq \varepsilon_0 > 0$. Thus $X-N_\varepsilon(C)$, which is disjoint from $\overline{C}$, contains some point in the support set of $\psi(\cdot)$, a contradiction to the definition of C.

(ii) follows easily from (i), and the proof is omitted. Q.E.D.

Discussion of the Conditions (B1)-(B3) of Theorem 4.

Under the conditions of Theorem 1, if $\overline{Q}$ is compact, then the measures for the stopped process are weakly bounded, and we can apply the invariance theorem to the stopped process. If the conditions of Theorem 1 hold for all $\lambda < \infty$, and each $\overline{Q}_\lambda$ is compact, then $\{m(t,\phi,\cdot)\}$ is weakly bounded. Usually, the function $k(\cdot)$ in Theorem (5) is the $k(\cdot)$ of Theorem 3. Furthermore, even if each Q_λ is not bounded, it may be that the measures for the process stopped on exit from Q_λ are weakly bounded. See Example 2 in [6].

Theorem 6. (B3) holds for a Feller process on any topological state space.

Proof. Let $\phi_n(\cdot) \overset{w}{\to} \phi(\cdot)$. We must show that

$$(*) \qquad \int f(x)m(t,\phi_n,dx) - \int f(x)m(t,\phi,dx) \to 0$$

for all $f(\cdot) \in C_X$. Write (*) as

$$\int [f(y)m(t,x,dy)](\phi_n(dx) - \phi(dx))$$

$$= \int h_t(x)[\phi_n(dx) - \phi(dx)] \to 0.$$

$m(t,x,\cdot)$ denotes the measure with initial condition x and $h_t(x) = E_x f(x_t)$ which is in C_X by the Feller property, and the convergence follows since $\phi_n(\cdot) \overset{w}{\to} \phi(\cdot)$. Q.E.D.

<u>Remark</u>. Theorem 6 implies that condition (B3) is not very restrictive.

<u>Theorem</u> 7. <u>Let</u>

$$(*) \qquad P_x\{|x_t - x| > \varepsilon\} \to 0$$

<u>as</u> $t \to 0$, <u>uniformly for</u> x <u>in any compact set</u>. <u>For each real</u> $T > 0$ <u>and compactum</u> $K \subset X$, <u>let the family</u> $\{m(t,x,\cdot),\ x \in K,\ t \le T\}$ <u>be weakly bounded</u>. <u>Then</u> (B2) <u>holds</u>.

<u>Proof</u>. Let $\{\phi_\alpha\}$ denote a weakly bounded set of measures. Then the second hypothesis implies that the family $\{m(t,\phi_\alpha,\cdot),\ t \le T, \text{ all } \alpha\}$ is weakly bounded (we omit the proof, which is not hard).

Write, for $t \ge 0$, $s \ge 0$, $s+t \le T$,

$$\left|\int f(x)[m(t+s,\phi_\alpha,dx) - m(t,\phi_\alpha,dx)]\right|$$

$$\le \int |E_x f(x_s) - f(x)| m(t,\phi_\alpha,dx)$$

$$= \int_{G'} |E_x f(x_s) - f(x)| m(t,\phi_\alpha,dx)$$

$$+ \int_{X-G'} |E_x f(x_s) - f(x)| m(t,\phi_\alpha,dx).$$

Choose compact G' to make the second term less than $\frac{\varepsilon}{2}$, for all α, $t \leq T$. Then, using the first hypothesis, choose $s_0 > 0$ so that $|E_x f(x_s) - f(x)| \leq \frac{\varepsilon}{2}$ for $s \leq s_0$ and $x \in G'$, thus proving the right continuity of $E_x f(x_t)$.

To prove left continuity, write, for $T \geq t-s \geq 0$, $s \geq 0$,

$$\left|\int f(x)[m(t,\phi_\alpha,dx) - m(t-s,\phi_\alpha,dx)]\right|$$

$$\leq \int_{G'} |E_x f(x_s)-f(x)|m(t-s,\phi_\alpha,dx) + \int_{X-G'} |E_x f(x_s)-f(x)|m(t-s,\phi_\alpha,dx).$$

Choose compact G' so that the second term is $\leq \frac{\varepsilon}{2}$ for $0 \leq t-s \leq T$, and all α, and then choose s_0 so that $|E_x f(x_s) - f(x)| \leq \frac{\varepsilon}{2}$ for $s \leq s_0$ and all $x \in G'$. Q.E.D.

III. EXAMPLES

Example 1. A relatively simple example is the diffusion process given by the Itô equation

$$dx_1 = x_2 dt$$

$$dx_2 = -g(x_1)dt - ax_2 dt - x_2 c dz$$

where

$$\int_0^t g(s)ds \to \infty \quad \text{as} \quad t \to \infty,\ sg(s) > 0,\ s \neq 0,\ g(0) = 0,$$

and $g(\cdot)$ satisfies a local Lipschitz condition. Let Q be a bounded open set in R^2. Then x_t can be defined up until the first exit time from Q, and the stopped process is a continuous Feller process and (B1)-(B3) hold. The function

$$V(x) = x_2^2 + 2\int_0^{x_1} g(s)ds$$

is in $\mathscr{D}(\tilde{A}_Q)$ and, for $x \in Q$,

$$\tilde{A}_Q V(x) = x_2^2(c^2 - 2a).$$

Let $c^2 \leq 2a$. Then

$$P_x\{\sup_{\infty > t \geq 0} V(x_t) \geq \lambda\} \leq V(x)/\lambda \to 0 \quad \text{as } \lambda \to \infty ,$$

and x_t can be uniquely defined on $[0,\infty)$ w.p.1., even though $g(\cdot)$ does not satisfy a global Lipschitz condition. It is a continuous Feller process and (B1)-(B3) hold.

Let $c^2 < 2a$. Then $x_{2t} \to 0$ w.p.1. and by Theorem 1, x_t tends in probability to the smallest invariant set whose support satisfies $x_{2t} = 0$, for all t. Thus $x_t \overset{P}{\to} 0$. This and the w.p.1. convergence of $V(x_t)$ implies that $x_t \to 0$ w.p.1.

Example 2. For the second example, we take a problem arising in the identification of the coefficients of a linear differential equation.

The system to be identified is the scalar input, scalar output asymptotically stable, reduced form, system

$$(1) \qquad \left(\frac{d}{dt^n} + \sum_{i=0}^{n-1} a_i \frac{d^i}{dt^i}\right)y = \left(\sum_0^m b_i \frac{d^i}{dt^i}\right)u, \; n > m,$$

where $u(t)$ is the input. We wish to know the a_i, b_i. The input $u(t)$ is $\Sigma\, c_i \bar{u}_i(t)$, where $\bar{u}(t)$ is a stationary Markov process. The "equation error" method of P. M. Lion ("Rapid Identification of Linear and Nonlinear Systems", Proc. 1966 Joint Automatic Control Systems Conference, University of Washington, Seattle) will be used. For this method, some estimate of the derivatives of a smoothed input and output are needed.

Let $H(s)$ denote a transfer function the degree of whose denomenator exceeds the degree of the numerator by at least n. For any real number c, define the "derivatives of the smoothed u,y as

$$y_k(s) = H(s)(s+c)^k y(s), \quad k = 0,\dots,n$$

$$u_k(s) = H(s)(s+c)^k u(s), \quad k = 0,\dots,m$$

and the equation error $\varepsilon(t)$ as

(2) $$\varepsilon(t) = y_n(t) + \sum_0^{n-1} \alpha_i y_i(t) + \sum_0^{m} \beta_i u_i(t)$$

where $\{\alpha_i, \beta_i\}$ are to be prescribed. Let the system $(y, y^{(1)}, \ldots, y^{(n-1)})$ be state variablized by the minimal order, (with asymptotically stable A^y) $\dot{x}^y = A^y x^y + B^y u$, $y = H^y x^y$, and write

$$y(s) = \frac{N(s)}{D(s)} u(s) + \sum_0^{n-1} \frac{Q_i(s)}{D(s)} x_i(0)$$

where the last term goes to zero exponentially.

Let us impose the following conditions:

(C1) $\bar{u}(t)$ is a right continuous stationary Feller strong Markov process with $E|\bar{u}(t)|^2 = M_0 < \infty$. Thus, the paths are Laplace transformable paths w.p.l. In particular, $\int_0^\infty e^{-kt}|\bar{u}(t)|dt < \infty$ w.p.l. for all $k > 0$.

(C2) $P_{\bar{u}}\{\sup_{\delta \geq h \geq 0} |\bar{u}(h) - \bar{u}| > \varepsilon\} \to 0$ as $\delta \to 0$ uniformly for the initial condition $\bar{u} = \bar{u}(0)$ in any compact region.

(C3) $E[\bar{u}(t+\tau)\bar{u}'(t)|\bar{u}(s), s \leq 0] \to \bar{R}(\tau)$, the covariance of the $\bar{u}(t)$ processes. Let $E\bar{u}(t) = 0$. (This condition is not essential.)

(C4) $S_u(\omega)$, the spectral density of $u(t)$, is nonzero over some interval.

There are real numbers $\{\alpha_i^o, \beta_i^o\}$ so that $\varepsilon(t) \equiv 0$ if all $x_i(0) = 0$. To see this write the Laplace transform of (2) where we have

$$\varepsilon(s) = y_n(s) + \sum_0^{n-1} \alpha_i^o y_i(s) + \sum_0^{m} \beta_i^o u_i(s)$$

$$= H(s)\left[(s+c)^n \frac{N(s)}{D(s)} + \sum_0^{n-1} \alpha_i^o (s+c)^i \frac{N(s)}{D(s)} + \sum_0^{m} \beta_i^o (s+c)^i\right] u(s) = 0$$

if

$$\frac{N(s)}{D(s)} = -\frac{\sum_0^m \beta_i^o (s+c)^i}{\sum_0^n \alpha_i^o (s+c)^i}, \qquad \alpha_n^o = \alpha_n = 1.$$

For $\{\alpha_i^o, \beta_i^o\}$ used in (2), $\varepsilon(t) \to 0$ exponentially. In fact, we suppose that the systems generating $y_i(t)$, $u_i(t)$ are connected to their inputs at $t = 0$, and that their initial conditions do not depend on the process $\bar{u}(t)$. Then $\varepsilon(t)$ is non-random. The condition can be relaxed to allow for random $y_i(0)$, $u_i(0)$, at some extra complication in the analysis.

The parameter adjustment procedure is

$$\dot{\alpha}_j = -\frac{k}{2}\frac{\partial \varepsilon^2}{\partial \alpha_j} = -k\,\varepsilon\, y_j$$

(3)

$$\dot{\beta}_j = -\frac{k}{2}\frac{\partial \varepsilon^2}{\partial \beta_j} = -k\,\varepsilon\, u_j.$$

Define the column vectors

$$z = (\alpha_0 - \alpha_0^o, \ldots, \alpha_{n-1} - \alpha_{n-1}^o, \ldots, \beta_m - \beta_m^o)$$

$$w = (y_0, \ldots, y_{n-1}, u_0, \ldots, u_m).$$

Then

(4) $$\dot{z} = -kw\varepsilon = -kw\{[y_0\alpha_0 + \ldots + y_{n-1}\alpha_{n-1} + \ldots + u_m\beta_m] + y_n + [y_0\alpha_0^o + \ldots + u_m\beta_m^o] - [y_0\alpha_0^o + \ldots + u_m\beta_m^o]\}$$

(5) $$= -kww'z + \tilde{\delta}_t$$

where $\tilde{\delta}_t = -kw[y_n + y_0\alpha_0^o + \ldots + u_m\beta_m^o]$.

We can assume that the $y_k(t)$, $u_k(t)$ are the outputs for asymptotically stable systems of the form $\dot{x}^{y_k} = A^{y_k}x^{y_k} + B^{y_k}y$, $y_k(t) = H^{y_k}x^{y_k}$, etc. Thus all y_k, u_k, y, u, z are state variabilized, and the composite state variabilization, namely $x(t)$, is a right continuous strong Markov process and Feller.

Furthermore, it is uniformly stochastically continuous in the sense of (C2).+

Let $E|z(0)|^2 < \infty$. Let $\Phi(t,s)$ denote the fundamental matrix solution of $\dot{z} = -kww'z$. Then $|\Phi(t,s)| \le 1$ and

$$|z(t)| \le |z(0)| + \int_0^\infty |\delta_s| ds$$

which, together with the bound $E|w_s||w_t| \le M < \infty$ for some M, yields that $E|z(t)|^2 \le M_1$ for some $M_1 < \infty$.

Next, let us introduce the Liapunov technique. Let

$$V = z'z.$$

Then

$$\text{(6)} \qquad \begin{aligned} \dot{V} &= -2k(z'w)^2 - k(z'w)[y_n + w'\alpha^o] \\ &\equiv -2k(z'w)^2 + \rho_t, \end{aligned}$$

where $\alpha^o = (\alpha_0^o,\ldots,\alpha_{n-1}^o, \beta_0^o,\ldots,\beta_m^o)$. Note that, since $V(z(t))$ is differentiable, $[E_{x_t} V(z(t+\Delta)) - V(z(t))]/\Delta \to \dot{V}(z(t))$ w.p.1. V does not involve the possibly non-differentiable components (namely $\overline{u}(t)$) of $x(t)$.

(a) $E|z'(t)w'(t)| \le E^{1/2}|z(t)|^2 E^{1/2}|w(t)|^2$ is uniformly bounded and $y_n(t) + w(t)\alpha^o$ is not random. Hence $\int_0^\infty |\rho_t| dt < \infty$ w.p.1. and $\int_0^\infty E|\rho_t| dt < \infty$.

(b) The $z(t)$ process is weakly bounded, since $E|z(t)|^2 \le M_1 < \infty$; hence the $\{x(t)\}$ process is weakly bounded.

(c) From (6) and (a), $z'(t)w(t)$ is square integrable on $\Omega \times [0,\infty)$

+E.g., let $\dot{x} = Ax + By$, where y satisfies (C2). Then we only need

$$P_{x,y}\{ \sup_{\delta \ge h \ge 0} |\int_0^h e^{A(h-\tau)} By(\tau) d\tau| > \varepsilon\} \to 0.$$

uniformly in (x,y) in compact intervals, as $\delta \to 0$.

and $\int_t^\infty (z'(\tau)w(\tau))^2 d\tau \to 0$ w.p.1. as $t \to \infty$. The components of $z(t)$ and $w(t)$ satisfy the uniform stochastic continuity condition. This, together with the weak boundedness and the convergence $\int_t^\infty (z'(\tau)w(\tau))^2 d\tau \to 0$ w.p.1. imply that $z'(t)w(t) \xrightarrow{P} 0$. (Indeed $z'(t)w(t)$ can be shown to converge to zero w.p.1.)

(d) Writing

$$\varepsilon - [y_n + \sum_0^{n-1} \alpha_i^o y_i + \sum_0^m \beta_i^o u_i] = z'w + y_n - y_n$$

we have that $\varepsilon(t) \xrightarrow{P} 0$, and is square integrable.

(e) The measures of $x(t)$ tend to an invariant set of measures. This invariant set must be consistent with $\varepsilon(t) = 0$ w.p.1. for each t. Thus, by (4), for the invariant set, $z(t) = \tilde{z}$, a random variable.

For the measures in the invariant set

$$0 = y_n(t) + \sum_0^{n-1} \tilde{\alpha}_i y_i(t) + \sum_0^m \tilde{\beta}_i u_i(t) \equiv \tilde{\varepsilon}(t) \tag{7}$$

for some set of random variables $\tilde{z} = (\tilde{\alpha}_0,\ldots,\tilde{\alpha}_{n-1}, \tilde{\beta}_0,\ldots,\tilde{\beta}_m) - (\alpha_0^o,\ldots,\beta_m^o)$. Write

$$0 = E[\tilde{\varepsilon}(t)\tilde{\varepsilon}(t+\tau) | \tilde{\alpha}_i, \tilde{\beta}_i, \text{ for all } i] = R_t(\tau).$$

Let $y(t)$ denote the components of $x(t)$ without $z(t)$. The probability law of the process $x(t)$ implies that $P_{y,z}\{y(t) \in A\} = P_y\{y(t) \in A\}$. This and (C3) imply that the limit $R(\tau)$ of $R_t(\tau)$, as $t \to \infty$, is the same as if the $\tilde{\alpha}_i, \tilde{\beta}_i$ were not random. Then, using the stationarity of the $\bar{u}(t)$, $y(t)$, $y_k(t)$, $u_k(t)$ processes, the Fourier transform of $R(\tau)$ is

$$0 = |\sum_0^n \tilde{\alpha}_j (i\omega + c)^j \frac{N(i\omega)}{D(i\omega)} + \sum_0^m \tilde{\beta}_j (i\omega+c)^j|^2 |H(i\omega)|^2 S_u(\omega). \tag{8}$$

[Consider $R(\tau)$ as the covariance of the output of an asymptotically stable linear system with input $u(t)$ - the transfer function of which is

$$\sum_0^n \tilde{\alpha}_j (s+c)^j H(s) \frac{N(s)}{D(s)} + \sum_0^m \tilde{\beta}_j (s+c)^j H(s).]$$

But (8) and (C4) imply that $\tilde{\alpha}_j = \alpha_j^o$, $\tilde{\beta}_j = \beta_j^o$, and the demonstration is complete.

BIBLIOGRAPHY

[1] H. J. Kushner, Stochastic Stability and Control, Academic Press, New York, 1967.

[2] H. J. Kushner, "The concept of invariant set for stochastic dynamical systems and applications to stochastic stability", in Stochastic Optimization and Control, H. Karreman, Editor, John Wiley and Sons, New York, 1968.

[3] H. J. Kushner, Introduction to Stochastic Control Theory, Holt, Rinehart and Winston, New York, 1971.

[4] M. Loeve, Probability Theory, 3rd Edition, Van Nostrand, Princeton, 1963.

[5] E. G. Dynkin, Markov Processes, Springer-Verlag, Berlin, 1965.

[6] H. J. Kushner, "On the stability of processes defined by stochastic difference-differential equations", Journal of Differential Equations, Vol. 4, No. 3, July 1968, pp. 424-443.

[7] J. P. LaSalle, "The extent of asymptotic stability", Proc. of the Nat. Acad. of Sciences, 46, (1960), p. 365.

[8] J. K. Hale, "Sufficient conditions for stability and instability of autonomous functional-differential equations", J. Diff. Eqns., 1, 1965.

[9] N. Dunford, J. T. Schwartz, Linear Operators, Part I, Interscience, Wiley, New York, 1958.

[10] I. I. Gikhman, A. V. Skorokhod, Introduction to the Theory of Random Processes, Saunders, Phila., 1969.

STABILIZATION OF LINEAR SYSTEMS WITH MULTIPLICATIVE NOISE

U.G. HAUSSMANN

The University of British Columbia
Vancouver, Canada

1. Introduction. Consider a system described by the stochastic differential equation

(1.1) $$dx = (Ax - Bu)dt - C(u)dw_1 + D(x)dw_2 + Edw_3$$

where x is an n-dimensional state vector, u an m-dimensional control vector, w_1, w_2, w_3 are independent Wiener processes of dimensions d_1, d_2, d_3 respectively,

$$C(u) = \sum_{i=1}^{m} C_i u_i, \qquad D(x) = \sum_{i=1}^{n} D_i x_i,$$

and where A, B, C_i, D_i, E are real constant matrices of appropriate dimensions. Assume $x(0)$ is a random variable independent of the increments of w_1, w_2, w_3, and let $u = \phi(x)$, where

(1.2) $$\|\phi(x) - \phi(y)\| \leqq k\|x - y\| .$$

Then (1.1) determines a diffusion process $X_\phi = \{x(t) : t \geq 0\}$.

Let $L(x,u) = x'Mx + u'Nu$ be a quadratic cost where M and N are positive definite. Here x' is the transpose of x. Suppose one is interested in minimizing the expected cost in the steady state. A steady state is given by an invariant probability distribution μ, i.e. if $x(0)$ has distribution μ then so does $x(t)$ for any $t \geqq 0$. Consequently of interest are those controls which give rise to a steady state and for which the expected cost is finite.

DEFINITION. $u = \phi(x)$ is a stabilizing control if

(i) (1.2) is satisfied for some constant k,

(ii) X_ϕ has at least one steady state,

(iii) any invariant measure μ has finite second moment, i.e.

$$E_\mu\{||x||^2\} \equiv \int_{R^n} ||x||^2 \mu(dx) < \infty .$$

The expected cost is then $\max_\mu E_\mu\{L(x,\phi(x))\}$, where μ is any invariant probability measure of the process X_ϕ. As pointed out by the author (c.f. [1]) for optimal controls all invariant measures yield the same cost.

2. <u>A Liapunov criterion for stability</u>. Let $\Gamma(P)$ and $\Delta(P)$ be the matrices with $ij^{\underline{th}}$ entry trace $[C_i'PC_j]$ and trace $[D_i'PD_j]$ respectively. Let L_u be the differential operator

$$L_u V(x) = \frac{1}{2}\{u'\Gamma(V_{xx})u + x'\Delta(V_{xx})x\} + \text{trace}[E'V_{xx}E]\} + (Ax - Bu)'V_x .$$

The subscript x denotes differentiation with respect to x. If $u = \phi(x)$ where ϕ satisfies (1.2), then L_ϕ is the differential operator of X_ϕ. From Zakai's work (c.f. [5]) the following result is readily deduced.

THEOREM 2.1. *If there exist two positive scalars λ and ρ, and two constant real matrices K and P with P non-negative definite, such that*

(2.1) $$L_\phi V(x) \leqq \lambda - \rho||x||^2 , \quad x \in R^n,$$

where $V(x) = x'Px$ and $\phi(x) = Kx$, then ϕ is a stabilizing control.

Moreover the author showed (c.f. [1], p.194) that the conditions of the theorem are also sufficient to initialize an algorithm due to Wonham (c.f. [3]) which yields an optimal control of the form $\phi^\circ(x) = K^\circ x$, where K° is given in terms of the solution of a matrix Riccati equation.

Wonham proved (c.f. [3], [4]) that if (A,B) is stabilizable, i.e. if there is a matrix K_1 such that $A - BK_1$ is stable, and if

$$\inf_K \left\| \int_0^\infty e^{t(A-BK)'}[K'\Gamma(I)K + \Delta(I)]e^{t(A-BK)}dt \right\| < 1,$$

then the conditions of theorem 2.1 are satisfied. There follows the conclusion that if the state and control dependent noise is sufficiently small then an optimal control exists.

However Wonham's conditions are not necessary. In fact the author showed (c.f. [1]) that if the system has a certain structure, i.e. the matrices A, B, C_i, D_i satisfy certain conditions, then the conditions of theorem 2.1 can be satisfied no matter how large the noise is. We shall now indicate the best such conditions for the case $D \equiv 0$. More details can be found in the author's forthcoming article [2].

3. A stability result. R^n decomposes into two complementary subspaces R^n_-, R^n_+ consisting of the asymptotically stable and asymptotically unstable modes of A respectively. From now on assume $D \equiv 0$. Then (1.1) decomposes into

$$dy = (A_+y - B_+u)dt - C_+(u)dw_1 + E_+dw_3 \tag{3.1}$$

$$dz = (A_-z - B_-u)dt - C_-(u)dw_1 + E_-dw_3 \tag{3.2}$$

where $x = y + z$, $y \in R^n_+$, $z \in R^n_-$, and A_- is stable. It can be shown that if the conditions of theorem 2.1 hold for equation (3.1) then they also hold for (1.1). If there is a control $u = K_+y$ such that $C_+(u) = 0$ and $(A_+ - B_+K_+)$ is stable, then it follows from the theory of the linear regulator without control dependent noise that (2.1) is satisfied for some P, λ, ρ, with $K = K_+$. Hence if the controllability space of (A_+, B_+), i.e. the column space of $[B_+, A_+B_+, \cdots, A_+^{n-1}B_+]$, with B_+ restricted to $\ker C_+$, the kernel of $C_+(\cdot)$, is all of R^n_+, then a stabilizing control exists no matter what C is.

If this space is not empty but is not all of R^n_+, take $\hat{A} = A_+ - B_+K_+$ where K_+ is such that $\hat{A}$ is stable on the controllability space of (A_+, B_+). Then (3.1) becomes

$$dy = (\hat{A}y - \hat{B}\hat{u})dt - \hat{C}(\hat{u})dw_1 + \hat{E}dw_3 \tag{3.3}$$

where $\hat{u}$ is in $R^m/\ker C_+$. Now $\hat{A}$ induces a decomposition of R^n_+ just as A did of R_n. This process can be continued until eventually either all of R^n is decomposed into subspaces on each of which a matrix of the form $A_o - B_oK_o$ is stable, or at some stage the corresponding controllability space is empty. In the former case a stabilizing control exists.

More precisely, if $\{F\}$ denotes the column space of the matrix F, define

$$M(V) \equiv \{Bu : u \in R^m, \{C(u)\} \subset V\} .$$

Let $R_o = R^n_-$ and let

$$R_i \equiv \text{span}\,\{R_{i-1}, \{x, Ax, \cdots, A^{n-1}x\} : x \in M(R_{i-1})\}, \quad i > 0.$$

THEOREM 3.1. <u>If $R_m = R^n$, then the conditions of theorem 2.1 are satisfied.</u>

Hence a stabilizing control exists without any conditions on the size of C.

4. <u>Instability</u>. It remains to discuss the second alternative, i.e. $R_m \neq R^n$. We shall show that in this case instability will occur if C is sufficiently large. Assume from now on that EE' is positive definite so that at most one invariant measure exists. Let $u = \phi(x)$ satisfy (1.2), and let X_ϕ be given by (1.1) (with $D \equiv 0$).

THEOREM 4.1. <u>Assume there exist two real valued functions V_1, V_2 such that</u>

(a) <u>for some</u> $r < \infty$, V_1 <u>and</u> V_2 <u>are defined and twice continuously differentiable for</u> $||x|| > r$,

(b) <u>there exists a sequence</u> $\{x_n\}$ <u>with</u> $||x_n|| \to \infty$ <u>such that</u> $V_1(x_n) \to \infty$ <u>as</u> $n \to \infty$,

(c) $V_2(x) \geqq 0$ <u>if</u> $||x|| > r$,

(d) $\limsup_{\rho\to\infty} \dfrac{\max\{V_1(x) : \|x\| = \rho\}}{\min\{V_2(x) : \|x\| = \rho\}} = 0,$

(e) *for* $\|x\| > r$, $L_\phi V_1(x) \geqq 0$ *and* $L_\phi V_2(x) \leqq \|x\|^2$.

Then no invariant measure of X_ϕ *has finite second moment.*

This result was originally established by Wonham (cf. [3]) ; however his proof used the partial differential equations approach. A probabilistic proof is given in [2].

With $V_1(x) = (x'Qx)^q$, $V_2(x) = \theta x'x$ for suitable constants θ, q, and matrix Q, (a) - (d) as well as the second part of (e) are satisfied. Another condition is required to ensure $L_\phi V_1 \geqq 0$.

THEOREM 4.2. *Assume* (i) *all the eigenvalues of* A *lie in the open right half-plane, and* (ii) $\ker C \subset \ker B$. *If there exists a positive definite matrix* P *such that*

$$\text{trace}\,[C(u)'QC(u)] > u'B'QP^{-1}QBu \tag{4.1}$$

for any μ *not in* $\ker B$, *where*

$$Q = \int_{-\infty}^{0} e^{tA'}Pe^{tA}\,dt,$$

then there is no stabilizing control.

Consider now the case $R_m \neq R^n$. Restricting the system (1.1) to the unstable modes of the corresponding matrix $\hat{A}$, i.e. to a certain complement of R_m, implies that (ii) of theorem 4.2 is satisfied. With $P = I$ (4.1) can be satisfied provided C(u) is sufficiently large. The final theorem results.

THEOREM 4.3. *Assume that* DD' *is positive definite and that* A *has no purely imaginary eigenvalues. If* $R_m \neq R^n$, *then for* C *sufficiently large there is no stabilizing control.*

References

[1] U.G. Haussmann, "Optimal Stationary Control with State and Control Dependent Noise", SIAM. J. Control, 9(1971), pp. 184-198.

[2] ____________, "Stability of Linear Systems with Control Dependent Noise", Ibid., forthcoming.

[3] W.M. Wonham, "Optimal Stationary Control of a Linear System with State-Dependent Noise", SIAM. J. Control, 5(1967), pp. 486-500.

[4] ____________, Random Differential Equations in Control Theory, Probabilistic Methods in Applied Mathematics, vol. II, A.T. Bharucha-Reid, ed., Academic Press, New York, 1969.

[5] M. Zakai, "A Lyapunov Criterion for the Existence of Stationary Probability Distributions for Systems Perturbed by Noise", SIAM. J. Control, 7(1969), pp. 390-397.

LYAPUNOV FUNCTIONS AND GLOBAL FREQUENCY DOMAIN STABILITY CRITERIA FOR A CLASS OF STOCHASTIC FEEDBACK SYSTEMS

JACQUES L. WILLEMS

University of Gent
Gent, Belgium

ABSTRACT

This paper deals with the stability of a particular class of stochastic systems; feedback systems are considered which have a feedback gain with a deterministic gain which may be nonlinear and/or time-varying and a stochastic component which is white noise. Lyapunov functions are constructed and criteria for global stability are derived similar to the results available for related deterministic feedback systems, such as the Routh-Hurwitz criterion, the Popov criterion, and the circle criteria.

1. INTRODUCTION

The difficult step in the application of Lyapunov theory to analyse the stability of deterministic systems as well as the stability of stochastic systems (Kushner 1967), is the construction of a suitable Lyapunov function. There is indeed no general systematic procedure for generating Lyapunov functions. For deterministic systems very interesting results, such as the Popov criterion and the circle conditions, have been obtained for a particular class of feedback systems containing a linear time-invariant forward path element and either a nonlinear or a time-varying feedback element (Zames 1966, J.C. Willems 1971a, J.L. Willems 1970). This is the motivation for considering a similar class of stochastic feedback systems, where the gain of the feedback element has a stochastic white noise component. It is the purpose of this paper to investigate whether or not the procedures for constructing Lyapunov functions for the deterministic case remain useful for the stability analysis in the stochastic case as well. In particular the path integral method (Brockett 1970) and the Kalman-Yacubovitch-Meyer lemma (Kalman 1963) are used to generate useful Lyapunov functions for the stochastic feedback system.

Section 2 deals with the case where the deterministic part of the feedback element is a constant gain; a Lyapunov function is obtained which proves a necessary and sufficient condition for mean-square stability; the white noise component of the feedback gain has a destabilizing effect on the mean-square stability properties. The criterion is also sufficient for Lyapunov stability with probability one, but not necessary. In Section 3 a nonlinear time-varying feedback element is considered; stability criteria are obtained which are closely related to the circle criterion for deterministic systems (Zames 1966), but which also show a destabilizing effect of the white noise gain component. A particularity of this analysis, which does not appear in deterministic stability theory, is that different solutions of the spectral factorization problem in the path integral method or of the Kalman-Yacubovitch-Meyer lemma yield non-equivalent stability criteria. Section 4 deals with the stability analysis when the deterministic feedback gain component is either nonlinear and time-invariant or linear and time-varying. Criteria similar to the well known Popov criterion are derived. Some possible extensions and generalizations are discussed in Section 5.

2. LINEAR TIME-INVARIANT FEEDBACK

In this section the stability is considered of a system consisting of a linear time-invariant forward path element with rational transfer function H(s) and a multiplicative feedback whose gain is the sum of a deterministic constant k and a stochastic white noise component. This system is described by the stochastic Itô differential equation

$$d\underline{x} = (\underline{A}_o\underline{x} - k\underline{b}\underline{c}\underline{x})dt - \underline{b}\underline{c}\underline{x}d\beta \tag{1}$$

where matrices and vectors are denoted by underlined symbols. The triple $\underline{A}_o, \underline{b}, \underline{c}$ is a realization of the transfer function $H(s) = \underline{c}(\underline{I}s-\underline{A}_o)^{-1}\underline{b}$, and β denotes a scalar Wiener process with independent increments and

$$E((\beta(t)-\beta(\tau))^2) = \sigma^2|t-\tau|, \quad E(d\beta(t)) = 0 \tag{3}$$

Let G(s) be the transfer function of the deterministic closed loop system

$$G(s) = \frac{q(s)}{p(s)} = \underline{c}(\underline{I}s-\underline{A}_o-k\underline{b}\underline{c})^{-1}\underline{b} = H(s)/(kH(s)+1)$$

Suppose that p(s) is strictly Hurwitz, and denote the polynomial p(s) by

$$p(s) = s^n + p_{n-1}s^{n-1} + \ldots + p_o, \quad q(s) = q_{n-1}s^{n-1} + \ldots + q_o$$

where, without loss of generality, p(s) is assumed a monic polynomial. Then (1) is equivalent to

$$p(D)y\,dt + q(D)y\,d\beta = 0 \tag{4}$$

where $D = d/dt$. In the sequel either equation (1) or equation (4) is used, whichever is more convenient. Both the state vectors $\underline{x}$ and $\underline{y} = [y \quad Dy \; \dots \; D^{n-1}y]^T$ are used; they are related by a linear transformation, and they are identical, if $\underline{A}$ has the companion form (Brockett 1970), $\underline{b} = [q_o \quad q_1 \; \dots \; q_{n-1}]^T$, and $\underline{c} = [0 \quad 0 \; \dots \; 0 \quad 1]$. For the stability analysis of a stochastic equation, such as (1), instead of the derivative of the Lyapunov function in deterministic stability theory, the sign definiteness of $LV(\underline{x})$ should be considered (Kushner 1967), where $LV(x)$ is the sum of the derivative for zero noise and

$$\frac{1}{2}\sigma^2(q(D)y)^2\frac{\partial^2 V}{\partial y_n^2}$$

where y_i denotes the ith component of the vector $\underline{y}$. Hence, if quadratic Lyapunov functions are used, it is the aim to find a Lyapunov function whose deterministic derivative contains a term proportional to $(q(D)y)^2$. This can be achieved by means of the path integral technique (Brockett 1970). Let the polynomial $h(s)$ be the solution of the set of linear equations

$$\frac{1}{2}[h(s)p(-s) + h(-s)p(s)] \equiv q(s)q(-s) \quad \text{for all } s \tag{5}$$

A unique solution exists if $p(s)$ has no zeros on the imaginary axis. Using the Lyapunov function

$$V(\underline{y}) = \int_{t(\underline{0})}^{t(\underline{y})} [\, p(D)z\; h(D)z - (q(D)z)^2 \,]\, dt \tag{6}$$

(where the notation is as introduced by Brockett), we obtain

$$LV(\underline{y}) = -(q(D)y)^2 + \frac{1}{2}\, h_{n-1}\sigma^2(q(D)y)^2 \tag{7}$$

Considering the path independence of the integral in (6) and evaluating this integral along solutions of $p(D)z = 0$, the positive definiteness of $V(\underline{y})$ is easily proved. By means of the Lyapunov theorems for stochastic systems (Kushner 1967) the following criterion is readily established :

<u>Criterion 1</u> The null solution of system (1) is mean-square stable in the large and stable in the large with probability one, if $p(s)$ is strictly Hurwitz, and

$$\sigma^2 h_{n-1}/2 \leq 1 \tag{8}$$

To prove asymptotic stability pole-shifting techniques can be applied, or the following Lyapunov function is used

$$V(\underline{y}) = \int_{t(\underline{0})}^{t(\underline{y})} \left[p(D)z(h(D)z+am(D)z) - (q(D)z)^2 - a\,(z^2+(Dz)^2+\ldots + (D^{n-1}z)^2) \right] dt$$

where m(D) is the solution of

$$\tfrac{1}{2}\left[p(D)m(-D)+p(-D)m(D)\right] = \sum_{i=0}^{n-1} (-1)^i\, D^{2i}$$

Suppose $\sigma^2 h_{n-1}/2 < 1$, and let a be a positive constant such that

$$\sigma^2(h_{n-1}+am_{n-1})/2 < 1$$

where m_i and h_i are the coefficients of D^i in the polynomials m(D) and h(D) respectively. Then

$$LV(\underline{y}) = -(q(D)y)^2 + \tfrac{1}{2}(q(D)y)^2(h_{n-1}+m_{n-1}a)\sigma^2 - a(y^2+(Dy)^2+\ldots+ (D^{n-1}y)^2)$$

is negative semi-definite and negative definite. This yields :

Criterion 2 The null solution of system (1) is asymptotically stable in the mean square and with probability one, and the asymptic stability is global, if p(s) is strictly Hurwitz, and $\sigma^2 h_{n-1}/2 < 1$. (9)

The conditions of criteria 1 and 2 on the stochastic gain component restrict the magnitude of σ^2; they can be transformed by the following interpretation of h_{n-1}. Since the integral in (6) is independent of path, let it be evaluated along the solution of p(D)z = 0 with initial conditions : z(0) = 0, Dz(0) = 0,..., $D^{n-2}z(0) = 0$, $D^{n-1}z(0) = 1$. Along this solution the integral is equal to $h_{n-1}/2$. This same trajectory is also the solution of the system equation

$$p(D)z = u(t)$$

with $u(t) = \delta(t)$ and initial state $z(0^-) = Dz(0^-) = \ldots = D^{n-1}z(0^-) = 0$. Let the output be

$$y(t) = q(D)z(t)$$

If w(t) is the impulse response of this system, that is the inverse Laplace transform of $G(s) = q(s)/p(s)$, then the above considerations show

$$h_{n-1}/2 = \int_0^{\infty} (w(t))^2 \, dt \tag{10}$$

Hence (9) is equivalent to

$$\sigma^2 \int_0^{\infty} (w(t))^2 dt = \sigma^2 \frac{1}{2\pi} \int_{-\infty}^{+\infty} |G(j\omega)|^2 \, d\omega < 1 \tag{11}$$

where Parseval's theorem is used. Explicit expressions for the conditions of criteria 1 and 2 can be found by writing the set of linear equations equivalent to (5) :

$$h_{n-1}p_{n-1} + h_{n-2}p_n = (-1)^{n-1} q_{n-1}^{\;2}$$

$$h_{n-1}p_{n-3} + h_{n-2}p_{n-2} + h_{n-3}p_{n-1} + h_{n-4}p_n = (-1)^{n-3} q_{n-1}q_{n-3} + (-1)^{n-2} q_{n-2}^{\;2} + (-1)^{n-1} q_{n-1}q_{n-3}$$

$$\cdots$$

$$h_o p_o = q_o^{\;2}$$

which yields the stability conditions of criteria 1 and 2 in terms of the Hurwitz determinants and an additional determinant :

$$D_1 > 0, \; D_2 > 0, \; \ldots, \; D_n > 0$$

where

$$D_i = \begin{vmatrix} p_{n-1} & p_n & 0 & \cdots & \\ p_{n-3} & p_{n-2} & p_{n-1} & \cdots & \\ & & & \vdots & \\ & & & & p_{n-i} \end{vmatrix}$$

are the Hurwitz determinants; moreover the stability condition requires

$$\sigma^2 D/2 \leq D_n$$

and asymptotic stability requires

$$\sigma^2 D/2 < D_n$$

where the determinant D is derived by replacing the elements in the first column of D_n by the elements $r_{2n-2}, r_{2n-4}, \ldots, r_o$, defined as follows

$$r_m = \sum_{j=0}^{m} (-1)^j q_j q_{m-j} \text{ for } 0 \leq m \leq n-1, \text{ and } r_m = \sum_{j=m-n+1}^{n-1} (-1)^j q_j q_{m-j}$$

$$\text{for } n \leq m \leq 2n-2$$

This agrees with the stability results obtained by Willems and Blankenship (1971) for mean square input-output stability. It also corresponds to the result obtained by Kleinman (1969), as is discussed by Willems and Blankenship (1971).

Remark 1 Using the same Lyapunov function we obtain that the null solution is not mean-square stable, if either p(s) has one or more zeros with positive real parts, or if p(s) is strictly Hurwitz and

$$\sigma^2 \int_0^\infty (w(t))^2 dt > 1$$

No conclusion is reached if p(s) has no zeros with positive real parts, but some zeros with zero real parts. Hence criterion 1 is almost a necessary condition for mean-square stability. However no conclusion can be reached concerning the necessity of the condition for stability with probability 1 ; indeed, Kushner (1967) has shown that quadratic Lyapunov functions cannot lead to necessary and sufficient conditions for almost sure stability. This shows that Kleinman's statement (1969) on the necessity of his stability criterion should be very carefully interpreted.

Remark 2 The criteria show the destabilizing effect of the stochastic element on mean square stability; this destabilizing effect increases with σ^2. This is also discussed by Kozin (1969) in his survey paper and also by Rabotnikov (1964).

Remark 3 The Lyapunov function used to prove criteria 1 and 2 can also be derived by means of the equation (with $\underline{A} = \underline{A}_0 - k\underline{bc}$) :

$$\underline{PA} + \underline{A}^T\underline{P} = -\underline{c}^T\underline{c}$$

which has a unique positive definite symmetric solution $\underline{P}$ if the eigenvalues of $\underline{A}$ have a negative real part and if $(\underline{A},\underline{b},\underline{c})$ is a minimal realization. Using the Lyapunov function $V(\underline{x}) = \underline{x}^T\underline{Px}$ for the stochastic equation (1), one obtains

$$LV(\underline{x}) = -(\underline{cx})^2 + \underline{b}^T\underline{Pb}\,(\underline{cx})^2\sigma^2$$

Since (J.L. Willems 1970)

$$\underline{P} = \int_0^\infty e^{\underline{A}^T t}\underline{c}^T\underline{c}e^{\underline{A}t} dt$$

we obtain the same stability criteria by virtue of

$$\underline{b}^T\underline{P}\underline{b} = \int_0^{\infty} (w(t))^2 dt$$

Remark 4 The same Lyapunov function can also be used for the stability analysis of the system described by the Itô equation

$$d\underline{x} = \underline{A}\underline{x} - \underline{b}\underline{c}\underline{x}\, f(\underline{x},t)\, d\beta \qquad (12)$$

or

$$p(D)y\, dt + f(y, Dy, \ldots, D^{n-1}y; t)\, q(D)y\, d\beta = 0 \qquad (13)$$

where the stochastic feedback gain is white noise noise multiplied by a nonlinear and/or time-varying coefficient. This system is considered by Blankenship (1971, 1972). The conditions for mean square stability in the large for the null solution are

(i) The polynomial p(s) is strictly Hurwitz or all eigenvalues of the matrix $\underline{A}$ have negative real parts.

(ii) For all $\underline{x}$ and t :

$$(f(\underline{x},t))^2 \sigma^2 \int_0^{\infty} (w(t))^2 dt \leq 1 \qquad (14)$$

asymptotic stability follows if (ii) is replaced by

(ii')

$$\sup_{\underline{x},t} (f(\underline{x},t))^2 \sigma^2 \int_0^{\infty} (w(t))^2 dt < 1 \qquad (15)$$

If $f(\underline{x},t)$ is constrained by

$$a \leq f(\underline{x},t) \leq b$$

then (14) requires

$$\sigma^2 \int_0^{\infty} (w(t))^2 dt \leq 1/a^2,\ 1/b^2.$$

Remark 5 The analysis can also be used to obtain stability conditions for non zero mean white noise. Let $E(d\beta(t)) = m\, dt$; let

$$d\beta(t) = m dt + d\beta_1(t)$$

where $d\beta_1$ is zero mean. If $w_1(t)$ is the inverse Laplace transform of the transfer function $G(s)(mG(s)+1)^{-1}$, then the stability conditions for (1) are

(i) The eigenvalues of $\underline{A}$-m$\underline{bc}$ have negative real parts, or, equivalently, the polynomial $p(s)+mq(s)$ is strictly Hurwitz.

(ii)

$$\sigma^2\int_0^{\infty}(w_1(t))^2dt < 1$$

Consider now the system described by (12) or (13) where the noise is non zero mean. Then (12) is equivalent to

$$d\underline{x} = (\underline{A}\underline{x}-f(\underline{x},t)\underline{bc}m)dt - f(\underline{x},t)\underline{bc}\, d\beta_1 \tag{16}$$

where the average system is nonlinear and/or time-varying. The stability of this system cannot be analysed by means of the technique of this section. The method of this section is restricted to cases where the average system is linear and time-invariant. The stability of (16) can be analysed by means of the technique of Section 3.

Remark 6 The method explained above can also be applied to systems where the average system is linear and time-varying. Consider the stochastic Itô equation

$$d\underline{x} = \underline{A}(t)\underline{x}dt-f(\underline{x},t)\underline{b}(t)\underline{c}(t)\underline{x}d\beta \tag{17}$$

The Lyapunov function is obtained from

$$\dot{\underline{P}}(t) + \underline{P}(t)\underline{A}(t) + \underline{A}(t)^T\underline{P}(t) = -\underline{c}(t)^T\underline{c}(t) \tag{18}$$

This yields the stability conditions for mean-square stability :

(i) The null solution of the deterministic differential equation $\dot{\underline{x}} = \underline{A}(t)\underline{x}$ is asymptotically stable in a uniform sense.

(ii) For all $\underline{x}$ and t

$$\sigma^2(f(\underline{x},t))^2 \int_{t_0}^{\infty}(w(t,\tau))^2d\tau \leq 1$$

For asymptotic stability, condition (ii) is to be replaced by

$$\sigma^2\sup_{\underline{x},t}\left\{f(\underline{x},t))^2\int_{t_0}^{\infty}(w(t,\tau))^2d\tau\right\} < 1$$

where $w(t,\tau)$ is the weighting pattern of the linear system

$$\dot{\underline{x}}(t) = \underline{A}(t)\underline{x}(t) + \underline{b}(t)u(t) \; ; \; y = \underline{c}(t)\underline{x}(t) .$$

3. THE STOCHASTIC CIRCLE CRITERIA

In this section systems are considered having a linear time-invariant forward path element and a feedback element of the multiplicative type whose gain is the sum of a nonlinear and/or time-varying gain and a stochastic white noise component. The system is described by the Itô equation

$$d\underline{x} = (\underline{A}\underline{x}-k(\underline{x},t)\underline{bcx})\,dt - \underline{bcx}\,d\beta \tag{19}$$

or

$$(p(D)y + k(y,Dy,\ldots,D^{n-1}y,t)q(D)y)dt + q(D)y\,d\beta = 0 \tag{20}$$

where β denotes a scalar Wiener process with independent increments and the statistics indicated in (3), where $G(s) = q(s)/p(s) = \underline{c}(\underline{I}s-\underline{A})^{-1}\underline{b}$ is the transfer function of the forward path element; the same symbols for the coefficients of the polynomials p(s) and q(s) are used as in Section 2. From the proof of the circle criteria for deterministic systems (Zames 1966, J.L. Willems 1970) the following criterion is derived :

Criterion 3 The stochastic feedback system (19) or (20) has a null solution which is mean square stable in the large or stable in the large with probability one, if the transfer function G(s) is positive real, and for all $\underline{x}$,t

$$k(\underline{x},t) \geqslant \sigma^2 q_{n-1}/2 = \sigma^2 w(0)/2 \tag{21}$$

The stability properties are asymptotic if, moreover, either G(s) is strictly positive real, or

$$\inf_{\underline{x},t} k(\underline{x},t) > \sigma^2 q_{n-1}/2 = \sigma^2 w(0)/2 \tag{22}$$

where w(t) denotes the inverse Laplace transform of G(s).

This criterion is proved by means of the following quadratic Lyapunov function

$$V(\underline{y}) = \int_{t(\underline{0})}^{t(\underline{y})} \left[q(D)z\;p(D)z - (r(D)z)^2\right] dt \tag{23}$$

where r(s) is the negative spectral factor of the even part of the product of p(s) and q(-s) :

$$r(s) = \sqrt{\tfrac{1}{2}}\,\left[p(s)q(-s)+p(-s)q(s)\right]^- \tag{24}$$

This Lyapunov function yields the desired result since

$$LV(\underline{y}) = -(r(D)y)^2 - (k(y,Dy,\ldots,t)-\sigma^2 q_{n-1}/2)(q(D)y)^2$$

The asymptotic stability is proved as in Section 2.

Remark 6 The criterion reduces to a well known stability criterion for deterministic feedback systems if the stochastic component of the feedback gain is absent. Criterion 3 is stricter than its deterministic counterpart; indeed $w(0)$ or q_{n-1} is positive if $G(s)$ is positive real. The condition on the deterministic part of the feedback gain gets more restrictive as σ^2 increases. The conditions of Criterion 3 require the passivity of the linear part of the open loop and a sufficient degree of passivity for the deterministic or average feedback element; the degree of passivity required depends on the statistics of the noise component of the feedback gain but also on the forward path element, since $w(0)$ appears in (22) and (23). In deterministic theory the passivity conditions of the forward path and the feedback path are uncoupled.

Remark 7 The criterion is equivalent to a corollary obtained by Willems and Blankenship (1971, Criterion 4) for systems where the deterministic component of the feedback gain is constant. (Note that a factor $\frac{1}{2}$ was omitted in that reference in the statements of corollaries 3 and 4). Here it is shown that the criterion is valid for a much larger class of stochastic systems.

Remark 8 The Lyapunov function can also be used for the system

$$d\underline{x} = (\underline{Ax} - k(\underline{x},t)\underline{bcx})\, dt - f(\underline{x},t)\underline{bcx}\, d\beta \tag{25}$$

The stability conditions are :

(i) $G(s)$ positive real.

(ii) For all $\underline{x}$ and t :

$$k(\underline{x},t) \geq \frac{\sigma^2}{2} q_{n-1} (f(\underline{x},t))^2 \tag{26}$$

For asymptotic stability it suffices that either of the conditions holds in the strict sense, as explained above.

Remark 9 Consider the case where the mean is non-zero. For system (19) with $E(d\beta) = m$, the function $G(s)/(mG(s)+1)$ is required to be positive real. For system (25), inequality (26) should be replaced by

$$k(\underline{x},t) \geq \frac{\sigma^2 w(0)}{2} (f(\underline{x},t))^2 - mf(\underline{x},t)$$

For $k(\underline{x},t) = 0$ the system described by (16) is obtained.

Using the same technique for Lyapunov function construction as in the proof of Criterion 3 the following related results are readily derived :

<u>Criterion 4</u> The stochastic system (19) has a null solution which is mean square stable in the large and stable in the large with probability one if
(i) $G(s)^{-1} - a$ is a positive real function, and
(ii) for all $\underline{x}$ and t :

$$k(\underline{x},t) \geq a + \sigma^2 w(0)/2 \tag{27}$$

for some (positive or negative) constant a. The stability properties are asymptotic if either of both conditions are true in the strict sense.

<u>Criterion 5</u> The null solution of the stochastic system (19) is mean square stable in the large and stable in the large with probability one if there exist constants k_1 and k_2 such that
(i) The function

$$\frac{k_1 G(s)+1}{k_2 G(s)+1}$$

is positive real.
(ii) For all $\underline{x}$ and t

$$\frac{k_1+k_2}{2} - b \leq k(\underline{x},t) \leq \frac{k_1+k_2}{2} + b \tag{28}$$

with

$$b = \left[\frac{k_1-k_2}{2} - \frac{\sigma^2}{2}\left(q_{n-1}(k_1+k_2)+2p_{n-1}-2r_{n-1}\right)\right]^{1/2} \tag{29}$$

where r_{n-1} is the coefficient of s^{n-1} in the polynomial r(s) :

$$r(s) = \sqrt{\tfrac{1}{2}}\,\Big[(k_1q(s)+p(s))(k_2q(-s)+p(-s))+(k_1q(-s)+p(-s))(k_2q(s)+p(s)\Big]^+$$

that is the <u>positive</u> spectral factor of the even part of the polynomial $(k_1q(s)+p(s))(k_2q(-s)+p(-s))$.

<u>Remark 10</u> The frequency domain conditions of Criteria 4 and 5 have an interesting interpretation in terms of circle conditions. Condition (i) of criterion 4, requires the frequency response $G(j\omega)$ for positive a to lie completely within the circle shown in Fig. 1. For negative a, the frequency response $G(j\omega)$ should lie completely outside the circle shown in Fig. 2, and should encircle the disk as many times as G(s) has poles with positive real parts (unstable open loop poles). This corresponds to a criterion obtained by Blankenship and Willems for a much smaller class of stochastic systems. The geometric interpretation of condition (i) of Criterion 5 is shown

Fig. 1

Fig. 2

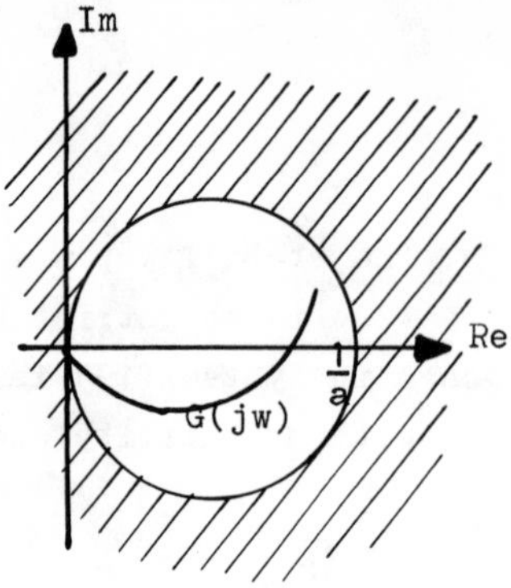

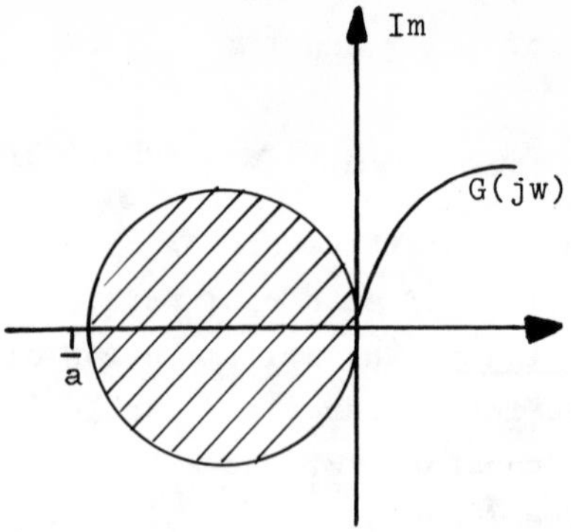

Frequency domain condition of Criterion 4.

Fig. 1 : a positive

Fig. 2 : a negative

Fig. 3

Frequency domain condition of Criterion 5

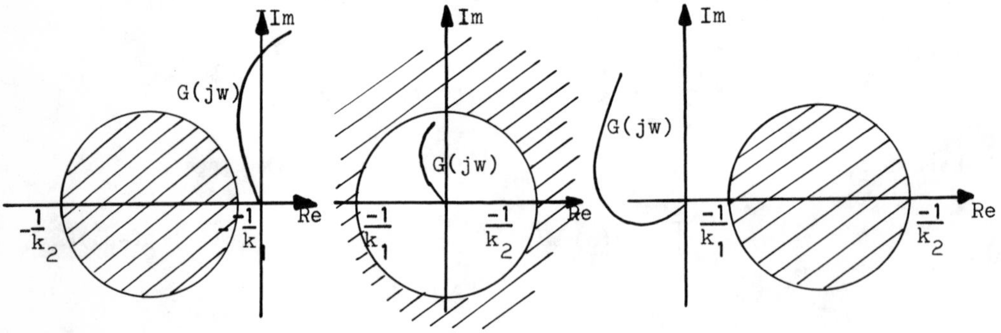

k_1, k_2 positive

k_1 positive, k_2 negative

k_1, k_2 negative

in Fig. 3.

Remark 11 The condition on the feedback element is different depending on whether the positive or negative spectral factor is used for generating the Lyapunov function. This is a particularity which does not appear for deterministic systems. This condition is least restrictive if the positive spectral factor is used. This same conclusion is obtained if the Kalman-Yacubovich-Meyer lemma is used for constructing the Lyapunov function as is shown below. The choice of the spectral factor has no effect on the stability condition if k_1 or k_2 are infinite, as in the cases considered in Criteria 3 and 4.

Remark 12 The same stability criteria and Lyapunov functions can also be obtained by means of the Kalman-Yacubovich-Meyer lemma and its generalizations instead of the path integral technique. Only Criterion 5 is discussed here. The positive real character of $(k_1G+1)/(k_2G+1)$ is sufficient for the existence of a negative definite symmetric matrix $\underline{P}$ satisfying the matrix inequality (Willems J.C. 1971b)

$$\underline{PA} + \underline{A}^T\underline{P} - (\underline{Pb}+m\underline{c}^T)(\underline{b}^T\underline{P}+m\underline{c}) + k_1k_2\underline{c}^T\underline{c} \geqslant \underline{0} \tag{30}$$

where $m = (k_1+k_2)/2$. The stability criteria are then obtained from the Lyapunov function $V(\underline{x}) = -\underline{x}^T\underline{Px}$. Here also the stability result depends on the choice of the solution $\underline{P}$ of the matrix inequality (30). Willems (1971b) has shown that the smallest $(-\underline{P})$ is obtained if the inequality (30) holds with the equality sign and if that particular solution is chosen such that the eigenvalues of $\underline{A}-\underline{b}(\underline{b}^T\underline{P}+m\underline{c})$ have negative real parts; this corresponds to the choice of the positive spectral factor as the solution of the factorization problem in the path integral method.

4. THE STOCHASTIC POPOV CRITERION

It is wellknown in deterministic stability theory that less restrictive stability results than the circle criteria can be obtained if the deterministic feedback gain is assumed to be either nonlinear and time-invariant or time-varying and linear. In this section it is shown that similar conclusions are obtained for the stochastic feedback system considered in Section 3. The case of a nonlinear time-invariant deterministic feedback is considered first, and a result similar to the Popov criterion is derived.

Consider the system considered in Section 3, but suppose that the deterministic feedback component is nonlinear, memoryless and stationary. The system is described by the Itô equation

$$d\underline{x} = (\underline{Ax} - f(\underline{cx})\underline{bcx})\, dt - \underline{bcx}\, d\beta \tag{31}$$

or

$$p(D)y\, dt + f(q(D)y)q(D)y\, dt + q(D)y\, d\beta = 0 \tag{32}$$

Criterion 6 The null solution of the stochastic system described by (31) or (32) is mean square stable in the large or stable in the large with probability one, if there exist a positive constant a such that
(i) the function

$$(1+as)q(s)/p(s)$$

is positive real, and
(ii) for all u :

$$f(u) \geq \frac{\sigma^2}{2}\left(aq_{n-2}p_{n-1} + q_{n-1} - 2r_n r_{n-1} + a\left(f(u) + u\frac{df(u)}{du}\right)q_{n-1}^2\right) \tag{33}$$

where r_n and r_{n-1} are the coefficients of D^n and D^{n-1} in the positive spectral factor of the even part of the polynomial $(1+as)q(s)p(-s)$.
The stability is asymptotic if either of the conditions holds in the strict sense.

The criterion is proved by means of the Lyapunov function

$$V(\underline{y}) = \int_{t(\underline{0})}^{t(\underline{y})} dt\left[(1+aD)q(D)y\; p(D)y - (r(D)y)^2 + aq(D)yf(q(D)y)q(D)y\right]$$

where

$$r(s) = \sqrt{\tfrac{1}{2}}\left[(1+as)q(s)p(-s) + (1-as)q(-s)p(s)\right]^+$$

Remark 13 The criterion reduces to the Popov criterion if σ^2 vanishes. Condition (33) is much simpler if q_{n-1} or w(0) vanishes, that is, if q(s) is of degree (n-2). Then (33) requires

$$f(u) \geq aq_{n-2}p_{n-1}\sigma^2/2$$

Criterion 6 reduces to the conditions of Criterion 3 if a = 0.

Next we consider the case of a linear deterministic feedback gain component. Then the system equation is

$$d\underline{x} = (\underline{Ax} - k(t)\underline{bcx})\, dt - \underline{bcx}\, d\beta \tag{34}$$

and the following stability criterion is obtained :

Criterion 7 The null solution of the stochastic system (34) is mean square stable in the large and stable in the large with probability one, if there exists a positive constant a such that the function $(1+as)q(s)/p(s)$ is positive real, and the gain $k(t)$ is nonnegative and satisfies (with the same notations as above)

$$k(t) \geq 2a\dot{k}(t) + \frac{\sigma^2}{2}(aq_{n-2}p_{n-1}+q_{n-1}-r_nr_{n-1}) \qquad (35)$$

for all t. It is asymptotically stable if either of both conditions is true in the strict sense.

Remark 14 Criterion 7 reduces to Criterion 3 if a = 0. Otherwise the condition on G(s) is less restrictive, but the condition (35) is stricter. By means of the techniques used in the stability analysis of deterministic feedback systems the criteria obtained in this section can be transformed and extended in various ways. Here Criteria 6 and 7 are given to illustrate the possible criteria one can obtain.

5. DISCUSSION

In this paper some stability criteria are derived for systems having a stochastic element in a feedback structure. The analysis has shown that the techniques for generating Lyapunov functions developed for the stability analysis of deterministic feedback systems also yield interesting stability results in the stochastic case. The analysis of this paper could be extended and generalized in various ways. Without conceptual problems, multivariable feedback systems can be dealt with; the path integral method is not well suited for this case, but the procedures indicated in Remark 6 and Remark 12 apply to multivariable feedback systems as well. Discrete systems could also be considered (Willems and Blankenship 1971); the analysis is even more straightforward, since the subtleties of Itô calculus disappear. It would be interesting to consider other types of noise (Blankenship 1972), and to see how the frequency domain conditions are affected. The weak point of the analysis of this paper is that only quadratic Lyapunov functions are generated, and all conditions are sufficient for mean square stability. It would be interesting to consider other types of Lyapunov functions in order to obtain less conservative conditions for stability with probability one, which would not necessarily imply mean square stability (Kushner 1967).

REFERENCES

Brockett, R.W., Finite Dimensional Linear Systems, New York : Wiley, 1970.

Blankenship, G.L., Stability of Uncertain Systems, Ph. D.thesis, M.I.T., Report ESL-R-448, June 1971.

Blankenship, G.L., Asymptotic properties of stochastic systems : a nonlinear integral equation, Technical Memo 24, Systems Research Centre, Case Western Reserve University, Cleveland, Ohio, 1972.

Kalman, R.E., Proceedings of the Nat. Academy of Science of the U.S.A., 49, pp. 201-205, 1963.

Kleinman, D.L., IEEE Trans. on Automatic Control, AC-14, 429-430, 1969.

Kozin, F., Automatica, 5, 95-112, 1969.

Kushner, H.J., Stochastic Stability and Control, New York : Academic Press, 1967.

Rabotnikov, Iu. L., Prikl. Mat. Mekh., 28, 935-940, 1964.

Willems, J.C., The Analysis of Feedback Systems, Cambridge, Mass. : M.I.T. Press, 1971(a).

Willems, J.C., IEEE Trans. on Automatic Control, AC-16, 621-634, 1971(b).

Willems, J.C., and Blankenship, G.L., IEEE Trans. on Automatic Control, AC-16, 292-299, 1971.

Willems, J.L., Stability Theory of Dynamical Systems, London : Nelson and New York : Wiley Interscience, 1970.

Zames, G., IEEE Trans. on Automatic Control, AC-11, 228-238 and 465-477, 1966.

STABILITY OF MODEL-REFERENCE SYSTEMS WITH RANDOM INPUTS

D.J.G. JAMES

Lanchester Polytechnic, Coventry, England

1. Introduction

In recent years model-reference adaptive control systems have proven to be one of the most popular methods in the growing field of adaptive control. The input to the system is also fed to a reference model, the output of which is proportional to the desired response; the outputs of the model and system are then differenced to form an error signal. Since this error signal is to be zero when the system is in its optimum state it is used as a demand signal for the adaptive loops which adjusts the variable parameters in the system to their desired values.

Various methods of synthesizing the adaptive loops have been proposed but the one that has proven most popular is that developed by Whitaker et.al. (1961) and referred to as the sensitivity or MIT rule. Here the performance criterion is taken as the integral of error squared and this leads to a rule that a particular parameter be adjusted according to the rule

$$\text{Rate of change of parameter} = -\text{Gain} \times (\text{error}) \times \frac{\partial(\text{error})}{\partial(\text{parameter})}$$

Although the MIT rule results in practically realizable,systems mathematical analysis of the adaptive loops, even for simple inputs, prove to be very difficult. A stability analysis for sinusoidal input has been previously considered by the author (James 1969, 1971); however, in practice, a more realistic input is a random one and the purpose of this paper is to investigate stability for such an input. Since the object is to pose the problem and illustrate the difficulties involved we shall limit our discussion to a first order MIT system with controllable gain.

2. Linear stochastic systems

All the stability problems considered in this paper reduce to one of investigating the stability of a system of linear differential equations with random coefficients. A vast amount of literature dealing with such systems has been published in recent years and various types of stability have been proposed (Kozin 1966, 1969). In this work we shall confine ourselves to three types of stability, namely, stability in mean, stability in mean square and almost sure asymptotic stability (a.s.a.s.) and will investigate such stability under two types of random coefficient variations, viz: (i) Gaussian white noise processes and (ii) Gaussian non-white processes.

2.1 Gaussian white noise processes

We shall be concerned with a system of equations of the form

$$\dot{\underset{\sim}{x}}(t) = \underset{\sim}{A}\,\underset{\sim}{x}(t) + \underset{\sim}{B}(t)\,\underset{\sim}{x}(t) + \underset{\sim}{\beta}(t) \qquad \forall t \in [0,\infty) \tag{1}$$

where $\underset{\sim}{A}$ is an $n \times n$ constant matrix, $\underset{\sim}{B} = (\beta_{ij}(t))$, $i, j = 1, 2, \ldots, n$ an $n \times n$ matrix and $\underset{\sim}{\beta} = (\beta_{i0}(t))$, $i = 1, 2, \ldots, n$ an n column vector of Gaussian white noise processes having properties

$$E\{\beta_{ij}(t)\} = 0, \quad E\{\beta_{ij}(t)\beta_{rs}(t+\tau)\} = 2D_{ijrs}\,\delta(\tau), \quad \begin{matrix} i,r = 1,2,\ldots,n \\ j,s = 0,1,2,\ldots,n \end{matrix}$$

Such a system has been studied extensively in the literature (Ariaratnam and Graeffe 1965, Caughey and Dienes 1962, Bogdanoff and Kozin 1962) and we shall confine ourselves here to a brief outline of the method of stability analysis.

The response of system (1) is a continuous n-dimensional Markov process and such processes are completely described by the Fokker-Planck equation

$$\frac{\partial p}{\partial t} = -\sum_{i=1}^{n} \frac{\partial}{\partial x_i}(A_i p) + \frac{1}{2}\sum_{i=1}^{n}\sum_{j=1}^{n} \frac{\partial^2 (B_{ij} p)}{\partial x_i \partial x_j} \tag{2}$$

appropriate to the system, where $p = p(\underset{\sim}{x}, t / \underset{\sim}{x}_0, t_0)$ and

$$A_i = \lim_{\delta t \to 0} \frac{E\{\delta x_i\}}{\delta t}, \quad B_{ij} = \lim_{\delta t \to 0} \frac{E\{\delta x_i \delta x_j\}}{\delta t}; \quad i,j = 1,2,\ldots,n$$

In stability investigation a knowledge of the moments of the system response is usually sufficient and a system of first order differential equations determining the moments of order N are readily obtained from (2) by multiplying throughout by $x_1^{K_1} x_2^{K_2} \cdots x_n^{K_n}$, $K_1 + K_2 + \cdots + K_n = N$, and integrating by parts over the entire state space. Necessary and sufficient conditions for stability in the mean and mean square are then readily obtained by applying the Routh-Hurwitz criteria to the system of differential equations obtained in the cases $N = 1, 2$ respectively.

2.2 Gaussian non-white processes

If the coefficient variations are Gaussian but non-white then the response of the system no longer forms a Markov process. However it is possible to construct linear time invariant filters which, with Gaussian white noise as input, will have the required non-white Gaussian coefficient variations as output. The response of the total system, which includes these linear filters, will then form a Markov process. Unfortunately, the additional state variables introduced by the filters render the system equations non-linear with the result that the moment equations, obtained from the appropriate Fokker-Planck, can no longer be solved recursively and one cannot obtain criteria for stability in the mean square.

We shall be concerned with linear systems of the form

$$\dot{\underset{\sim}{x}}(t) = [\underset{\sim}{A} + \sum_{i=1}^{R} f_i(t)\underset{\sim}{G}_i]\underset{\sim}{x}(t), \; R < n^2, \; \forall t \in [t_0, \infty) \tag{3}$$

where $\underset{\sim}{A}$ is an $n \times n$ constant stability matrix, $f_i(t)$ are stationary ergodic processes and $\underset{\sim}{G}_i$ constant matrices.

Sufficient conditions guaranteeing a.s.a.s. of (3) have been obtained by many authors and the most recent improvements in the stability criteria obtained appear to be those due to Infante (1968) and Man (1970). By applying the results of the extremal properties of the eigenvalues of pencils of quadratic form Infante showed that a sufficient condition for a.s.a.s. in the large of system (3) is that

$$\sum_{i=1}^{R} \tfrac{1}{2} E\{|f_i(t)|\}[\lambda_{max}(\underset{\sim}{G}_i' + \underset{\sim}{B}\underset{\sim}{G}_i\underset{\sim}{B}^{-1}) - \lambda_{min}(\underset{\sim}{G}_i' + \underset{\sim}{B}\underset{\sim}{G}_i\underset{\sim}{B}^{-1})] \leq -\lambda_{max}(\underset{\sim}{A}' + \underset{\sim}{B}\underset{\sim}{A}\underset{\sim}{B}^{-1}) \tag{4}$$

where $\lambda_{max}(\cdot)$ denotes the largest eigenvalue of the matrix $(\cdot)$ and $\underset{\sim}{B}$ is a symmetric positive definite matrix. By simultaneously reducing two quadratic forms to diagonal from Man extended the development of Infante to obtain the sufficient condition for a.s.a.s. in the form

$$\sum_{i=1}^{R} E\{|f_i(t)|\}[\lambda_{max}(\underset{\sim}{G}_i\underset{\sim}{P} + \underset{\sim}{P}\underset{\sim}{G}_i)\underset{\sim}{Q}^{-1}] < 1 \tag{5}$$

where $\underset{\sim}{P}$ and $\underset{\sim}{Q}$ are positive definite constant matrices satisfying

$$\underset{\sim}{A}'\underset{\sim}{P} + \underset{\sim}{P}\underset{\sim}{A} = -\underset{\sim}{Q}$$

3. Gain adaption model reference system

Consider a model and system to be governed respectively by the equations

$$\begin{aligned} T\dot{\theta}_m(t) + \theta_m(t) &= K\,\theta_i(t) \\ T\dot{\theta}_s(t) + \theta_s(t) &= K_v K_c\,\theta_i(t) \end{aligned} \tag{6}$$

where the time constant T and model gain K are constant and known, but the process gain K_v is unknown and possibly time varying. The problem here is to determine a suitable adaptive loop to control K_c so that $K_v K_c$ eventually equals the model gain K. The MIT rule gives

$$\dot{K}_c = -\,G e\,(\partial e/\partial K_c) = B e\,\theta_m(t) \tag{7}$$

where $B = GK_v/K$, and this leads to the scheme of fig.1.

4. Random Input

4.1 Stability analysis

If to the system of fig.1 a general random signal $\alpha(t)$ is applied at time $t=0$, when $\theta_m(t), \theta_s(t)$ are zero and $K_v K_c \neq K$ and if subsequently K_v remains constant but K_c is adjusted according to (7) then the system equations (6) become

$$\begin{aligned} T\dot{\theta}_m(t) + \theta_m(t) &= K\alpha(t) \\ T\dot{e}(t) + e(t) &= x(t)\,\alpha(t) \\ \dot{x}(t) &= -B K_v\, e(t)\,\theta_m(t) \end{aligned} \tag{8}$$

where $x(t) = K - K_v K_c(t)$.

Despite the recent progress in stochastic stability theory, methods of investigating the stability of (8), where the system is not asymptotically stable when the noise terms are equated to zero, are not forthcoming. However, digital simulation of the system, which will be discussed in section 4.2, suggests that stability boundaries exist for such an input. In order to have a first look at the problem, we shall assume that

$$\alpha(t) = \sum_{r=0}^{N} A_r\,\delta(t - \tau r) \tag{9}$$

where $\tau \gg T$ and $A_r (r = 0, 1, 2, \ldots, N)$ are random variables drawn from an amplitude probability distribution $p(y)$.

Substituting (9) in (8) and solving within the time interval $r\tau < t < (r+1)\tau$ furnishes the following recurrence relationship for $x(r\tau)$

Fig.1 <u>First order system - MIT gain adaption</u>

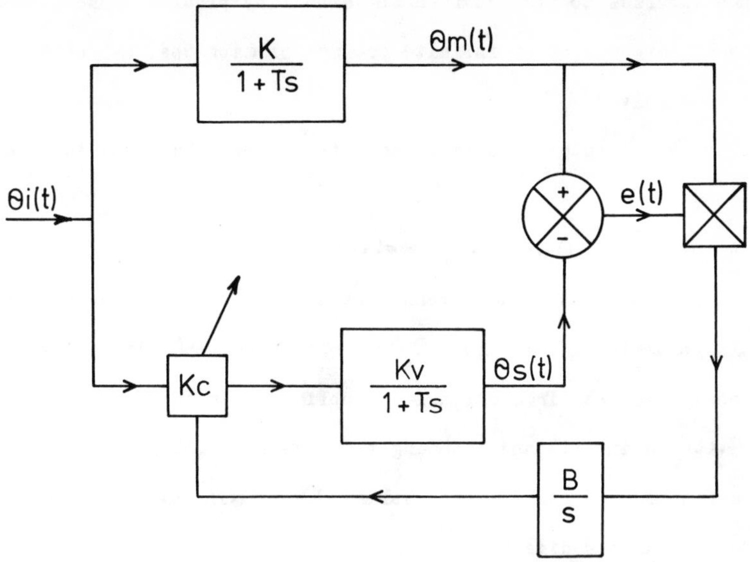

$$x(\overline{r+1}\,\tau) = \left[1 - \frac{KK_v BA_r^2}{2T}\right] x(r\tau)$$

which on successive application leads to

$$x(r+1\,\tau) = x(0)\prod_{s=0}^{r}\left[1 - \frac{KK_v BA_s^2}{2T}\right]$$

Thus in order to investigate the stability of the system we must examine, by letting $r \to \infty$ the convergence of the infinite product

$$P = \prod_{r=0}^{\infty}\left[1 - aA_r^2\right], \quad a = \frac{KK_v B}{2T} > 0.$$

By considering a large number of terms and their distribution and by considering the logarithm of the product of these terms we are led to consider the integral

$$I = \int_0^{\infty} \log|1 - ay^2|\, p(y)\, dy \tag{10}$$

If $I > 0$ then P diverges and the system is unstable; if $I < 0$ then P diverges to zero, $x(r\tau) \to 0$ as $\tau \to \infty$, and the system is stable. If y has a Gaussian distribution with zero mean and variance σ^2 then (10) leads to the stability criterion $a\sigma^2 < 2.5$, a result which has been confirmed by direct digital simulation of the infinite product .

If one attempts to develop the above approach a step further and allow the effects of the impulses to overlap then the stability problem reduces to one of examining the convergence of an infinite product of matrices, a problem which, as yet, has not been solved.

In order to avoid the difficulties encountered when the input is purely random let us take

$$\theta_i(t) = R + \alpha(t) \tag{11}$$

where R is a constant and $\alpha(t)$ a random variable. Since the system of fig.1 is stable for all parameter values when $\theta_i(t) = R$ (Parks 1966) it follows that the stability theory discussed in section 2 is applicable in this case.

Taking $\theta_i(t)$ as in (11) and assuming that the adaption is switched on when $\theta_m(t)$ has reached its steady-state value $(KR + \beta(t))$ the system equations (8) may be written in the non-dimensional form

$$\frac{d}{d\tau}\begin{bmatrix}\xi_1(\tau)\\ \xi_2(\tau)\end{bmatrix} = \begin{bmatrix}-1 & -1\\ \pi_1 & 0\end{bmatrix}\begin{bmatrix}\xi_1(\tau)\\ \xi_2(\tau)\end{bmatrix} + \begin{bmatrix}0 & -\pi_2\alpha_1(\tau)\\ \pi_1\pi_3\beta_1(\tau) & 0\end{bmatrix}\begin{bmatrix}\xi_1(\tau)\\ \xi_2(\tau)\end{bmatrix} \tag{11}$$

where $\pi_1 = BKK_vR^2T$, $\pi_2 = \sigma_i^2/R$, $\pi_3 = \sigma_o^2/KR$, σ_i^2, σ_o^2 being the variances of $\alpha(t)$ and $\beta(t)$ respectively; $\xi_1 = e(t)/(KR)$, $\xi_2 = (K_vK_c(t) - K)/K$, $\tau = t/T$ and $\alpha_1(\tau), \beta_1(\tau)$ are the normalized form of $\alpha(\tau)$ and $\beta(\tau)$ respectively.

Since $\beta(t)$ is a filtered form of $\alpha(t)$ white noise assumptions on the coefficients of (11) may not be made so that a Fokker-Planck type analysis is not suitable. Before applying criteria (4) and (5) for a.s.a.s. we require first to obtain a relationship between σ_i and σ_o.

If $G_i(\omega)$ and $G_o(\omega)$ are the power spectral densities of $\alpha(t)$ and $\beta(t)$ respectively then

$$G_o(\omega) = K^2 G_i(\omega)/(1 + T^2\omega^2) \tag{12}$$

Since in the digital simulation of the system the numerical procedure commits us to a small step length we shall take

$$G_i(\omega) = \sigma^2 h/\pi = 3\sigma_i^2 h/(2\pi) \tag{13}$$

where h, σ defined as in section 4.2. Substituting (13) in (12) and applying the Wiener-Khintchine relationship gives

$$\sigma_0^2 = \int_0^\infty G_0(\omega)\,d\omega = \frac{\sigma^2 h K^2}{2\pi}$$

so that $\pi_3^2 = \frac{1}{4}\pi_4\pi_5^2$, where π_4 and π_5 are the non-dimensional parameters h/T and σ/R respectively.

By taking $\underset{\sim}{B}$ to be the most general positive definite matrix (Infante 1968) and then optimizing, criterion (4) gives the system (11) is a.s.a.s. if

$$\pi_5^2 < \frac{3\pi}{16} \cdot \frac{4\pi_1 - 1}{\pi_1^2\left[1 + (3\pi_4/4)^{1/2}\right]^2} \tag{14}$$

Taking $\underset{\sim}{Q}$ to be the identity matrix criterion (5) gives that system (11) is a.s.a.s. if

$$\pi_5 < \frac{\sqrt{3\pi}}{-1 + \sqrt{1 + (1+\pi_1)^2} + \sqrt{.75\pi_4}\left[\pi_1 + \sqrt{\pi_1^2 + (8+\pi_1)^2}\right]} \tag{15}$$

4.2 <u>Digital simulation</u>

The basic numerical procedure generates a sequence of normally distributed random numbers (μ, σ). By spacing these numbers at interval h and joining them by straight lines, a segmentally linear function is obtained that approximate white noise; the degree of approximation being dependent on h. It can be shown from the autocorrelation function (James 1972) that, for small values of ωh, the power spectral density of the function is given by

$$G(\omega) = \frac{\sigma^2 h}{\pi}\left[1 - \frac{(\omega h)^2}{6}\right]$$

Thus, by making h sufficiently small, it follows that the function produced by the numerical procedure approximates white noise $\alpha(t)$ having time domain statistical properties

$$E\{\alpha(t)\} = 0 \quad , \quad E\{\alpha(t)\,\alpha(t+\tau)\} = \sigma^2 h\,\delta(\tau)$$

Using the above numerical subroutine to generate the random variable $\alpha(t)$ the system of fig.1, with input given by (11), is simulated on a digital computer; two methods of numerical integration being employed, as a check, using in turn the Runge-Kutta and Crank-Nicolson procedures. In order to obtain the variations of the ensemble moments with time the system equations are solved m times simultaneously with each set having a different $\alpha(t)$. The mean square values, over the m outputs, of both $e(t)$ and $K - K_v K_c(t)$ are printed out at each instant of time and a decision taken as to whether the system is stable or unstable in the mean square.

Stability boundaries, in parameter space $\pi_1 - \pi_5$, obtained using both the numerical methods of integration, are shown in fig.2. Also shown in fig.2 is the theoretical stability boundary corresponding to criterion (14); the boundary corresponding to criterion (15) is slightly inferior and is therefore not included in the figure.

Although we have been unable to obtain theoretical results when the input is purely random nevertheless the system was simulated digitally under such conditions. Stability boundaries, obtained using both numerical procedures, in the $\pi_6 - \pi_7$ plane, where π_6 and π_7 are the non-dimensional parameters $\pi_6 = \sigma^2/K_v^2$, $\pi_7 = BKK_v^3T$, are shown in fig.3. Also plotted in fig.3 is the rectangular hyperbola $\pi_6\pi_7 = 175$ and approximately this represents the mean of the two boundaries obtained numerically. Thus, a form of stability criterion one would like to obtain theoretically is $\pi_6\pi_7 < 175$. Substituting for π_6 and π_7 this condition suggests, on preceeding to the limit $h \to 0$, that the system is unstable, for all parameter values, when the input is pure white noise.

5. Random variations in environment

So far we have assumed K_v constant. However, in practice, the adapting parameter may be required to change not only in response to the system input signal but also to changes in the environment so that a stability analysis when K_v is time varying is also essential. We shall consider the system of fig.1 to be subjected to a step input of magnitude R at time $t = 0$ and subsequently let K_v vary with time according to

$$K_v(t) = S + \alpha(t) \tag{16}$$

where S is a constant and $\alpha(t)$ a stationary gaussian white noise process with statistics $E\{\alpha(t)\} = 0$, $E\{\alpha(t)\alpha(t+\tau)\} = 2D\,\delta(\tau)$.

Substituting (16) in the system equations then white noise assumptions on the coefficients may be made and the corresponding Fokker-Planck equation is

$$\frac{\partial p}{\partial \tau} = \frac{\partial}{\partial \xi_1}(\xi_1 p) - \frac{\partial p}{\partial \xi_1} - \pi_8 \xi_1 \frac{\partial p}{\partial \xi_2} + \pi_9 \xi_2^2 \frac{\partial^2 p}{\partial \xi_1^2} + \xi_2 \frac{\partial p}{\partial \xi_1} \tag{17}$$

where $\pi_8 = BKR^2ST$, $\pi_9 = D/(TS^2)$, $\xi_1 = e/(KR)$, $\xi_2 = SK_c/K$ and $\tau = t/T$

By considering the first order moment equations, deduced from (17), it is readily shown that the system is stable in the mean for all parameter values and that both $e(t)$ and $K - K_v(t) K_c(t)$ tend to zero, in the mean, as required. Similarly, by considering the second order moment equations, deduced from (17), it can be shown that the system is stable in the mean square if and only if

$$\pi_8 \pi_9 < 1 \tag{18}$$

Further, when condition (18) holds, the asymptotic values, as $\tau \to \infty$, of the second moments are obtained in the form

$$\lim_{\tau \to \infty} \langle \xi_1^2(\tau) \rangle = \frac{\pi_9}{1 - \pi_8 \pi_9}, \quad \lim_{\tau \to \infty} \langle \xi_1(\tau)\xi_2(\tau) \rangle = 0, \quad \lim_{\tau \to \infty} \langle \xi_2^2(\tau) \rangle = \frac{1}{1 - \pi_8 \pi_9}$$

Using the same procedure as described in section 4.2 the system equations, with K_v given by (16), are simulated on a digital computer. Stability boundaries obtained using both numerical procedures are shown in fig.4 together with the stability boundary corresponding to (18).

6. Conclusions

It is apparent from the problems considered in this paper that a great deal of further research is required in analysing the stability of linear differential equations with random coefficients. If the system equations may be written in the form $\dot{\underset{\sim}{x}}(t) = [\underset{\sim}{A} + \underset{\sim}{F}(t)]\underset{\sim}{x}$, where $\underset{\sim}{A}$ is a constant stability matrix and the non vanishing elements of $\underset{\sim}{F}(t)$ are random processes, then two different cases have been considered.

(i) When the elements of $\underset{\sim}{F}(t)$ are Gaussian white noise processes the Fokker-Planck equation has been used to obtain necessary and sufficient conditions for stability in the mean square. The results suggest that the stability boundaries obtained using this method agree favourably with those obtained by direct simulation of the system.

(ii) When the elements of $\underset{\sim}{F}(t)$ are Gaussian non-white processes the Fokker-Planck equation approach is essentially a non-linear problem for which there is, to date, no satisfactory solution. The stability criteria based on Liapunov's second method, which constitute sufficient conditions for a.s.a.s., have proven to be highly conservative when compared with results obtained by digital simulation.

Fig.2 **Stability boundaries for step plus random input**

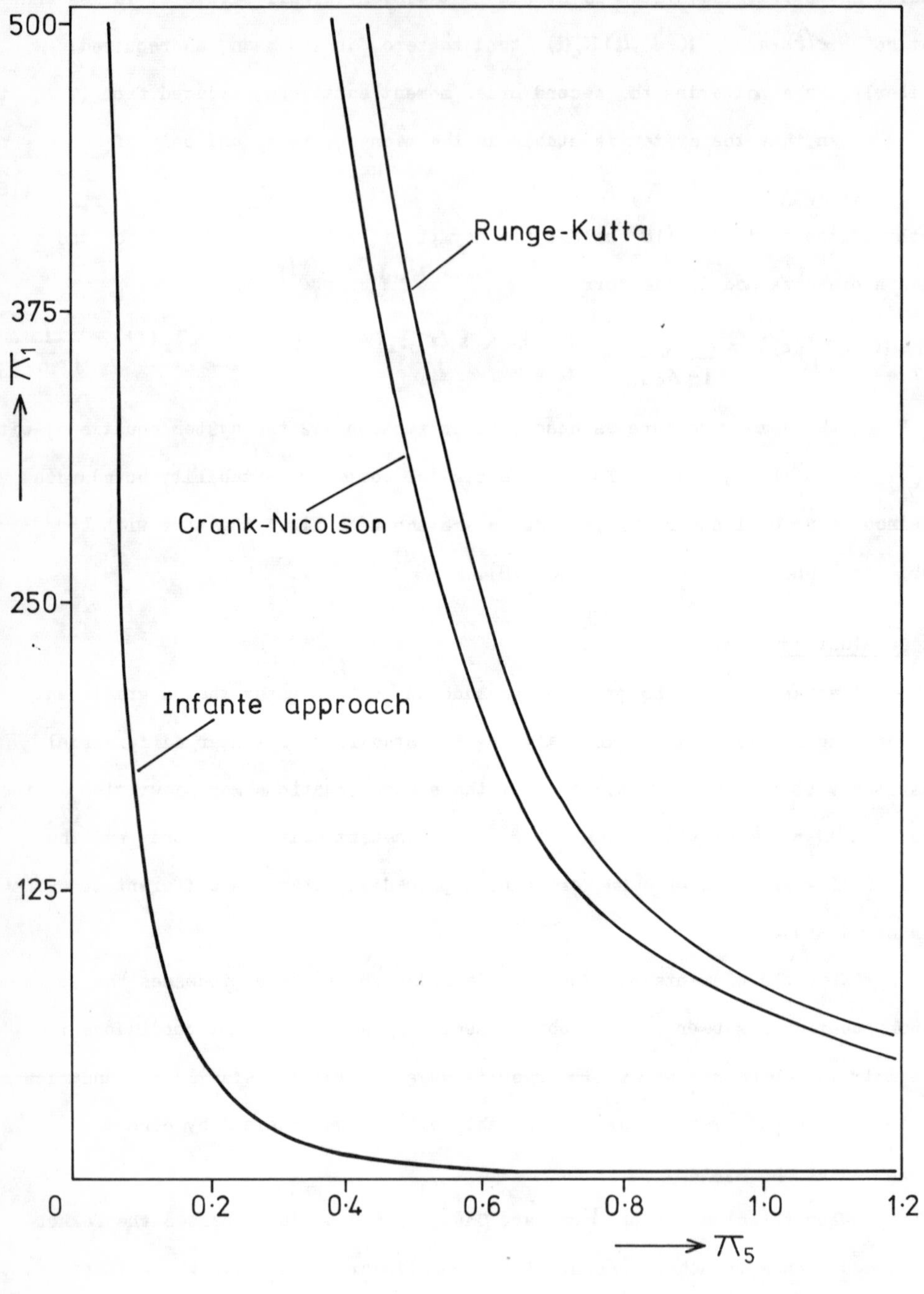

Fig.3 **Stability boundaries for random input**

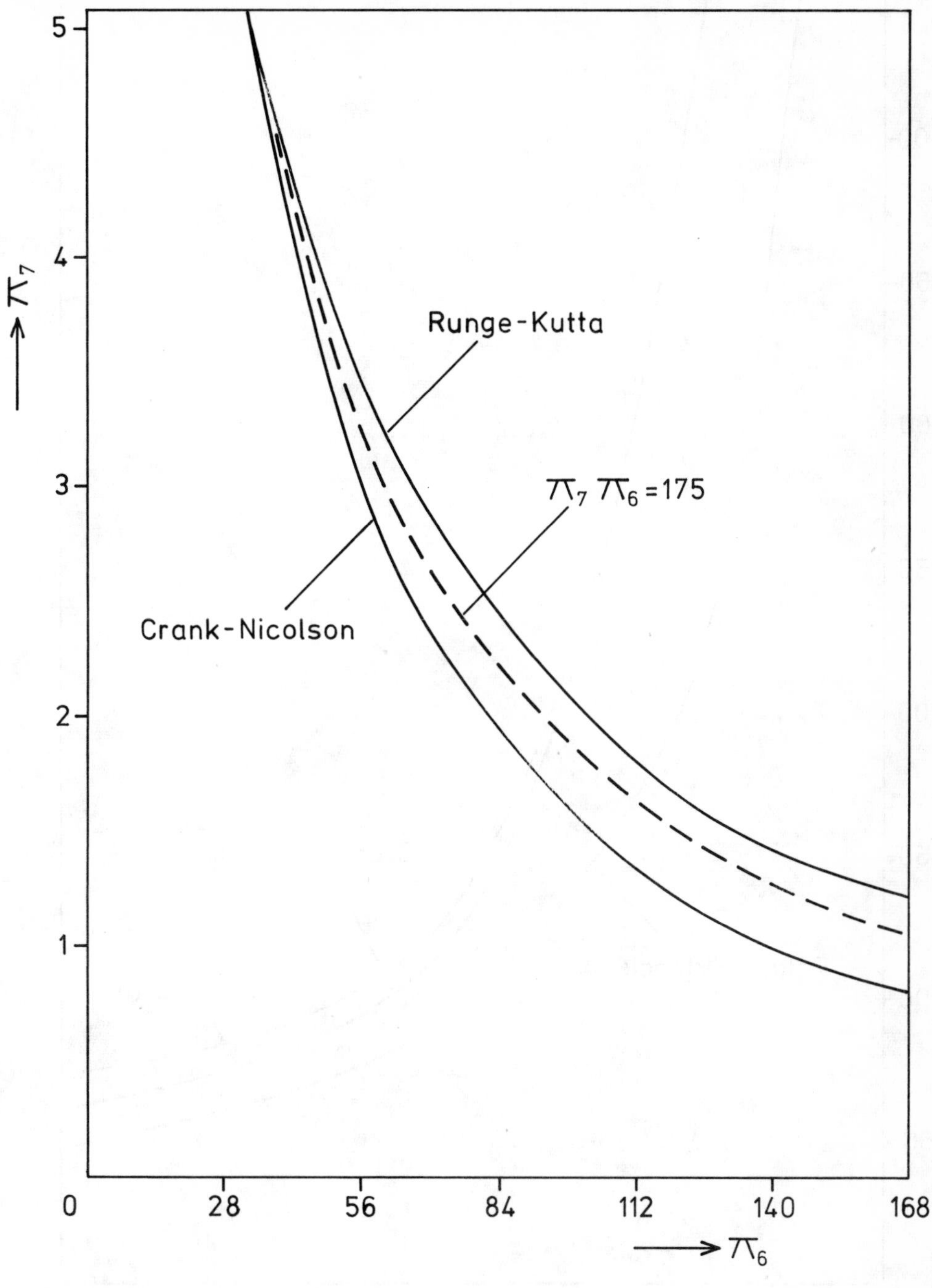

Fig.4 **Stability boundaries when K_V varies randomly**

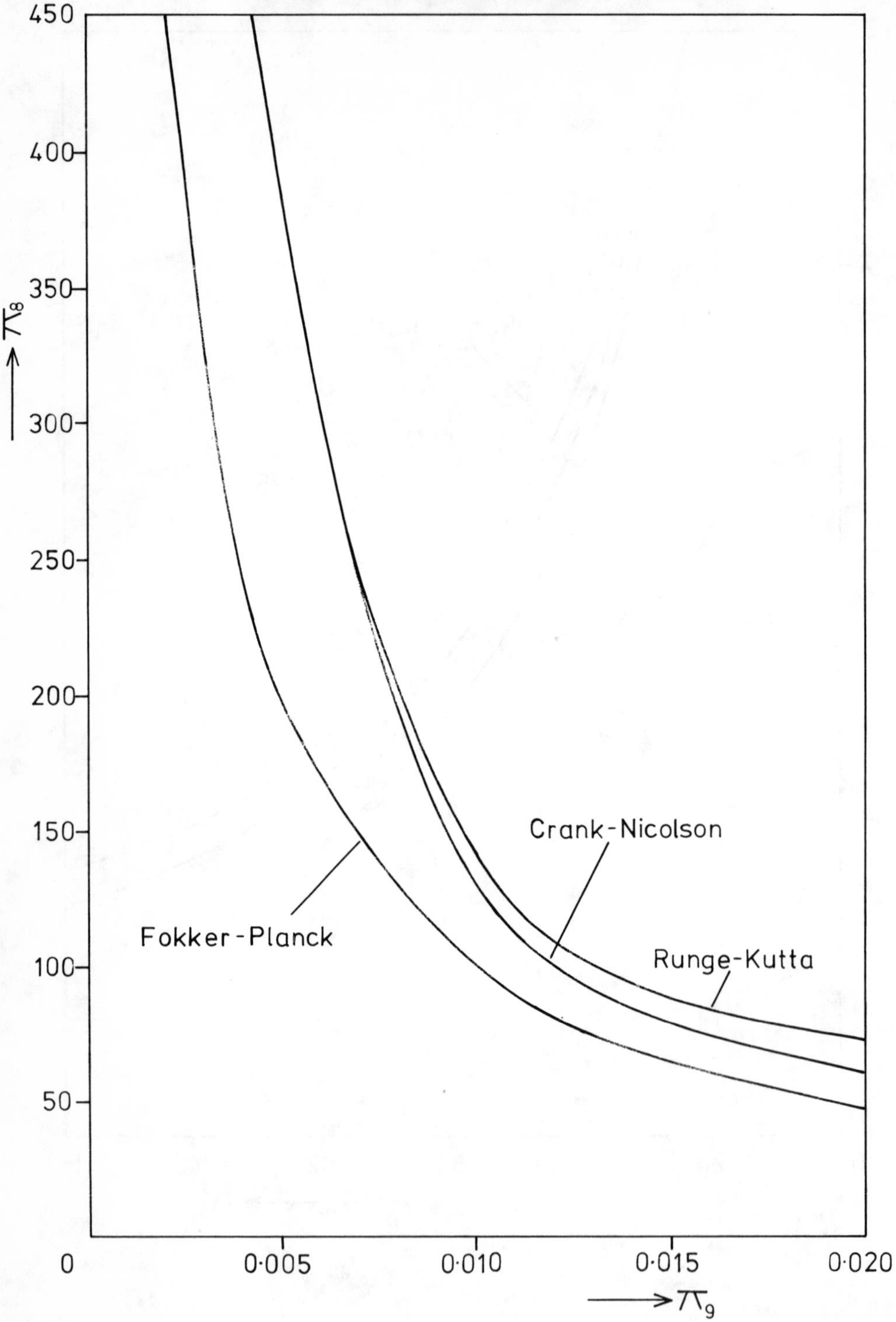

The digital simulation results for the case when the input is purely random suggests that an outstanding problem is that of obtaining stability criteria for the linear system $\dot{\underset{\sim}{x}}(t) = \underset{\sim}{A}(t)\,\underset{\sim}{x}(t)$, where the time-varying elements of $\underset{\sim}{A}(t)$ are correlated Gaussian non-white processes and the system is not asymptotically stable when the time-varying elements of $\underset{\sim}{A}(t)$ are made identically zero.

ACKNOWLEDGMENT

The author would like to express his gratitude to Dr P. C. Parks, University of Warwick, for his helpful discussions concerning the work described in this paper.

REFERENCES

ARIARATNAM, S.T., and GRAEFE, P.W.U., 1965, Int.J.Control, 1, 239; 1965, Int.J.Control, 2, 161; 1965, Int.J.Control, 2, 205.

BOGDANOFF, J.J., and KOZIN, F., 1962, J.Acoust.Soc.Am., 34, 1065.

CAUGHEY, T.K., and DIENES, J.K., 1962, J.Math. and Phys., 41, 300.

INFANTE, E.F., 1968, A.S.M.E.Jour.App.Mech., 5, 7.

JAMES, D.J.G., 1969, Int.J.Control, 9, 311; 1971, Amer.Inst.Aero.Astro., 9, 950; 1972, Int.J.Control, to be published.

KOZIN, F., 1966, Paper 3A, 3rd IFAC Congr., London; 1969, Automatica, 5, 95.

MAN, F.T., 1970, A.S.M.E.Jour.App.Mech., 37, 541.

PARKS, P.C., 1966, IEEE Trans.Aut.Cont., AC-11, 362.

WHITAKER, M.P., OSBURN, P.V., and KEZER, A., 1961, Inst.Aero.Sci., paper 61-39.

REGIONS OF INSTABILITY FOR A LINEAR SYSTEM WITH RANDOM PARAMETRIC EXCITATION

W. WEDIG

Universität Karlsruhe (TH), Karlsruhe, Germany

INTRODUCTION

The dynamic stability of elastic structures under fluctuating loads has been extensively investigated. In simple cases, this study leads to a linear second order differential equation, in which the external so-called parametric excitation appears as a time dependent coefficient. A technical example for such stability problems is a uniform pin-ended column, subjected to a time varying axial force.

The stability properties of this system for periodic excitations are well-known by the Mathieu equation. According to the corresponding Strutt diagram [1] we know those ranges of values for the excitation frequencies and amplitudes, for which the position of equilibrium of the system becomes unstable. The first two, and most important, instability regions of this diagram are situated near twice the natural frequency of the system and near the natural frequency itself.

If the external excitation fluctuates in a stochastic manner, the stability equation mentioned above contains a stochastic coefficient. The zero position of the system is then said to be stable in mean square, when all second order moments of its state coordinates remain bounded, if the time tends to infinity [2].

This stability definition has been applied in the case of Gaussian white noise [3], which has the mean value zero and a flat spectrum. The corresponding stability condition may be graphically illustrated (see fig. 1), if we record the critical spectral density suitably multiplied by the natural frequency of the system. Comparing this figure with the stability map mentioned above, we find the same threshold value. But, in contrast to harmonic excitations, we have here a constant stability limit and we find no instability regions, the limits of which vary above the excitation frequencies in a similar manner. It is physically obvious, that such stability limits must exist for non-white noise coefficients of the stability equation. These limits are to be derivated in the following.

STATE EQUATIONS

If we select the initially mentioned column under end thrust as a simple example of stability problems, the corresponding stability equation may be taken in the form

$$\ddot{y} + 2\omega_1 D\dot{y} + \omega_1^2[1+\varepsilon x(t)]y = 0. \tag{1}$$

The unknown time function $y(t)$ of this equation then describes the transverse motion of the column, ω_1 is its natural frequency, ε is a dimensionless parameter of the external excitation $\varepsilon x(t)$ and D is a dimensionless coefficient as well, by means of which we are assuming a viscous damping in the material of the column.

For such mechanical problems we can restrict our interest to lightly damped and weakly excited systems, so that ε and D are small in comparison to one $(\varepsilon, D \ll 1)$. It is then possible to make the exponential substitution

$$y(t) = T(t)\exp(-\omega_1 Dt) \tag{2}$$

for $y(t)$ and to convert the original equation (1) into the shorter form

$$\ddot{T} + \nu_1^2[1+\varepsilon_1 x(t)]T = 0,$$

$$\varepsilon_1 = \varepsilon/(1-D^2), \qquad \nu_1 = \omega_1\sqrt{1-D^2},$$

containing the small parameter ε_1 and the damped natural frequency ν_1 of the system. This stability equation can be written as a set of state equations

$$T = z_1 + z_2, \quad \dot{T} = i\nu_1(z_1 - z_2),$$

$$\dot{z}_1 = +i\nu_1[z_1 + \varepsilon_1 x(z_1+z_2)/2], \tag{3a}$$

$$\dot{z}_2 = -i\nu_1[z_2 + \varepsilon_1 x(z_1+z_2)/2], \tag{3b}$$

where z_1 and z_2 are the generalized coordinates of the system.

LOW-PASS PROCESSES

We shall first of all discuss a low-pass process $x(t) = z_3(t)$. Such processes can be obtained by passing Gaussian white noise $\xi(t)$ through a low-pass filter

$$\dot{z}_3 + \omega_g z_3 = \omega_g \xi(t),$$

having an arbitrary limit frequency ω_g. After a short time, the filter's response is stationary and its spectral density has the well-known form

$$S_x(\omega) = S_o \omega_g^2/(\omega^2+\omega_g^2).$$

Herein the coefficient S_o is the constant spectrum of $\xi(t)$, so that its auto-correlation function is given by $E[\xi(t_1)\xi(t_2)] = S_o \delta(t_1 - t_2)$.

Since the input $\xi(t)$ of the filter is Gaussian white noise, its response $x(t) = z_3(t)$ as well as the state coordinates z_1 and z_2 of (3) form Markov processes. Therefore, the conditional probability density p of all coordinates z_1, z_2 and z_3 is given by the corresponding Fokker-Planck equation

$$\frac{\partial p}{\partial t} = -\sum_{i=1}^{3} \frac{\partial}{\partial z_i}(a_i p) + \frac{1}{2}\sum_{i,j=1}^{3} \frac{\partial^2}{\partial z_i \partial z_j}(b_{i,j} p), \qquad a_i = \lim_{\Delta t \to 0} \frac{1}{\Delta t} E[\Delta z_i],$$
$$p = p(\underset{\approx}{z}, t; \underset{\approx}{z}_o, t_o), \qquad \underset{\approx}{z} = (z_1, z_2, z_3)^T, \qquad b_{i,j} = \lim_{\Delta t \to 0} \frac{1}{\Delta t} E[\Delta z_i \Delta z_j],$$

in which the coefficients a_i and $b_{i,j}$ are the incremental moments and may be computed by the method indicated in [4].

At present, there is no general method of solving the Fokker-Planck equation available. For our purposes, however, a knowledge of the expectation values of all state coordinates is sufficient. According to the definition of these moments

$$M_{k_1,k_2,k_3} = E[z_1^{k_1} z_2^{k_2} z_3^{k_3}] = \iiint_{-\infty}^{+\infty} z_1^{k_1} z_2^{k_2} z_3^{k_3} p \, dz_1 dz_2 dz_3,$$

we have to multiply both sides of the Fokker-Planck equation by powers of the state coordinates and to integrate it over the entire state plane.

As a result of this procedure, we then obtain the differential equations of the moments, which are best written in matrix form [5].

$$\frac{1}{2\nu_1}\dot{M}_k = A_k M_k + \eta_g \sigma_x^2 k(k-1) M_{k-2} + \varepsilon_1 R M_{k+1}, \tag{4}$$

$$M_k = \begin{Bmatrix} E[z_1^2 z_3^k] \\ E[z_1 z_2 z_3^k] \\ E[z_2^2 z_3^k] \end{Bmatrix}, \quad A_k = \begin{bmatrix} i-k\eta_g & 0 & 0 \\ 0 & -k\eta_g & 0 \\ 0 & 0 & -i-k\eta_g \end{bmatrix}, \quad R = \frac{i}{2}\begin{bmatrix} 1 & 1 & 0 \\ -1/2 & 0 & 1/2 \\ 0 & -1 & -1 \end{bmatrix}.$$

They contain the dimensionless limit frequency $\eta_g = \omega_g/2\nu_1$ and the variance $\sigma_x^2 = \omega_g S_0/2$ of the low-pass process. The matrices A_k and R of (4) are noted above. The index k of the moment vector M_k is a positive integer and indicates the power of the low-pass process. If k is equal to zero, the components of the equation's vector are the second order moments of the system (3), whose behaviour we have to examine for the stability decision. It becomes evident, that these moments are weakly coupled by ε_1 with higher order moments, so that we actually have a sequence of linear differential equations with constant coefficients.

APPROXIMATE SOLUTION

A sufficiently general solution of the homogeneous equations (4) is the exponential function

$$M_k(t) = C_k \exp(2\nu_1 \varrho t) \tag{5}$$

with an amplitude vector C_k and the unknown eigenvalue ϱ adequately multiplied by $2\nu_1$. When we introduce this function into the moments' equations (4), we get an infinite system of homogeneous algebraic equations.

$$\begin{bmatrix} A_0-\varrho E & \varepsilon_1 R & & \\ & A_1-\varrho E & \varepsilon_1 R & \\ 2\eta_g\sigma_x^2 E & & A_2-\varrho E & \\ & 6\eta_g\sigma_x^2 E & & \\ & & 12\eta_g\sigma_x^2 E & \\ & & & \end{bmatrix} \begin{Bmatrix} C_0 \\ C_1 \\ C_2 \\ C_3 \\ C_4 \end{Bmatrix} = 0.$$

For ε_1 equals zero, the latter can exactly be solved. In this case we obtain three times the infinite magnitude of eigenvalues

$$\varrho_{n,k}^{(0)} = -k\eta_g + in,$$

and the corresponding amplitude vectors, as noted below.

$$\underset{\sim}{C}^{(0)}_{n,k+2l} = \sigma_x^{2l} \frac{(k+2l)!}{2^l k!\, l!} \begin{bmatrix} n(n+1)/2 \\ (1-n)(1-n) \\ n(n-1)/2 \end{bmatrix}, \qquad \underset{\sim}{C}^{(0)}_{n,k+2l+1} = 0,\ n=0,\pm 1,\ \ k,l=0,1,\ldots$$

The number n has the three values $n = 0,\pm 1$ and determines the three roots of the diagonal matrices A_k, which we can obtain for each positive integer k. It should be remarked, that the components of the amplitude vectors are unbounded for increasing values of k.

In case ε_1 is not equal to zero, it is reasonable to seek an approximate solution by a perturbation method expanding the unknown eigenvalue ϱ and the corresponding amplitude vector of (5) in power series of the small parameter ε_1.

$$\varrho = \varrho^{(0)} + \varepsilon_1 \varrho^{(1)} + \varepsilon_1^2 \varrho^{(2)} + \ldots, \qquad \underset{\sim}{C}_k = \underset{\sim}{C}_k^{(0)} + \varepsilon_1 \underset{\sim}{C}_k^{(1)} + \varepsilon_1^2 \underset{\sim}{C}_k^{(2)} + \ldots. \tag{6}$$

The comparison of its coefficients results in the recurrence formula

$$\varepsilon_1^i: \quad (\varrho^{(0)} \underset{\sim}{E} - \underset{\sim}{A}_k) \underset{\sim}{C}_k^{(i)} + \sum_{j=0}^{i-1} \varrho^{(i-j)} \underset{\sim}{C}_k^{(j)} = \eta_g \sigma_x^2 k(k-1) \underset{\sim}{C}_{k-2}^{(i)} + \underset{\sim}{R} \underset{\sim}{C}_{k+1}^{(i-1)}, \tag{7}$$

which can be solved step by step.

It can easily be shown, that the first eigenvalue has the greatest real part. Thus, only the approximations of this eigenvalue need to be noted here.

$$\varrho^{(0)} = 0, \quad \varrho^{(2)} = (\varkappa_1 + \bar{\varkappa}_1)\sigma_x^2/8, \quad \varrho^{(1)} = \varrho^{(3)} = 0,$$
$$\varrho^{(4)} = [i(\varkappa_1 \bar{\varkappa}_1 + 1)(\varkappa_1^2 - \bar{\varkappa}_1^2) - (\varkappa_2 \varkappa_1^2 + \bar{\varkappa}_2 \bar{\varkappa}_1^2)]\sigma_x^4/32.$$

In particular, the solutions of an odd degree of approximation are vanishing and the others are calculated up to the fourth approximation. These results contain only positive real parts, because the values of

$$\varkappa_p = 1/(\eta_g + i/p), \qquad \bar{\varkappa}_p = 1/(\eta_g - i/p) \qquad (p = 1,2)$$

are complex, respectively conjugate complex.

STABILITY IN MEAN SQUARE

As mentioned above, the first eigenvalue has the greatest real part. It therefore determines the increasing behaviour of all second order moments of the state equations (3) and it determines, together with the initially introduced exponential function (2), the behaviour of the moment

$$E[y^2(t)] = E[(z_1+z_2)^2]\exp(-2\omega_1 D t)$$

of the original stability equation (1) as well. Consequently, the zero position of the system (1) will be stable in mean square

$$\lim_{t\to\infty} E[y^2(t)] \leq k = \text{const.},$$

if the value of this root is not greater than the damping coefficient D.

$$2\nu_1\varrho - 2\omega_1 D \leq 0.$$

In this stability condition we now have to introduce the approximate solution of the first eigenvalue. If we restrict our solution to those terms up to the order ε_1^4, we obtain a quadratic equation for the critical variance $(\varepsilon_1\sigma_x)^2$ of the parametric excitation $\varepsilon x(t)$ [6].

$$\eta_g\left[\frac{(\varepsilon_1\sigma_x/2)^2}{\eta_g^2+1}\right]^2 \frac{\eta_g^4+3+11\eta_g^2/2}{(\eta_g^2+1)(\eta_g^2+1/4)} + \eta_g\frac{(\varepsilon_1\sigma_x/2)^2}{\eta_g^2+1} + O(\varepsilon_1^6) \leq \frac{D}{\sqrt{1-D^2}}.$$

Upon computation of this variance, we have the critical spectral density as well

$$\varepsilon_1^2\omega_1 S_x = 2\frac{\omega_1\omega_g(\varepsilon_1\sigma_x)^2}{\omega_g^2+\omega^2}$$

and we can plot it for the various values of its limit frequency ω_g in a stability map (fig. 1). We see that the envelope of all critical spectral densities defines the limit of an instability region, the lowest point of which is situated near twice the natural frequency of the system. This threshold value is the already known stability condition, having been derived for Gaussian white noise [3] and stochastic coefficients with weakly varying spectral densities [7].

Low-pass processes, however, do not endanger the stability of the system as much. The smaller the bandwiths, the greater the maximum of their spectral densities may be, without destabilizing the system. In

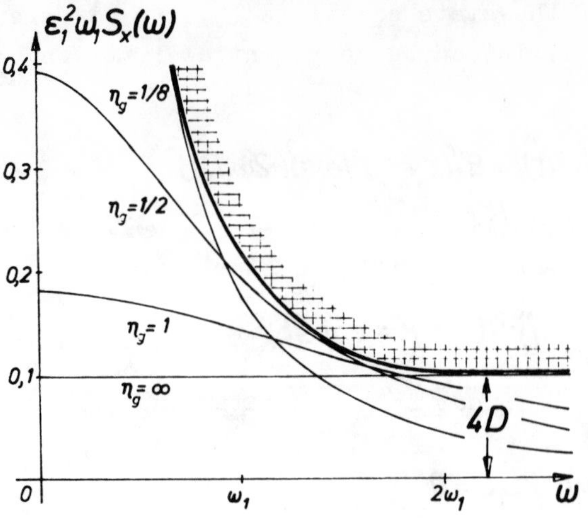

Figure 1: Stability map of low-pass processes

the case of small limit frequencies, an increasing spectrum of the process will first touch, and then cross, the stability limit in the upper proximity of twice the natural frequency of the system

CONVERGENCE OF THE POWER SERIES

From the mathematical point of view, these results are only ensured, if we can determine under which conditions the power series (6) are convergent. For this purpose, it would be necessary to know the general term of the expansions. It is, of course, impossible to calculate it since the recurrence formula (7) is too complicated.

In order to simplify this formula, we must first transform the moments's equations by taking the real state coordinates of the position $y_1 = T$ and of the dimensionless velocity $y_2 = \dot{T}/\nu_1$ of the system. We then add up two of these equations, so that we obtain a single differential equation

$$\left[\frac{1}{2\nu_1}\frac{d}{dt} + k\eta_g\right]E[(y_1^2+y_2^2)x^k] = \eta_g\sigma_x^2 k(k-1)E[(y_1^2+y_2^2)x^{k-2}] - \varepsilon_1 E[y_1 y_2 x^{k+1}], \qquad (8)$$

for the sum of the quadratic moments

$$E[(y_1^2+y_2^2)x^k] = D_k \exp(2\nu_1 \varrho t),$$

the amplitude D_k and the eigenvalue ϱ of which we shall investigate in the following.

The mixed moment, remaining on the right side of the equation (8), is smaller than half the absolute sum of the quadratic moments.

$$|E[(y_1^2+y_2^2)x^{k+1}]|/2 \geq E[y_1 y_2 x^{k+1}].$$

If on the right-hand side we now increase by means of this relation, and on the left-hand side decrease by the eigenvalue, we then obtain the simpler recurrence formula

$$k\eta_g \bar{D}_k = \eta_g \sigma_x^2 k(k-1)\bar{D}_{k-2} + \varepsilon_1 \bar{D}_{k+1}/2$$

for a majorant of the amplitudes, the indices of which are greater than zero. The coefficient $\bar{D}_0$ and the eigenvalue $\bar{\varrho}$ itself are given by

$$\bar{D}_0 = 1, \qquad \bar{\varrho} = \varepsilon_1 \bar{D}_1/2.$$

Applying the perturbation method, we can exactly calculate the first two terms of the expansions

$$\bar{D}_k = \bar{D}_k^{(0)} + \varepsilon_1 \bar{D}_k^{(1)} + \varepsilon_1^2 \bar{D}_k^{(2)} + \dots, \qquad \bar{D}^{(0)} = 1,\ \bar{D}_0^{(i)} = 0\ (i=1,2\dots)$$

and must then estimate the higher terms once again.

$$\bar{D}_{2l+1}^{(1)} = (1;2;l+1)\sigma_x^{2(l+1)}/2\eta_g, \qquad \bar{\bar{D}}_{2l}^{(2n)} = (1;2;l)l\sigma_x^{2l}(\sigma_x/\eta_g)^{2n}\prod_{j=2}^{n}\left[1+\frac{l-1}{(2j-1)2j}\right],$$
$$\bar{D}_{2l}^{(0)} = (1;2;l)\sigma_x^{2l}, \qquad \bar{\bar{D}}_{2l+1}^{(2n+1)} = (1;2;l+1)\frac{1}{6}(l+3)\sigma_x^{2l}(\sigma_x/\eta_g)^{2(n+1)}\prod_{j=2}^{n}\left[1+\frac{l}{2j(2j+1)}\right],$$
$$l = 0,1,2\dots,\quad n = 1,2,3\dots.$$

The respective inequalities of this second majorant may be proved by comparing the coefficients of the same powers of l. For this purpose we introduce this second majorant into the corresponding recurrence formula and obtain two inequations for the amplitudes $\bar{\bar{D}}_{2l+1}^{(2n+1)}$ and $\bar{\bar{D}}_{2l}^{(2n)}$. The most important part of the first inequality leads to

$$2l^2\prod_{j=2}^{n}\left[1+\frac{l-1}{(2j-1)2j}\right] \geq 2l^2\prod_{j=2}^{n}\left[1+\frac{l-2}{(2j-1)2j}\right] + \frac{2l^2}{12}\prod_{j=2}^{n-1}\left[1+\frac{l}{2j(2j+1)}\right],$$
$$\prod_{i=1}^{n-1}[l+1+2i(2i+3)] - \prod_{i=1}^{n-1}[l+2i(2i+3)] \geq \frac{n}{2}\prod_{i=2}^{n-1}[l+2i(2i+1)], \tag{9}$$
$$l = 0,1,2\dots,\quad n = 3,4,5\dots$$

and shall be proved in the following.

The first two coefficients of l^p can easily be calculated and result for the powers $p=n-1$ and $p=n-2$ in

$$l^{n-1}(n\geq 1):\ 1-1\geq 0, \qquad\qquad l^{n-2}(n\geq 2):\ n-1\geq n/2.$$

In the case of $p=n-3$ we obtain a single sum on the right-hand side of (9) and a double sum on the left-hand side. Its first term may be neglected and its two last terms can be reduced to a single sum, so that we are able to compare the sums on both sides of the inequality.

$$l^{n-3}(n\geq 3):\ \sum_{i,j=1(i<j)}^{n-1}[1+2i(2i+3)+2j(2j+3)]\geq (n-2)\sum_{i=1}^{n-1}2i(2i+3)\geq \frac{n}{2}\sum_{i=2}^{n-1}2i(2i+1).$$

Similarly, we obtain the inequality of the coefficients of a general power of l.

$$l^{n-p}(n\geq p):(n-p+1)\sum_{\substack{i_1,i_2,\dots i_{p-2}=1\\(i_1<\dots<i_{p-2})}}^{n-1}2i_1(2i_1+3)\dots 2i_{p-2}(2i_{p-2}+3)\geq \frac{n}{2}\sum_{\substack{i_1,\dots i_{p-2}=2\\(i_1<\dots<i_{p-2})}}^{n-1}2i_1\dots 2i_{p-2}(2i_{p-2}+1).$$

It is obvious, that this last inequality is fulfilled, if $(n-p+1)$ is smaller than $n/2$. For increasing values of p its left-hand side becomes greater. Consequently, it must be proved only for p equals n.

$$l^0\cdot\sum_{\substack{i_1,\dots i_{p-2}=1\\(i_1<\dots<i_{p-2})}}^{p-1}2i_1(2i_1+3)\dots\quad 2i_{p-2}(2i_{p-2}+3)\geq \frac{n}{2}\sum_{\substack{i_1,\dots i_{p-2}=2\\(i_1<\dots<i_{p-2})}}^{p-1}2i_1(2i_1+1)\dots\ 2i_{p-2}(2i_{p-2}+1).$$

When we calculate the sums on both sides of this inequality, we obtain on the right-hand side a single finite product and on the left-hand side $(p-1)$ finite products.

$$\prod_{i=1(i\neq 1)}^{p-1}2i(2i+3)+\prod_{i=1(i\neq 2)}^{p-1}2i(2i+3)+\dots+\prod_{i=1(i\neq p-1)}^{p-1}2i(2i+3)\geq \frac{p}{2}\prod_{i=2}^{p-1}2i(2i+1).$$

The first two terms of this sum suffice to verify the inequality.

$$(2p+1)/5+(2p+1)/14+\dots \qquad\qquad \geq p/2, \quad p=3,4,\dots n.$$

The other inequality may be proved in a similar manner.

By means of these results we arrive at a closed solution, which is an upper bound of the amplitudes as well as of the eigenvalue itself.

$$\bar{\bar{D}}_{2l+1}=(1;2;l+1)\sigma_x^{2(l+1)}(\varepsilon_1/2\eta_g)\left[1+\sum_{n=1}^{\infty}\frac{1}{3}(l+3)(\varepsilon_1\sigma_x/\eta_g)^{2n}\prod_{j=2}^{n}1+\frac{l}{2j(2j+1)}\right],$$

$$\bar{\bar{\varrho}} = \frac{1}{4\eta_g}(\varepsilon_1\sigma_x)^2 \sum_{n=0}^{\infty}(\varepsilon_1\sigma_x/\eta_g)^{2n}, \qquad (\varepsilon_1\sigma_x/\eta_g)^2 < 1.$$

It is evident, that, if the variance of the stochastic coefficient divided by η_g^2 is smaller than one, both power series are convergent. In the case of broadband processes, the limit frequency of which approaches infinity, there is no convergence problem.

BANDPASS PROCESSES

Let us now dispense with the purely mathematical aspect and regard other excitation processes of the system (1), which are more interesting for their technical application. We especially want to consider a bandpass process, which can be obtained by passing white noise $\xi(t)$ through a bandpass filter. In the time domain, this filter is represented by two uncoupled differential equations of the first order.

$$\dot{z}_3 + (\omega_g + i\omega_e)z_3 = +i\omega_g\xi(t),$$
$$\dot{z}_4 + (\omega_g - i\omega_e)z_4 = -i\omega_g\xi(t).$$

Since all coefficients of these equations are conjugate complex, their stochastic solutions are conjugate complex as well, so that their sum

$$x(t) = z_3(t) + z_4(t)$$

is only a real random excitation. The corresponding spectral density

$$S_x(\omega) = \frac{4\omega_g^2\omega_e^2 S_o}{[\omega_g^2 + (\omega+\omega_e)^2][\omega_g^2 + (\omega-\omega_e)^2]}$$

is represented in the figure above.

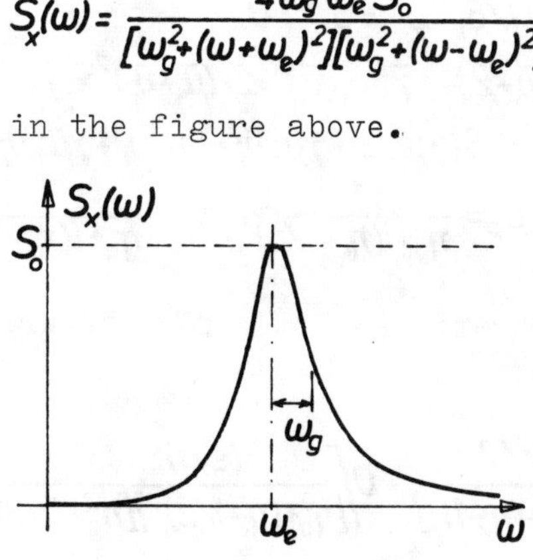

Figure 2: Spectrum of a bandpass process

It is obvious, that the middle excitation frequency is determined by the imaginary part of the filter equations' coefficients and the bandwidth of this process is determined by the coefficient's real part.

As mentioned above, its destabilizing effect may be investigated in a similar manner. Applying the Fokker-Planck equation, we obtain similar differential equations for the moments of the system (3)

$$\frac{1}{2\nu_1}\dot{\underset{\sim}{M}}_{k,l} = \underset{\sim}{A}_{k,l}\underset{\sim}{M}_{k,l} + \varepsilon_1 \underset{\sim}{R}(\underset{\sim}{M}_{k+1,l} + \underset{\sim}{M}_{k,l+1}) + \eta_g \frac{\sigma^2}{2}[-k(k-1)\underset{\sim}{M}_{k-2,l} + 2kl\underset{\sim}{M}_{k-1,l-1} - l(l-1)\underset{\sim}{M}_{k,l-2}],$$

$$\underset{\sim}{M}_{k,l} = \begin{bmatrix} E[z_1^2 z_3^k z_4^l] \\ E[z_1 z_2 z_3^k z_4^l] \\ E[z_2^2 z_3^k z_4^l] \end{bmatrix}, \quad \underset{\sim}{A}_{k,l} = \begin{bmatrix} i-\eta_{k,l} & 0 & 0 \\ 0 & -\eta_{k,l} & 0 \\ 0 & 0 & -i-\eta_{k,l} \end{bmatrix}, \quad \underset{\sim}{R} = \frac{i}{2}\begin{bmatrix} 1 & 1 & 0 \\ -1/2 & 0 & 1/2 \\ 0 & -1 & -1 \end{bmatrix},$$

$$\eta_{k,l} = k(\eta_g + i\eta_e) + l(\eta_g - i\eta_e), \quad \eta_g = \omega_g/2\nu_1, \quad \eta_e = \omega_e/2\nu_1, \quad \sigma^2 = \omega_g S_o$$

and approximately calculate its solution

$$\underset{\sim}{M}_{k,l} = \underset{\sim}{C}_{k,l}\exp(2\nu_1 \rho t)$$

by the perturbation method.

$$\rho = \rho^{(0)} + \varepsilon_1 \rho^{(1)} + \varepsilon_1^2 \rho^{(2)} + \dots, \qquad \underset{\sim}{C}_{k,l} = \underset{\sim}{C}_{k,l}^{(0)} + \varepsilon_1 \underset{\sim}{C}_{k,l}^{(1)} + \varepsilon_1^2 \underset{\sim}{C}_{k,l}^{(2)} + \dots,$$

$$\varepsilon_1^i: \quad [\rho^{(0)}\underset{\sim}{E} - \underset{\sim}{A}_{k,l}]\underset{\sim}{C}_{k,l}^{(i)} + \sum_{j=0}^{i-1}\rho^{(i-j)}\underset{\sim}{C}_{k,l}^{(j)} = \underset{\sim}{R}(\underset{\sim}{C}_{k+1,l}^{(i-1)} + \underset{\sim}{C}_{k,l+1}^{(i-1)}) + $$
$$+ [-k(k-1)\underset{\sim}{C}_{k-2,l}^{(i)} + 2kl\underset{\sim}{C}_{k-1,l-1}^{(i)} - l(l-1)\underset{\sim}{C}_{k,l-2}^{(i)}]\eta_g\sigma^2/2.$$

If in this calculation we also retain only terms up to the order of ε_1^4

$$\rho^{(0)} = 0, \quad \rho^{(2)} = i\eta_e\sigma^2[\mu_1\upsilon_1 - \bar{\mu}_1\bar{\upsilon}_1]/8, \quad \rho^{(1)} = \rho^{(3)} = 0,$$

$$\rho^{(4)} = \frac{\sigma^4}{256}[8\eta_g\gamma\bar{\gamma}\mu_1\bar{\mu}_1(\mu_1\upsilon_1 + \bar{\mu}_1\bar{\upsilon}_1 + 2) - \gamma\bar{\gamma}[(\mu_1 + \bar{\upsilon}_1)^2\varkappa_2 + (\bar{\mu}_1 + \upsilon_1)^2\bar{\varkappa}_2] + 2(\gamma^2\mu_2\mu_1^2 +$$
$$+ \bar{\gamma}^2\bar{\mu}_2\bar{\mu}_1^2 + \gamma^2\upsilon_2\upsilon_1^2 + \bar{\gamma}^2\bar{\upsilon}_2\bar{\upsilon}_1^2) - 8(\gamma\mu_1^2\upsilon_1^2 + \bar{\gamma}\bar{\mu}_1^2\bar{\upsilon}_1^2 + \gamma\mu_1^3\upsilon_1^3 + \bar{\gamma}\bar{\mu}_1^3\bar{\upsilon}_1^3)],$$

$$\gamma = \frac{1}{\eta_g + i\eta_e}, \quad \varkappa_p = \frac{1}{\eta_g + i/p}, \quad \mu_p = \frac{1}{\eta_g + i(\eta_e + 1/p)}, \quad \upsilon_p = \frac{1}{\eta_g + i(\eta_e - 1/p)}, \qquad p = 1,2,$$

we once again obtain a quadratic equation for the critical variance of the bandpass excitation.

$$\frac{D}{\sqrt{1-D^2}} \geq \frac{\eta_g\eta_e(\varepsilon_1\sigma)^2/2}{[\eta_g^2 + (\eta_e+1)^2][\eta_g^2 + (\eta_e-1)^2]} + O\left[\frac{(\varepsilon_1\sigma)^4}{[\eta_g^2 + (\eta_e+1/2)^2][\eta_g^2 + (\eta_e-1/2)^2]}\right].$$

STABILITY MAP OF BANDPASS PROCESSES

The next figure represents its corresponding critical spectral densities for various values of the middle excitation frequency ω_e and a relatively great value of the limit frequency ω_g. In this case we see that we have two threshold values and two instability regions.

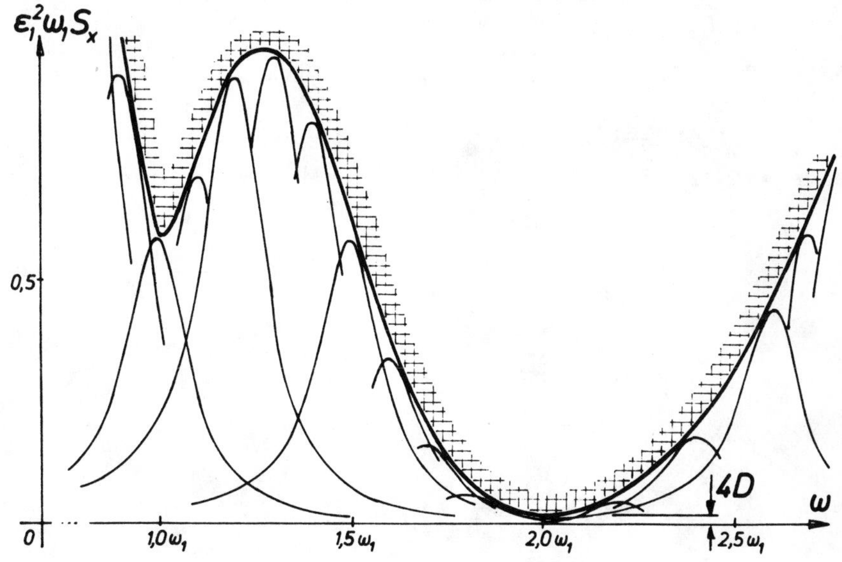

Figure 3: Stability map of bandpass processes

The first and most important region is situated near twice the natural frequency of the system and its lowest point has the already known value $4D$. On both sides of this point the stability limit ascends. For this reason, the greater the distance of the middle excitation frequency from this resonance frequency, the smaller will be the destabilizing effect of the bandpass process.

Near the natural frequency itself of the system there is a second instability region, having a second threshold value. Its dip depends upon the limit frequency ω_g of the bandpass process and becomes all the more deep, the smaller the bandwidth of the process is. Similar to the Strutt-diagram of the Mathieu equation, instability regions of higher order may be found if higher terms of the power series are used.

REFERENCES

[4] Ariaratnam, S.T., and Graefe, P.W.U. Int. Journ. of Control. 1, 293 (1965), 2, 161 (1965), 3, 205 (1965)

[3] Ariaratnam, S.T., Dynamic Stability of Structures. Hrsgg. von Herrmann, G., 255, New York (1967)

[2] Bertram, J.E., and Sarachik, P.E. Trans. Inst. Radio Engr. P.G.I., Z-5, 260 (1959)

[1] Bolotin, W.W., Kinetische Stabilität elastischer Systeme. Berlin (1961)

[5] Kozin, F., J. Math. Phys. 42, 59 (1963)

[6] Wedig, W., ZAMM, 52, T 77 (1972)

[7] Weidenhammer, F., Ing.-Arch. 33, 404 (1964)

ANALYTICAL STUDY ON n-th ORDER LINEAR SYSTEM WITH STOCHASTIC COEFFICIENTS

Takayoshi NAKAMIZO
Department of Mechanical Engineering
Defense Academy, Japan

Yoshikazu SAWARAGI
Department of Applied Mathematics and Physics, Kyoto University, Japan

1. INTRODUCTION

During the past decade, the stability analysis of physical systems with stochastic parameters has been extensively developed [1-10]. From the applied point of view, it is naturally of interest to evaluate how stochastic parameters affect the dynamic behaviors. In the present paper, analysis will be made of n-th order linear differential equation with stochastic coefficients, although the treatment may be extended to the more general case. The special class of systems considered here is described by

$$\frac{d^n x}{dt^n} + \sum_{k=1}^{n} [a_k + \dot{\alpha}_k(t)] \frac{d^{k-1}x}{dt^{k-1}} = a_o + \dot{\alpha}_o(t) \tag{1}$$

where a_i, $i=0,1,\ldots,n$ are constants, and $\dot{\alpha}_i(t)$ are stationary white Gaussian noises. Alternatively Eq.(1) can be written in the precise form*

$$\begin{cases} dx_i = x_{i+1}dt & i= 1,2,\ldots, n-1 \\ dx_n = -\sum_{k=1}^{n} [a_k dt + d\alpha_k(t)]x_k + a_o dt + d\alpha_o(t) \end{cases} \tag{2}$$

in which all the non-zero coefficients $\alpha_k(t)$ are Brownian motion processes with incremental covariances such that

$$\varepsilon\{d\alpha_i(t)d\alpha_j(t)\} = q_{ij}dt, \quad i,j = 0,1,\ldots,n \tag{3}$$

It is pertinent, on account of its convenience in analysis, to introduce the so-called frozen system defined by

$$\begin{cases} \dot{z}_i = z_{i+1} & i= 1,2,\ldots, n-1 \\ \dot{z}_n = -\sum_{k=1}^{n} a_k z_k + a_o + \dot{\alpha}_o(t) \end{cases} \tag{4}$$

* Some care must be needed when using Eq.(2) as a model of a physical process governed by Eq.(1) where $\dot{\alpha}_k(t)$ are not ideal white noises. In reality $\dot{\alpha}_k(t)$ is a sample function of a Gaussian process with a spectrum which is flat up to very high frequency. Then the best model equation for such a case may be expressed by replacing a_k in Eq.(2) by $(a_k - \frac{1}{2}q_{kn})$.

where $\dot{\alpha}_o(t)$ can be viewed as a white noise with $\varepsilon\{\dot{\alpha}_o(t)\dot{\alpha}_o(\tau)\} = q_{oo}\delta(t-\tau)$. Here we use z_k to denote the phase variables of the frozen system to avoid confusion.

2. EVALUATION OF RESPONSE

2.1 Moment Equation The continuous state trajectory of the system (2) forms a Markov process in the space R^n. Thus for the scalar function V(x), it follows that[6]

$$\frac{d\varepsilon\{V(x)\}}{dt} = \varepsilon\{LV(x)\} \tag{5}$$

where L denotes the differential generator.

First Moment $m_i(t)$ Putting $V(x) = x_i$, then Eq.(5) yields

$$\begin{cases} \dfrac{dm_i(t)}{dt} = m_{i+1}(t) & i=1,2,\ldots, n-1 \\ \dfrac{dm_n(t)}{dt} = -\sum\limits_{k=1}^{n} a_k m_k(t) + a_o \end{cases} \tag{6}$$

which can be solved by the standard method. Note that the first order moment is clearly equivalent to the solution of Eq.(4) with $\dot{\alpha}_o(t) = 0$.

Second Moment $m_{ij}(t)$ It can be easily shown from Eq.(5) that the covariance $m_{ij}(t) = \text{cov}\{x_i(t), x_j(t)\}$ satisfies the following equation:

$$\begin{cases} \dfrac{dm_{ij}(t)}{dt} = m_{i+1,j}(t) + m_{i,j+1}(t) & i,j=1,2,\ldots, n-1 \\ \dfrac{dm_{in}(t)}{dt} = m_{i+1,n}(t) - \sum\limits_{k=1}^{n} a_k m_{ik}(t) & i=1,2,\ldots, n-1 \\ \dfrac{dm_{nn}(t)}{dt} = -2\sum\limits_{k=1}^{n} a_k m_{kn}(t) + \sum\limits_{k=1}^{n}\sum\limits_{\ell=1}^{n} q_{k\ell} m_{k\ell}(t) + q_{oo} + f(t) \end{cases} \tag{7}$$

where f(t) contains only the first moments as

$$f(t) = \sum_{k=1}^{n}\sum_{\ell=1}^{n} q_{k\ell} m_k(t) m_\ell(t) - 2\sum_{k=1}^{n} q_{ok} m_k(t) \tag{8}$$

2.2 Solution of Eq.(7) The evaluation of covariances can be made by solving Eq.(7) for unknowns $m_{ij}(t)$. In practice this may be rather complex. It is however possible by Laplace transform technique to obtain somewhat compact form of the solution, which enables us not only to evaluate directly the steady state results, but also to establish the moment stability. Assume that the system is at rest in the negative range of time (this is not essential). Using the notation $\overline{m}_{ij}(s) = \mathcal{L}[m_{ij}(t)]$ and $\overline{f}(s) = \mathcal{L}[f(t)]$, the Laplace transform of Eq.(7) becomes

$$\begin{cases} s\overline{m}_{ij}(s) - \overline{m}_{i+1,j}(s) - \overline{m}_{i,j+1}(s) = 0 \\ s\overline{m}_{in}(s) - \overline{m}_{i+1,n}(s) + \sum_{k=1}^{n} a_k \overline{m}_{ik}(s) = 0 \\ s\overline{m}_{nn}(s) + 2\sum_{k=1}^{n} a_k \overline{m}_{kn}(s) = \frac{1}{s} q_{oo} + \overline{f}(s) + \sum_{k=1}^{n}\sum_{\ell=1}^{n} q_{k\ell}\overline{m}_{k\ell}(s) \end{cases} \tag{9}$$

By Cramer rule, we can formally obtain

$$\overline{m}_{ij}(s) = [\frac{1}{s} q_{oo} + \overline{f}(s) + \sum_{k=1}^{n}\sum_{\ell=1}^{n} q_{k\ell}\overline{m}_{k\ell}(s)]\frac{\Delta_{ij}(s)}{\Delta(s)} \tag{10}$$

in which $\Delta(s)$ represents the $n(n+1)/2$ dimensional determinant with the coefficients of $\overline{m}_{ij}(s)$ $(i \geq j)$ in the LHS of Eq.(9) as elements, and $\Delta_{ij}(s)$ is a cofactor of ij-element of the last row of $\Delta(s)$. It is not difficult to show that, if we denote the impulse response function of the frozen system by $w(t)$ [13]

$$\frac{\Delta_{ij}(s)}{\Delta(s)} = L[w^{(i-1)}(t)w^{(j-1)}(t)] \equiv \overline{w}_{ij}(s) \tag{11}$$

Thus Eq.(10) becomes

$$\overline{m}_{ij}(s) = \overline{z}_{ij}(s) + \overline{w}_{ij}(s)[\overline{f}(s) + \sum_{k=1}^{n}\sum_{\ell=1}^{n} q_{k\ell}\overline{m}_{k\ell}(s)] \tag{12}$$

bearing also in mind the fact that the first term of Eq.(10) expresses the covariance of the frozen system;

$$\overline{z}_{ij}(s) = L[\mathrm{cov}\,\{z_i(t), z_j(t)\}]$$

Note that Eq.(12) is again a set of linear simultanious equation for unknowns $\overline{m}_{ij}(s)$. It can be easily seen that the solution of Eq.(13) is given by [13]

$$\overline{m}_{ij}(s) = \overline{z}_{ij}(s) + \frac{\overline{w}_{ij}(s)}{D(s)}[\overline{f}(s) + \sum_{k=1}^{n}\sum_{\ell=1}^{n} q_{k\ell}\overline{z}_{k\ell}(s)] \tag{15}$$

where

$$D(s) = 1 - \sum_{k=1}^{n}\sum_{\ell=1}^{n} q_{k\ell}\overline{w}_{k\ell}(s) \tag{14}$$

Since the covariance for the frozen system is given by

$$\overline{z}_{ij}(s) = \frac{1}{s} q_{oo}\overline{w}_{ij}(s)$$

Eq.(13) is simply expressed as

$$\overline{m}_{ij}(s) = G(s)\,\overline{z}_{ij}(s) \tag{15}$$

where

$$G(s) = \frac{1}{D(s)}[1 + \frac{s}{q_{oo}}\overline{f}(s)] \tag{16}$$

It should be noted that $D(s)=0$ is the characteristic equation associated with the second moment equation.

2.3 Steady State If the stochastic system is stable in the appropriate sense, then the second moment will tend to certain finite value asymptotically after a sufficiently long time. Such asympototic value is given by

$$m_{ij} = \lim_{t\to\infty} m_{ij}(t) = \lim_{s\to 0} s\bar{m}_{ij}(s)$$

$$= \frac{k}{D(0)} z_{ij} \tag{17}$$

where $z_{ij} = \lim_{t\to\infty} z_{ij}(t)$ and

$$k = 1 - \frac{2a_o q_{o1}}{a_1 q_{oo}} + \frac{a_o^2 q_{11}}{a_1^2 q_{oo}} \tag{18}$$

The steady state value of covariance is equal to k/D(0) times that of the frozen system.

2.4 Correlation Function and Spectral Density By a similar method, it can be shown that the correlation function matrix is given by

$$\phi_{\underline{xx}}(t,\tau) = \varepsilon\{[\underline{x}(t) - \underline{m}(t)][\underline{x}(\tau) - \underline{m}(\tau)]'\}$$

$$= e^{A(t-\tau)}M(\tau)1(t-\tau) + M(t)e^{A'(t-\tau)}1(\tau-t) \tag{19}$$

where

$$\underline{x}(t) = [x_1(t)\ x_2(t)\ \ldots\ x_n(t)]'$$

$$\underline{m}(t) = \varepsilon\{\underline{x}(t)\}$$

$$M(t) = \varepsilon\{[\underline{x}(t) - \underline{m}(t)][\underline{x}(t) - \underline{m}(t)]'\}$$

$$A = \begin{bmatrix} 0 & 1 & 0 & \ldots & 0 \\ 0 & 0 & 1 & \ldots & 0 \\ \vdots & \vdots & \vdots & & \vdots \\ -a_1 & -a_2 & -a_3 & \ldots & -a_n \end{bmatrix}$$

In particular the steady state correlation is expressed as

$$\phi_{\underline{xx}}(\tau) = \frac{k}{D(0)} \phi_{\underline{zz}}(\tau) \qquad \text{for all } \tau \tag{20}$$

where $\phi_{\underline{zz}}(\tau)$ represents the correlation matrix of output vector of the frozen system.

The spectral density is given by taking the Fourier transform of Eq.(20). Hence

$$\Phi_{\underline{xx}}(\omega) = \frac{k}{D(0)} \Phi_{\underline{zz}}(\omega) \tag{21}$$

It is worth noting that the spectral density for the stochastic system has the same shape as that for the frozen system: however the magnitude is increased by the factor $k/D(0)$.

3. MOMENT STABILITY

There are two types of moment stability that have widely appeared in the literature, namely i) mean stability and ii) mean square stability. A stochastic system is said to possess the mean (or the mean square) stability, if all the first (or second) moments of output variables are bounded for all $t \geq 0$.

It is easily observed that the necessary and sufficint condition to guarantee the mean stability for the stochastic system is that the frozen system frozen system is stable. Clearly this problem includes no difficulty, because the standard Routh-Hurwitz criterion will provide such a test. Here we will study the boundness of covariances. The boundness of covariances will be determined by the nature of the roots of characteristic equation

$$D(s) = 1 - \sum_{k=1}^{n} \sum_{\ell=1}^{n} q_{k\ell}\overline{w}_{k\ell}(s) = 0 \tag{22}$$

provided that the frozen system is stable. It can be stated that the stochastic linear system possesses the mean square stability, if and only if all the roots of Eq.(22) lie in the left half of s-plane. This fact enables us to apply the Routh-Hurwitz criterion to Eq.(22). However such direct task will require the complicated computation. This section is devoted to establishment of a practical criterion for the mean square stability.

To establish it, we need a lemma:

Lemma(Bergen[2]) Let $G(s)$ be a Laplace transform of the corresponding non-negative function $g(t) \geq 0$, for $t \geq 0$. Then all the roots of the equation $G(s) = 1$ have negative real parts, if and only if $G(0) \geq 1$.

If we denote

$$G(s) = \sum_{k=1}^{n} \sum_{\ell=1}^{n} q_{k\ell}\overline{w}_{k\ell}(s)$$

then the corresponding time function is

$$g(t) = \sum_{k=1}^{n} \sum_{\ell=1}^{n} q_{k\ell} w^{(k-1)}(t) w^{(\ell-1)}(t)$$

Since the covariance matrix $Q = (q_{ij})$ is always non-negative definite, the quadratic form $g(t)$ is also non-negative. From lemma, we can immediately obtain the stability condition as

$$D(0) = 1 - \sum_{k=1}^{n} \sum_{\ell=1}^{n} q_{k\ell} \bar{w}_{k\ell}(0) > 0 \tag{23}$$

It is of interest to note the boundness of covariances depends only on the constant term of the characteristic equation. Thus all we have to do is to compute D(0) analytically. Different methods for evaluating D(0) present different kinds of stability criteria.

I. By definition

$$\bar{w}_{k\ell}(0) = \int_0^\infty w^{(k-1)}(t) w^{(\ell-1)}(t) dt$$

Invoking the Parseval theorem gives

$$\bar{w}_{k\ell}(0) = \frac{1}{2\pi} \int_{-\infty}^{\infty} (j\omega)^{k-1} (-j\omega)^{\ell-1} \left| \bar{w}(j\omega) \right|^2 d\omega \tag{24}$$

where $\bar{w}(s)$ is the transfer function of the frozen system, i.e.,

$$\bar{w}(s) = \frac{1}{s^n + a_n s^{n-1} + \ldots + a_2 s + a_1} \tag{25}$$

The integral must vanish for $k+\ell$=odd. Here we arrive at:

Frequency Domain Stability Criterion: Assume that the frozen system is stable. Then the linear stochastic system of Eq.(1) is mean square stable if and only if

$$1 - \sum_{k=1}^{n} \sum_{\ell=1}^{n} (-1)^{m-k} q_{k\ell} K_m > 0$$

where

$$K_m = \frac{1}{2\pi} \int_{-\infty}^{\infty} \left| (j\omega)^{m-1} \bar{w}(j\omega) \right|^2 d\omega$$

II. There is an alternative way to compute D(0). This will lead us to the Nevelson & Khasminskii criterion. After performing some determinantal calculations, we can obtain as*

$$\begin{aligned} \Delta(0) &= 2^n H_n \\ \Delta_{ij}(0) &= (-1)^{n-k} 2^{n-1} H_{ij} \end{aligned} \tag{26}$$

where

$$H_n = \begin{vmatrix} a_n & a_{n-2} & a_{n-4} & \cdots & 0 \\ 1 & a_{n-1} & a_{n-3} & \cdots & 0 \\ 0 & a_n & a_{n-2} & \cdots & 0 \\ \vdots & \vdots & \vdots & & \vdots \\ 0 & 0 & 0 & \cdots & a_1 \end{vmatrix}, \quad H_{ij} = \begin{vmatrix} e_n & e_{n-1} & e_{n-2} & \cdots & e_1 \\ 1 & a_{n-1} & a_{n-3} & \cdots & 0 \\ 0 & a_n & a_{n-2} & \cdots & 0 \\ \vdots & \vdots & \vdots & & \vdots \\ 0 & 0 & 0 & \cdots & a_1 \end{vmatrix}$$

in which

*This is not difficult, but somewhat lengthy. So it is omitted here (cf. [12,13])

$$e_m = \begin{cases} 1 & \text{for} \quad i + j = 2m \\ 0 & \text{for} \quad i + j \neq 2m \end{cases}$$

It should be noted that H_n is equal to the n-th order Hurwitz determinant for the frozen system, and that H_{ij} is resulted from H_n by replacing its first row by a row vector $(e_n, e_{n-1}, \ldots, e_1)$. Thus we can compute as

$$\overline{w}_{ij}(0) = (-1)^{n-k} \frac{H_{ij}}{H_n} \tag{27}$$

The above fact gives :

Algebraic Criterion: Assume that the frozen system is stable. Then the linear stochastic system of Eq.(1) is stable in the mean square sense, if and only if

$$H_n > \frac{1}{2} \sum_{k=1}^{n} \sum_{\ell=1}^{n} (-1)^{n-k} q_{k\ell} H_{k\ell} \quad (k+\ell=\text{even})$$

which is very relevant to Nevelson & Khasminskii criterion [14]

Some other discussions on stability will be includes in appendices.

4. EQUIVALENT SYSTEM

4.1 Derivation of Equivalent System Consider a system described by the differential equation with constant coefficients of the form

$$\frac{d^n x_e}{dt^n} + \sum_{k=1}^{n} b_k \frac{d^{k-1} x_e}{dt^{k-1}} = b_o + \dot{\beta}(t) \tag{28}$$

where $b_k \geq 0$, and $\dot{\beta}(t)$ is a Gaussian white noise with $\varepsilon\{\dot{\beta}(t)\dot{\beta}(\tau)\} = q\delta(t-\tau)$. In some applications, it is of interest to use the representation of Eq.(28) instead of Eq.(1). The question to be asked here is how the coefficients b_k and the noise covariance q should be selected. The criterion for the selection taken here that

$$\begin{aligned} \varepsilon\{\underline{x}_e(t)\} &= \varepsilon\{\underline{x}(t)\} \\ \text{cov}\,\{\underline{x}_e(t), \underline{x}_e(t)\} &= \text{cov}\,\{\underline{x}(t), \underline{x}(t)\} \end{aligned} \tag{29}$$

It is easily verified that Eq.(29) holds for the same initial condition if we choose

$$\begin{aligned} b_k &= a_k \qquad k = 0,1, \ldots , n \\ q &= q_{oo} - 2\sum_{k=1}^{n} q_{ok} m_k(t) + \sum_{k=1}^{n} \sum_{\ell=1}^{n} q_{k\ell}[m_{k\ell}(t) + m_k(t)m_\ell(t)] \end{aligned} \tag{30}$$

It is also proved that the constant coefficient system with Eq.(30) is equivalent in the sense that it will have the same correlation matrix and hence the same cross spectral densities (in the steady state). In this sense, the system defined by Eq.(28) is said to be equivalent to the original system defined by Eq.(1).

Application to State Estimation Briefly the problem is to estimate the n state vector $\underline{x}(t)$ of the stochastic linear system described by Eq.(1) provided that the m vector valued observation:

$$d\underline{y}(t) = H\underline{x}(t)dt + Rd\underline{\gamma}(t) \tag{31}$$

is available. Here H and R are mxn and mxs matrices respectively, and $\underline{\gamma}(t)$ is the s vector independent Wiener process.

The equivalent system to Eq.(1) is given by Eq.(28) with Eq.(30):

$$d\underline{x}_e(t) = A\underline{x}_e(t)dt + \underline{u}[a_o dt + d\beta(t)] \tag{32}$$

where $\underline{u} = (0,0, \ldots ,1)'$. Similarly the equivalent system to Eq.(31) can be expressed by

$$d\underline{y}_e(t) = H\underline{x}_e(t)dt + Rd\underline{\gamma}(t) \tag{33}$$

Thus the equivalent Kalman filter can be immediately obtained based on Eqs.(32) and (33) as

$$d\hat{\underline{x}}_e(t) = A\hat{\underline{x}}_e(t)dt + \underline{u}a_o dt + K[d\underline{y}_e(t) - H\hat{\underline{x}}_e(t)dt] \tag{34}$$

where

$$\begin{aligned} K &= PH'R^{-1} \\ \dot{P} &= AP + PA' - PH'R^{-1}HP + \underline{uu}'q \end{aligned} \tag{35}$$

Since the actual observation process is given by Eq.(31), the process $\underline{y}_e(t)$ in Eq.(34) should be replaced by $\underline{y}(t)$ to obtain the physical realizable filter. Thus

$$d\hat{\underline{x}}(t) = [A\hat{\underline{x}}(t) + \underline{u}a_o]dt + K[d\underline{y}(t) - H\hat{\underline{x}}(t)dt] \tag{36}$$

It is also not difficult to prove that the estimate $\hat{\underline{x}}(t)$ governed by Eq.(36) is the optimal linear estimate in the minimal variance sense. The use of the equivalent system provides with a simple method for determining the optimal linear estimate of the stochastic linear system. In appendix A.3 extension will be made to the more genral case.

5. CONCLUSION

Some analytical results have been presented here for the single input-single output linear stochastic system. The description begins with the derivation of moment differential equation from which the second moment of output variables is presented in somewhat compact form. Then the correlation matrix can be expressed in a similar way. It turns out that the output spectral density has the same shape as that for the frozen system: however the magnitude is increased by the constant factor.

An explicit form of the condition is obtained for quaranteeing the mean square stability. The resultant criterion gives a simple method for estabilishing the stability of the sigle input-single output system. Thus the required computation labour can be lessened considerablly. However it is very doubtful whether some simple criterion does exist for the general vector case. An alternative attempt to the feedback stochastic system has been made by Sawaragi et al [15].

Finally the equivalent deterministic system is derived in the sense that it has the same second order statistics as ones for the original stochastic system, and its application to the state estimation from the noisy observation is presented. Some of results presented here may be straightforwardly extended to the more general case, as indicated in appendix.

APPENDICES

A.1 *Stochastic Liapunov Method* Consider the free system as

$$\begin{cases} dx_i = x_{i+1}dt \\ dx_n = -\sum_{k=1}^{n} [a_k dt + d\alpha_k(t)] x_k \end{cases} \tag{37}$$

An equiblium solution of Eq.(37) is evidently $\underline{x}=0$, whose stability is in question. Roughly speaking the equiblium solution is stable with probability one, if there exists a Liapunov function $V(\underline{x}) > 0$ which satisfies $LV(\underline{x}) = -k(\underline{x}) < 0$ (cf. Kushner[10]).

Let the Liapunov function be of the quadratic form

$$V(\underline{x}) = \underline{x}'P\underline{x}\ , \qquad P = (p_{ij})$$

where P is an nxn symmetric positive definite matrix. Then

$$LV(\underline{x}) = (A\underline{x})'V_{\underline{x}} + p_{nn}\underline{x}'Q\underline{x} \tag{38}$$

the first term on the RHS of which is related to the stability of the frozen system. Assume that the frozen system is stable, then the matrix P can be chosen so that

$$(A\underline{x})'V_{\underline{x}} = -\underline{x}'Q\underline{x} \tag{39}$$

because the covariance matrix Q is non-negative definite. It follows from Eqs.(37) and (38) that

$$LV(\underline{x}) = (p_{nn} - 1)\underline{x}'Q\underline{x} \tag{40}$$

from which the stability condition is obtained as $p_{nn} > 1$. Thus the problem is only to evaluate the element p_{nn}. From Eq.(39)

$$a_i p_{jn} + a_j p_{in} - p_{i-1,j} - p_{i,j-1} = q_{ij} \qquad (41)$$

$$(i,j = 1,2,\ldots,n)$$

which is a set of equation for $n(n+1)/2$ unknown p_{ij}. By a direct computation, the p_{nn} can be solved as (cf. Nakanizo[12])

$$p_{nn} = \frac{1}{2H_n} \sum_{k=1}^{n} \sum_{\ell=1}^{n} (-1)^{n-k} q_{k\ell} H_{k\ell} \qquad (42)$$

Thus we arrive at the same algebraic criterion as presented in 2. It should be noted that the stability of second moments implies the stability with probability one.

A.2 *Stability of r-th moments** It follows from Eq.(5) that

$$\varepsilon\{V(\underline{x})\} - \varepsilon\{V(\underline{x}_o)\} = \int_o^t \varepsilon\{LV(\underline{x})\}dt$$

If $\varepsilon\{V(\underline{x})\} < 0$, then

$$\varepsilon\{V(\underline{x})\} \leq \varepsilon\{V(\underline{x}_o)\}$$

Choose the Liapunov function as

$$V(x) = \{\tilde{V}(x)\}^{\frac{r}{2}} = \{x'Px\}^{\frac{r}{2}} \qquad (43)$$

There are k_1, $k_2 > 0$ such that

$$k_1\|\underline{x}\|^2 \geq \tilde{V}(\underline{x}) \geq k_2\|\underline{x}\|^2$$

Hence

$$\varepsilon\{\|\underline{x}\|^r\} \leq \alpha\varepsilon\{\|\underline{x}_o\|^r\}$$

where $\alpha = k_1^{\frac{1}{2}r}/k_2^{\frac{1}{2}r}$. We can therefore conclude that, if $LV(\underline{x}) < 0$, th the r-th moments are bounded. Using Eq.(43)

$$LV(x) = \frac{1}{2}r\tilde{V}^{\frac{1}{2}r-1}\{x'(A'P+PA)x + p_{nn}x'Qx + (r-2)\tilde{V}^{-1}(\sum_k p_{nk}x_k)^2 x'Qx\} \qquad (44)$$

Since the frozen system is assumed to be stable, then the matrix P can be chosen so that

$$A'P + PA = -Q$$

Note that

$$(\sum_k p_{nk}x_k)^2 \leq p_{nn}\tilde{V}$$

Hense Eq.(44) becomes

* This section mostly follows Nevelson & Khasminskii paper[14]

$$\varepsilon\{LV(\underline{x})\} \le \frac{1}{2} r\, \varepsilon\{(\tilde{V}^{\frac{1}{2}r-1})\underline{x}'Q\underline{x}\}[(r-1)p_{nn} - 1] \qquad (45)$$

from which the sufficient condition becomes

$$(r-1)\, p_{nn} < 1 \qquad (46)$$

where p_{nn} is given previously by Eq.(42). Thus we arrive at the following stability condition:

Sufficient Condition for r-th order moment stability:

Assume that the linear stochastic system is stable in the (r-1)th moment sense. Then the system has the r-th moment stability, if

[*Algebraic Criterion*]

$$H_n > \frac{r-1}{2} \sum_{\substack{k=1 \\ (k+\ell=\text{even})}}^{n} \sum_{\ell=1}^{n} (-1)^{n-k} q_{k\ell} H_{k\ell}$$

[*Frequency Domain Criterion*]

$$\frac{1}{r-1} > \sum_{\substack{k=1 \\ (k+\ell=2m)}}^{n} \sum_{\ell=1}^{n} (-1)^{m-k} q_{k\ell} K_m$$

A.3 *Application of Equivalent System to State Estimation* A linear stochastic system may in general be described by

$$dx_i(t) = \sum_{j=1}^{n} [a_{ij}dt + dw_{ij}(t)]x_j + dw_{io}(t) \qquad (47)$$
$$(i = 1,2,\ldots,n)$$

where a_{ij} are non-random coefficients, and $w_{ij}(t)[i=1,\ldots,n:j=0,1,\ldots,n]$ are the standard Brownian motion processes. It is supposed that there is a set of noisy observations

$$dy_i(t) = \sum_{j=1}^{n} [h_{ij}dt + dv_{ij}(t)]x_j + dv_{io}(t) \qquad (48)$$
$$(i = 1,2,\ldots,m)$$

where h_{ij} are constant, and $v_{ij}(t)[i=1,\ldots,n:j=0,1,\ldots,m]$ are Brownian motion processes. Equations (47) and (48) can be written in vector-matrix form

$$d\underline{x}(t) = A\underline{x}(t)dt + dW(t)\underline{x} + d\underline{w}_o(t) \qquad (49)$$

$$d\underline{y}(t) = H\underline{x}(t)dt + dV(t)\underline{x} + d\underline{v}_o(t) \qquad (50)$$

The incremental covariances of Brownian motion processes are given by

$$\varepsilon\{dw_{ij}(t)dw_{k\ell}(t)\} = q_{ijk\ell}dt$$
$$(i,k=1,\ldots,n:\ j,\ell=0,1,\ldots,n)$$
$$\varepsilon\{dv_{ij}(t)dv_{k\ell}(t)\} = r_{ijk\ell}dt$$
$$(i,k=1,\ldots,m:\ j,\ell=0,1,\ldots,m)$$

and

$$\varepsilon\{dw_{ij}(t)dw_{k\ell}(t)\} = s_{ijk\ell}dt \quad (i=1,\ldots,n;\ j=0,1,\ldots,n;\ k=1,\ldots,m;\ \ell=0,1,\ldots,m)$$

The present problem is to find the minimal variance linear estimate of the state $\underline{x}(t)$ provided that the process $\{y(\tau),\ t_o\leq\tau\leq t\}$ is available as observation. This problem was solved by using the Wiener-Hopf equation (McLane[11]). In this section, an advantageous use will be made of the equivalent system to solve the above problem.

The equivalent system may be defined in the sense that

$$\begin{aligned}\varepsilon\{\underline{x}(t)\} &= \varepsilon\{\underline{x}_e(t)\}\\ \varepsilon\{\underline{y}(t)\} &= \varepsilon\{\underline{y}_e(t)\}\end{aligned} \tag{52}$$

and

$$\begin{aligned}\text{cov}\{\underline{x}(t),\underline{x}(t)\} &= \text{cov}\{\underline{x}_e(t),\underline{x}_e(t)\}\\ \text{cov}\{\underline{y}(t),\underline{y}(t)\} &= \text{cov}\{\underline{y}_e(t),\underline{y}_e(t)\}\\ \text{cov}\{\underline{x}(t),\underline{y}(t)\} &= \text{cov}\{\underline{x}_e(t),\underline{y}_e(t)\}\end{aligned} \tag{53}$$

Thus Eqs.(49) and (50) may be equivalent to

$$\begin{aligned}d\underline{x}_e(t) &= A\underline{x}_e(t)dt + d\underline{u}_1(t)\\ d\underline{y}_e(t) &= H\underline{x}_e(t)dt + d\underline{u}_2(t)\end{aligned} \tag{54}$$

where $\underline{u}_i(t)$ $(i=1,2)$ are Brownian motion processes with

$$\begin{aligned}\varepsilon\{d\underline{u}_1(t)d\underline{u}_1'(t)\} &= D_{11}dt\\ \varepsilon\{d\underline{u}_1(t)d\underline{u}_2'(t)\} &= D_{12}dt\\ \varepsilon\{d\underline{u}_2(t)d\underline{u}_2'(t)\} &= D_{22}dt\end{aligned} \tag{55}$$

Here the matrices D_{11}, D_{12} and D_{22} have respective components as

$$\begin{aligned}(D_{11})_{ij} &= q_{iojo} + \sum_{k=1}^{n}(q_{iojk}+q_{joik})m_k(t) + \sum_{k=1}^{n}\sum_{\ell=1}^{n} q_{ikj\ell}[m_{k\ell}(t)+m_k(t)m_\ell(t)]\\ (D_{12})_{ij} &= s_{iojo} + \sum_{k=1}^{n}(s_{iojk}+s_{joik})m_k(t) + \sum_{k=1}^{n}\sum_{\ell=1}^{n} s_{ikj\ell}[m_{k\ell}(t)+m_k(t)m_\ell(t)]\\ (D_{22})_{ij} &= r_{iojo} + \sum_{k=1}^{n}(r_{iojk}+r_{joik})m_k(t) + \sum_{k=1}^{n}\sum_{\ell=1}^{n} r_{ikj\ell}[m_{k\ell}(t)+m_k(t)m_\ell(t)]\end{aligned} \tag{56}$$

where $m_k(t)=\varepsilon\{x_k(t)\}$, $m_{k\ell}(t)= \varepsilon\{x_k(t)x_\ell(t)\}$. By the direct application of the Kalman theory to the equivalent system (54), the linearized filter is readily found as

$$d\hat{\underline{x}}_e(t) = A\hat{\underline{x}}_e(t)dt + K(t)[d\underline{y}_e(t) - H\hat{\underline{x}}_e(t)dt] \tag{57}$$

where

$$K = (PH' + D_{12})D_{22}^{-1}$$

$$\dot{P} = (A - D_{12}D_{22}^{-1}H)P + P(A - D_{12}D_{22}^{-1}H)'$$

$$- PH'D_{22}^{-1}HP - D_{12}D_{22}^{-1}D_{12}' + D_{11} \qquad (59)$$

To obtain the physical realizable linear filter, the equivalent observation process $\underline{y}_e(t)$ should be replaced by the actual process $\underline{y}(t)$. Hence the linear optimal filter is given by

$$d\hat{\underline{x}}(t) = A\hat{\underline{x}}(t)dt + K(t)[d\underline{y}(t) - H\hat{\underline{x}}(t)dt] \qquad (60)$$

It is not difficult to prove that Eq.(60) gives the linear optimal unbias estimate for the stochastic system in the minimal variance sense.

REFERENCES

[1] Ariaratnam, T., and Graefe, P.W.U., *Int. J. Control,* 1, 239-250 (1965)

[2] Bergen, Y., *AIEE Trans. Appl. and Indus.,* 142-145, July (1961)

[3] Bertram, J.E., and Sarachik, P.E., *IRE Trans. Information Theory,* IT-5, 239-250 (1959)

[4] Caughey, T.K. and Dienes, J.K., *J. Math. and Phys.,* 41, 300-318 (1962)

[5] Caughey, T.K., and Gray, A.H., *J. Appl Mech.,* 365-372 (1965)

[6] Cumming, I.J., *Int. J. Control,* 5, 85-90 (1967)

[7] Gray, A.H., *J. Acoust. Soc. Am.,* 37, 235-239 (1965)

[8] Infante, E.F., *J. Appl. Mech.,* 35, 7-12 (1968)

[9] Kozin, F., *J. Math. and Phys.,* 43, 59- (1963)

[10] Kushner, H.J., *Stochastic Stability and Control,* Academic Press, New York (1967)

[11] McLane, P.J., *Int. J. Control,* 10, 41-51 (1969)

[12] Nakamizo, T., *IEEE Trans. Automatic Control,* Ac-14, 584-585 (1969)

[13] Nakamizo, T., *Memoirs Defense Academy,* VIII, 857-885 (1969)

[14] Nevelson, N.B., and Khasminskii, R.Z., *J. Appl. Math. and Mech.,* 30, (1966)

[15] Sawaragi, Y., Nakamizo, T., and Kikuchi, H., *Bulletin of the JSME,* 13,1419-1425, (1970)

STABILITY OF THE LINEAR STOCHASTIC SYSTEM

F. KOZIN

Department of Electrical Engineering, Polytechnic Institute of Brooklyn, U.S.A.

I. Introduction

In this lecture we shall be concerned with the linear stochastic system. As in the case of deterministic systems, the linear system plays the fundamental role in the modeling of phenomena whose dynamical evolution is governed by probabilistic laws. Thus, knowledge concerning the behavior of the solution processes to linear stochastic equations is of prime importance in the study of stochastic systems. We wish to direct our attention to linear differential equations with coefficients that may be stochastic processes, in the present lecture.

In applications linear differential equations with stochastic coefficients occur in a wide variety of problems. For the structual engineer, beams and plates subjected to random forces at the boundaries generates such equations. The appliedmechanist studies this class of problems under the heading of Dynamic Stability. For the control engineer, linearized equations of the dynamics of satellites in orbit provide a basic example generating such studies. For the electrical engineer, parametric amplifier dynamics and for the communications engineer the stability of time varying channels lead to these equations. The study of the scattering phenomena in a random medium, as well as certain chemical and biological problems also involve linear differential equations with stochastic coefficients.

Generally speaking, in all problems above, it is usually assumed that the coefficient processes are known. Thus, the problem becomes that of determining the properties of the solution process as they depend upon the system parameters and the properties of the coefficient processes. In particular, one is interested in statistical properties as characterized by moments, correlationsand distributions as well as sample behavior properties as characterized by bounds, first passages,exceedences, axis crossings, and last but not least, stability.

The study of the stability of stochastic systems has taken a long time in developing and has involved the efforts of many. The recent origins of the study of stability of stochastic systems commences with Rosenbloom [1] and Stratonovich [2] in the early 1950's , then leads to Bertram and Sarachik [3], Kats and Krasovskii [4] as well as Samuels [5] in the early 1960's and continues with Caughey and his coworkers [6, 7], Kushner [8], Khazminskii and his coworkers [9,10] Infante [11], the present author [12, 13], as well as many other fine contributions to the topic [see e.g., [14]] .

The study of stochastic systems has very naturally split into two branches. These are, stochastic differential equations with Gaussian white noise coefficients (the Itô Differential Equations) and stochastic differential equations with non-white noise coefficients.

The greatest advances in the study of stability of stochastic systems have been accomplished for the Ito differential equations. This is primarily due to the fact that the solution processes of Ito differential equations are Markov diffusion processes, and the extensive tools and techniques of diffusion process theory are available to study the behavior of the solution. However, the extensive use of diffusion process theory has required the reader to possess greater maturity in stochastic process theory, which to a certain extent has slowed its development as a practical tool for studying stochastic phenomena. In the non-white coefficient case, the techniques have been somewhat more direct, but the advances have not been as great.

In either case the results that have been obtained in the past are basically sufficient conditions for stability. These conditions have been obtained in terms of the system parameters as well as the statistical properties of the coefficient processes.

Although at the beginning of the modern development of the subject most researchers were concerned with the stability of the moments of the solution processes, it was later realized that sample stability is the significant property to ascertain, and most research has been directed to almost sure stability properties during the past 7 or 8 years. For the stability problem, it is our personal conviction that the moment behavior of the solution process is of importance only insofar as it implies or negates the desired sample stability behavior. We are led to this conviction by the fact that we observe a sample solution, when we test a real system subjected to random excitation. Thus, stability must relate to what the stability behavior of the system will be each time we observe it in operation. Furthermore, there is a certain ambiguity to moment stability that we shall discuss later.

The succession of sufficient conditions for stability of linear stochastic systems that have been obtained in the past, although important to the development of the subject, have often been too conservative to be of practical use. Therefore, it appears to be necessary to have some idea of the true extent of of stability for the linear stochastic system. Those techniques that can bring us closer to the true stability boundaries are, therefore, quite important to study at this stage in the development of the subject. It is exactly this feature that has motivated a number of research activities during the past few years, and it is to this point that we will direct our attention in the present lecture.

In the next section we shall present definitions of some of the typical stability concepts that have been studied relative to stochastic systems. In Section III we discuss those systems with non-white coefficients, for which the ordinary calculus can be applied to analyze their stability properties. In Section IV, we shall discuss those systems with Gaussian white noise coefficients for which the Itô calculus must be applied. We will close this paper with the Discussion in Section V.

Our objective here is to convey information and, hopefully, generate new insights for the reader. We shall present the material as we see it and concentrate on those points that have been of interest to us. Although, we may not be successful in coming up to our objective we are truly sincere in our desire to instill an interest in the topic on the part of the reader.

Much of the material that we present here today has been obtained in joint study with our students, R. Mitchell, S. Prodromou, C.M. Wu, G, N. Sarma, S. Alter, and A. Belle Isle.

II. Stability

The common stability properties of stochastic systems that have been studied in the literature have generally been related to Lyapunov stability - [see [14] for a survey of the subject]. Recognizing that stability in the Lyapunov sense is merely a uniform convergence with respect to the initial conditions, various concepts of stability for stochastic systems can be immediately defined by invoking one of the usual modes of convergence of probability theory. That is, convergence in probability, convergence in the mean and almost sure convergence.

In what follows, $x(t; x_o, t_o)$ will denote the n-dimensional vector solution at time, t, with initial state x_o at time t_o, $\|x\|$ will denote a suitable norm, such as an absolute value or quadratic norm, and we shall be testing the stability of the equilibrium solution $x \equiv 0$.

The definitions of Lyapunov stability for deterministic systems that we shall be transcribing to the stochastic case are stated as follows:

Definition I Lyapunov Stability

The equilibrium solution is said to be stable if, given $\epsilon > o$, there exists $\delta(\epsilon, t_o) > 0$ such that for $\|x_o\| < \delta$, it follows that

$$\sup_{t \geqslant t_o} \|x(t; x_o, t_o)\| < \epsilon \tag{2.1}$$

Definition 2 Asymptotic Lyapunov Stability

The equilibrium solution is said to be asymptotically stable, if it is stable and if there exists δ such that $\|x_o\| < \delta$ implies

$$\lim_{t \to \infty} \|x(t; x_o, t_o)\| = 0 \quad . \tag{2.2}$$

If (2.1) holds for any t_o, the stability is said to be uniform, and if (2.2) holds for any x_o, the equilibrium solution is asymptotically stable in the large.

In order to make the transition to stochastic stability concepts, we merely have to employ the convergence concepts of probability theory in Definition I. However, we must recognize that it is the random variable

$\sup_{t \geq t_o} \| x(t; x_o, t_o) \|$ whose convergence is being tested relative to the parameter x_o, where $x(t; x_o, t_o)$ now represents a sample solution to a stochastic differential equation. We wish to emphasize the fact that this random variable depends upon the behavior of the sample solution on the entire half line (t_o, ∞).

The stochastic versions of Definitions I and II are given in the definitions below.

Definition Ip. Lyapunov Stability in Probability

The equilibrium solution is stable in probability if given $\epsilon, \epsilon' > 0$, there exists $\delta(\epsilon, \epsilon', t_o)$ such that $\|x_o\| < \delta$ implies

$$P\left\{ \sup_{t \geq t_o} \|x(t; x_o, t_o)\| > \epsilon' \right\} < \epsilon \quad . \tag{2.3}$$

Definition I_m. Lyapunov Stability in the Mean

The equilibrium solution is stable in the mean if the expectation exists and given $\epsilon > 0$, there exists $\delta(\epsilon, t_o)$ such that $\|x_o\| < \delta$ implies

$$E\left\{ \sup_{t \geq t_o} \|x(t; x_o, t_o)\| \right\} < \epsilon \ . \tag{2.4}$$

Definition $I_{a.s.}$ Almost Sure Lyapunov Stability

The equilibrium solution is said to be almost surely stable if

$$P\left\{ \lim_{\|x_o\| \to o} \sup_{t \geq t_o} \|x(t; x_o, t_o)\| = 0 \right\} = 1 \ . \tag{2.5}$$

Almost sure Lyapunov stability states that the equilibrium solution is stable for almost all sample systems. This is the same as saying the Definition I holds with probability one.

Asymptotic stability can be extended to the stochastic case in a similar way.

Definition II_p. Asymptotic Stability in Probability

The equilibrium solution is said to be asymptotically stable in probability if I_p holds and if there exists $\delta > 0$ such that $\|x_o\| < \delta$ implies

$$\lim_{T \to \infty} P\left\{ \sup_{t \geq T} \| x(t; x_o, t_o) \| > \epsilon \right\} = 0 \qquad (2.6)$$

for any $\epsilon > 0$.

Definition II_m. Asymptotic Stability in the Mean

The equilibrium solution is said to be asymptotically stable in the mean if I_m holds and if there exists $\delta > 0$ such that $\|x_o\| < \delta$ implies

$$\lim_{T \to \infty} E\left\{ \sup_{t \geq T} \|x(t; x_o, t_o)\| \right\} = 0. \qquad (2.7)$$

Definition $II_{a.s.}$ Almost Sure Asymptotic Stability

The equilibrium solution is said to be almost surely asymptotically stable if $I_{a.s.}$ holds and

$$P\left\{ \lim_{T \to \infty} \sup_{t \geq T} \|x(t; x_o, t_o)\| = 0 \right\} = 1 \quad . \qquad (2.8)$$

The definitions above, as we have stated before, are direct transitions to the stochastic setting of Lyapunov stability. These stability concepts are concerned with sample behavior on the half line.

In the early stages of the development of this subject, most studies were concerned with the stability of various statistics of the solution process at a given time, rather than of the samples. This is probably due to the fact that it is easier to study statistical behavior than to study sample behavior.

Two typical stability concepts related to the moments are as follows; [5], [15] .

Definition III. Lyapunov Stability of the Mean

The equilibrium solution is said to possess stability of the mean if the expectation exists and

$$\lim_{\|x_o\| \to 0} E\left\{ \|x(t; x_o, t_o)\| \right\} = 0 \qquad (2.9)$$

for all $t \geq t_o$.

Definition IV. Exponential Stability of the Mean

The equilibrium solution is said to possess exponential stability of the mean if the expectation exists and if there exists constants α, β, δ, all greater than zero such that $\|x_o\| < \delta$ implies

$$E\left\{ \|x(t;x_o,t_o)\| \right\} < \beta \|x_o\| \exp[-\alpha(t-t_o)] \tag{2.10}$$

for all $t > t_o$.

Although the stability definitions III and IV above do not appear to be as strong a restriction on the solution process as given in I_p-$I_{a.s.}$, II_p-$II_{a.s.}$, there are significant implications in III and IV for sample stability behavior [16], [17]. However, stability of the moments alone does not always provide a satisfactory intuitive basis upon which to judge the stability characteristics of the system. This can easily be illustrated by the simple first order linear Ito differential equation,

$$dx = axdt + \sigma x \, dB \quad , \tag{2.11}$$

where a, σ are constants, and the B-process is the zero mean Wiener process with $E\left\{ B^2(t) \right\} = t$, [14].

The solution process to (2.11) with initial condition $x(o) = x_o$, is obtained via the Ito calculus as

$$x(t) = x_o \exp[(a-\sigma^2/2)t + \sigma B(t)] \quad , \tag{2.12}$$

with probability one.

Furthermore, the nth moments of (2.12) are easily shown to be

$$E\left\{ x^n(t) \right\} = x_o^n \exp\left[(a-\sigma^2/2)nt + \frac{\sigma^2 n^2}{2} t \right] . \tag{2.13}$$

From (2.13), we find that there is exponential stability of the nth moment if and only if

$$a < \frac{\sigma^2}{2}(n-n^2) \quad . \tag{2.14}$$

Thus, for $a < 0$, the first moment is exponentially stable, but higher moments are unstable. For $a < -\sigma^2$, the first and second moments are exponentially stable, and higher moments are unstable, etc., etc. It seems difficult to associate a physical meaning to the system behavior, knowing only that the first moment is stable but the second moment is unstable, or that the first N moments are stable and all higher moments are unstable. To make matters even more interesting, it is well known that the sample functions of the Brownian motion grow no faster than $\sqrt{t \log \log t}$, with probability one. Therefore, the stability of the sample solutions (2.12) are determined by the algebraic sign of $a - \sigma^2/2$ only. That is, $a < \frac{\sigma^2}{2}$ is necessary and sufficient for the equilibrium solution to be asymptotically stable with probability one. The reader should notice that for $0 < a < \sigma^2/2$, the sample solutions possess asymptotic stability and yet all moments will diverge exponentially as seen from (2.13), (2.14).

Hence, we see that the moment behavior can be quite different from the sample behavior regarding stability characteristics.

It is basically for this reason that more effort has been directed, in recent years, to sample stability properties.

We shall be concerned in this lecture with almost sure sample stability properties as defined by $I_{a.s.}$, $II_{a.s.}$.

III. The Second Order Linear Stochastic Differential Equation

The stability properties of the equilibrium solution of linear systems described by stochastic differential equations of the form

$$\frac{dx}{dt} = (A + F(t)\,)\,x \tag{3.1}$$

have been the topic of study of a number of authors in recent years. Under the assumption that A is a stability matrix and $F(t)$ is a matrix whose non-identically zero elements are stationary ergodic processes, sufficient conditions guaranteeing almost sure sample asymptotic stability have been obtained, for example, in references [6], [11], [12]. The sufficiency conditions are usually in terms of the real parts of the eigenvalues of A and the second moments of $F(t)$.

In [11], Infante applied a clever approach based upon properties of quadratic forms, that can be traced to Wintner [see [18], p. 48], to obtain the best sufficient conditions to date for systems of the type described by (3.1). Briefly, if we let $\|x\|_P = (x'Px)$, where P is a symmetric, positive definite constant matrix it is easy to verify that any solution to (3.1) satisfies

$$\log \|x(t)\|_P - \log \|x(o)\|_P = \int_0^t \frac{x'[A + F(s))'\,P + P(A + F(s))]\,x}{x'Px}\,ds. \tag{3.2}$$

But, by the min max properties of pencils of quadratic forms, it follows that

$$\begin{aligned} \Omega(t) &= \max_x \frac{x'[\,(A + F(t))'P + P(A + F(t))\,]\,x}{x'Px} \\ \omega(t) &= \min_x \frac{x'\,[(A + F(t))'\,P + P(A + F(t))\,]\,x}{x'Px} \end{aligned} \tag{3.3}$$

where $\Omega(t)$, $\omega(t)$ are respectively the maximum and minimum eigenvalues of the matrix

$$[\,(A + F(t))'P + P(A + F(t))]\,P^{-1}\ .$$

It immediately obtains from (3.2), (3.3) that

$$\|x(t)\|_P < \|x(o)\|_P \exp \int_o^t \Omega(s)ds \tag{3.4}$$

Infante, then chooses the matrix P to acheive the largest possible stability region, after using the fact that F(t) in (3.1) is a matrix of ergodic processes to imply that

$$\lim_{t \to \infty} 1/t \int_o^t \Omega(s)ds = E\left\{\Omega\ (s)\right\} \tag{3.5}$$

with probability one.

Therefore the sufficient condition for sample asymptotic stability with probability one is

$$E\left\{\Omega(s)\right\} < 0 .$$

For the simple second order oscillator

$$x + 2\beta x + [\ c + f(t)\ ]\ x = 0 \tag{3.6}$$

where $\left\{ f(t);\ t \in [0, \infty)\right\}$ is a zero mean stationary ergodic process with finite second moments, one is led to the sufficient condition

$$-2\beta + DE\left\{\ |\ \ell - f(t)\ |\ \right\} < 0 \tag{3.7}$$

where $D^{-1} = (\ \ell + c - \beta^2)^{1/2}$, and ℓ is to be determined to yield the largest sufficiency region. [see [19] for a slightly different approach to the derivation of (3.7)]. Applying the Schwartz inequality to (3.7), using the fact that $E\left\{ f(t)\right\} = 0$, yields

$$E\left\{f^2(t)\right\} < 4\beta^2 D^{-2} - \ell^2 \ . \tag{3.8}$$

The best sufficiency region is obtained by maximizing the right hand side of (3.8). From the definition of D, we find $\ell_{max} = 2\beta^2$, which produces Infante's sufficient condition

$$E\left\{\ f^2(t)\ \right\} < 4\beta^2 c\ , \tag{3.9}$$

as obtained in [11] for $c = 1$. The region generated by (3.9) is shown in Fig. 1. A characteristic of the approach taken up to this point, is that only the second moment statistic is employed in defining the sufficient condition.

No further detailed statistical structure is assumed. However, as we stated in the introduction, our quest is to use any information available to try to approach the true stability region for the systems that we are studying. With this end in mind, we shall treat (3.7) somewhat differently from the way it has been studied in the past. We shall assume that we know the first distrbution function P(f) of the stationary f-process. Hence, we can write

$$E\left\{ \left| \ell - f(t) \right| \right\} = \int_{-\infty}^{\infty} \left| \ell - f \right| dP$$

$$= \int_{\ell}^{\infty} f dP + \ell (1-2P_{\ell}) - \int_{-\infty}^{\ell} f dP \tag{3.10}$$

$$= 2 \int_{\ell}^{\infty} f dP + \ell (1-2P_{\ell}) ,$$

where $P_{\ell} = P\left\{ f(t) \geqslant \ell \right\}$, and using the fact that the f-process is zero mean. Upon substituting (3.10) into (3.7) obtains the sufficient condition

$$\left[\ell (1-2P_{\ell}) + 2 \int_{\ell}^{\infty} f dP \right] D < 2\beta . \tag{3.11}$$

We note that (3.11) implies $\frac{\ell}{2} \in \left[\beta^2 - \beta\sqrt{c} ,\ \beta^2 + \beta\sqrt{c} \right]$, and that ℓ remains to be chosen in an optimum way.

Clearly, the inequality (3.11) depends upon the nature of the distribution function P . The region generated by (3.11) is obtained through simple numerical procedures.

As an example, we shall assume the f-process in (3.6) to be gaussian with first density function,

$$p(f) = \frac{1}{\sqrt{2\pi}\,\sigma} e^{-f^2/2\sigma^2} . \tag{3.12}$$

It follows that the integrals

$$\left.\begin{aligned} &\int_{\ell}^{\infty} \frac{1}{\sqrt{2\pi}\,\sigma}\, e^{-f^2/2\sigma^2}\, df \\ &\int_{\ell}^{\infty} \frac{1}{\sqrt{2\pi}\,\sigma}\, f e^{-f^2/2\sigma^2}\, df \end{aligned}\right\} \tag{3.13}$$

must be calculated.

For fixed values of β, c and each value of ℓ lying within the constraints $2\beta^2 \pm 2\beta\sqrt{c}$, the integrals (3.13) are calculated for increasing values of σ and substituted into (3.11). Hence, for a given value of ℓ, the maximum value of σ, denoted by $\sigma(\ell)$, is obtained by increasing σ until (3.11) fails to be satisfied. Then, the value

$$\sigma_{max} = \max_{\ell} \sigma(\ell)$$

is obtained for a fixed (β, c), for ℓ varying within its constraints.

The region shown in Fig. 1 was generated for $c = 1$, and ℓ varying over $2\beta^2 \pm .4k\beta$, where $k = 0, 1, 2, 3, 4$, and σ was increased in steps of 0.1. The increase in this region over that given by the sufficient condition (3.9) is quite substantial. Indeed the curve grows roughly like $\beta^{3.3}$ for $\beta > 2$. Figure 1 also includes the sufficiency curve generated by (3.11) for the periodic coefficient

$$f(t) = A\cos(\omega t + \theta),$$

where A, ω are fixed amplitude and frequency and θ is a random phase uniformly distributed over the interval $[0, 2\pi]$.

For the stochastic differential equation

$$x + [2\beta + f(t)]\,x + cx = 0, \tag{3.14}$$

Infante obtained the sufficient condition

$$E\left\{f^2(t)\right\} < \frac{4\beta^2 c}{\beta^2 c}, \tag{3.15}$$

which generates the curve shown in Fig. II for $c = 1$. However, using the procedure above based upon the assumed knowledge of the first density function of the f-process, we obtain the sufficient condition

$$\left[(1-2P_\ell)\, \ell + 2 \int_\ell^\infty f dP \right] G < 2\beta \tag{3.16}$$

where P_ℓ is as before and

$$G = \left[1 - \frac{(\ell + 2\beta)^2}{4c} \right]^{-1/2} .$$

For $c = 1$, the region generated by (3.16) is also compared with (3.15) in Fig. II. Here the increase in the region is even more dramatic since the curve grows like $\beta^{1.4}$ for $\beta > 2$, but (3.15),possesses a horizontal asymptote. The periodic case is also illustrated in Fig. II. For further details concerning the present development, see [19].

What has been accomplished in this section? Basically, we have seen that the sufficiency conditions that have previously appeared in the literature for linear systems with stationary ergodic coefficients apparently are quite conservative when compared with the yet to be determined, true stability region.

Simply by adding slightly more detail concerning the statistical structure of the coefficient process, we have achieved a rather dramatic advance over previous results. The succession of stronger sufficiency conditions beginning with [12],[6] and [11] and now the conditions (3.11), (3.16) bring with them the question "where does it finally end?". What is the true stability boundary for the second order system (3. 6) with a stationary ergodic Gaussian coefficient? To this question, unfortuantely, we have no answer at the present time.

But, let us return to the equality (3. 2). Upon dividing by t, we obtain

$$\frac{\log \|x(t)\|_P - \log \|x(o)\|_P}{t} = \frac{1}{t} \int_o^t \frac{x'[(A+F(s))'P + P(A+F(s))]\, x}{x'PX}\, ds . \tag{3.17}$$

If it can be established that the quotient on the left hand side of (3.17) remains negative as t approaches infinity, with probability one, then it must follow that

$$\lim_{t \to \infty} \left\| x(t) \right\| = 0 \qquad (3.18)$$

with probability one, yielding the almost sure asymptotic stability of the equilibrium solution of (3.1), since (3.18) implies Lyapinov stability for linear systems.

If the quotient remains positive as t approaches infinity, with probability one, the equilibrium solution is unstable.

Hence, the algebraic sign of the limit, as t approaches infinity, of the integral on the right hand side of (3.17) becomes the necessary and sufficient condition for almost asymptotic stability. That is, if the limit is negative, we have stability and if the limit is positive, we have instability. One must, therefore, establish the existence of the limit with probability one, and then evaluate this limit. At this time, there does not appear to be an easy way to accomplish this for the systems that we have been investigating in this section.

However, quite recently Khaz'minskii has recognized that such a limit can be studied for linear Ito stochastic differential equations [20]. The next section is devoted to applications of Khaz'minskii's results.

IV. Linear Itô Differential Equations

In this section we shall be concerned with linear stochastic Itô differential equations, which can be written in differential form for $i = 1, \ldots, \ell$, as

$$dx_i = \sum_{j=1}^{\ell} b_i^j x_j \, dt + \sum_{r=1}^{n} \sum_{j=1}^{\ell} \sigma_{ir}^j x_j \, dB_r , \tag{4.1}$$

where b_i^j, σ_{ir}^j are constants and the B_r are mutually independent Wiener processes for which $E\{B_r(t)\} = 0$ and $E\{[B_t(t) - B_r(s)]^2\} = |t-s|$.

It is well known [21] that the unique, solution process to the stochastic system (4.1) is a Markov diffusion process, with an associated generator $\mathcal{L}$, defined by

$$\mathcal{L}u = (Bx, \text{ grad } u) + \frac{1}{2} \sum_{i,j=1}^{\ell} a_{ij}(x) \frac{\partial^2 u}{\partial x_i \partial x_j} , \tag{4.2}$$

where

$$a_{ij}(x) = \sum_{k,s=1}^{\ell} \sum_{r=1}^{n} \sigma_{ir}^k \sigma_{jr}^s x_k x_s$$

and

$$B = (b_i^j) .$$

Upon applying the Itô differential formula [22] to the function $\log \|x\|$, where x is the solution process to the system (4.1) and $\|x\|$ is the Euclidean norm $(x,x)^{1/2}$, one obtains the expression in differentials,

$$d \log \|x\| = \mathcal{L} \log \|x\| \, dt + \sum_{r=1}^{n} (\sigma(r)\lambda, \lambda) dB_r(t) \tag{4.3}$$

where, $\lambda = x/\|x\|$, $\sigma(r) = (\sigma_{ir}^j)$, $i, j = 1, \ldots, \ell$, and $\mathcal{L}$ is given by (4.2).

In particular,

$$\mathcal{L} \log \|x\| = Q(\lambda) = (B\lambda, \lambda) + \frac{1}{2} \sum_{i=1}^{\ell} a_{ij}(\lambda) - \sum_{i,j=1}^{\ell} a_{ij}(\lambda)\lambda_i \lambda_j . \tag{4.4}$$

If we substitute (4.4) into (4.3), integrate the resulting equation, and divide by t, one obtains

$$\frac{\log \|x(t)\| - \log \|x(o)\|}{t} = \frac{1}{t} \int_o^t \Big[(B\lambda(s), \lambda(s)) + \frac{1}{2} \sum_{i=1}^{\ell} a_{ij}(\lambda(s))$$

$$- \sum_{i,j=1}^{\ell} a_{ij}(\lambda(s))\lambda_i(s)\lambda_j(s) \Big] ds \tag{4.5}$$

$$+ \frac{1}{t} \int_o^t \sum_{r=1}^{n} (a(r)\lambda(s), \lambda(s)\, dB_r(s) ,$$

which is the analogue to the formula (3. 17) of Section III.

Thus it follows that the stability properties that we are attempting to determine will be implied by the limit of the integrals on the right hand side of (4. 5) as t approaches infinity.

Although, we were stopped at this point for stochastic systems of the type studied in Section III, Khaz'minskii has shown the way through the dilemma for systems of the type studied in the present section, [see [20], and [23], chapter 6]. He recognized that the vector $\lambda = x/\|x\|$, where x is the solution process to (4. 1), itself satisfies an Itô differential equation. That is, λ satisfies an equation of the form,

$$d\lambda = \Lambda_1(\lambda)dt + \Lambda_2(\lambda)dB \ , \tag{4.6}$$

Hence, the λ process is a Markov process defined on the surface of the sphere $\|\lambda\| = 1$. Furthermore, this process is ergodic under certain conditions. We note, however, that the ergodic properties of the λ - process are determined by its singularities. In particular, the ergodic properties are determined by the nature of the λ - process in the neighborhood of its singularities. A singularity of a Markov process is defined as a point at which the diffusion component vanishes. For the λ-process, given by (4. 6), the singularities are the solutions to $\Lambda_2(\lambda) = 0$.

Since the second integral on the right hand side of (4. 5) approaches zero with probability one as t approaches infinity [23] it follows from (4.4), (4.5) that

$$\lim_{t \to \infty} \frac{\log \|x(t)\| - \log \|x(o)\|}{t} = E\left\{Q(\lambda)\right\} \tag{4.7}$$

with probability one, in the ergodic case.

This is exactly the formula we want since it yields the necessary and sufficient condition for almost sure sample asymptotic stability in terms of the expectation $E\left\{Q(\lambda)\right\}$. If the expectation is negative, the desired stability property follows. If the expectation is positive, the samples are unstable with probability one. We must evaluate the arithmetic sign of the expectation.

As a simple example, consider the first order linear Itô equation (2.11). For this case the generator is simply

$$\mathcal{L} = ax \frac{d}{dx} + \frac{\sigma^2}{2} x^2 \frac{d^2}{dx^2} \ .$$

One easily finds from (4.4),

$$Q(\lambda) = (a - \frac{\sigma^2}{2}) \lambda^2 \ .$$

Recalling that $\lambda^2 = 1$, we immediately obtain the well known conditions

$$E\left\{Q(\lambda)\right\} = (a - \frac{\sigma^2}{2}) \quad \begin{cases} > 0, & a > \sigma^2/2 \\ < 0, & a < \sigma^2/2 \end{cases}$$

for almost sure sample stability. For this simple example, we did not require any knowledge of the invariant measure of the λ-process with respect to which the expectation is defined. However, for the second order systems that we study in this section, the situation is not so simple.

We are concerned with the second order systems

$$\left.\begin{aligned} dx_1 &= x_2 dt \\ dx_2 &= -x_1 dt - \sigma x_1 dB \end{aligned}\right\} \quad (a)$$

$$\left.\begin{aligned} dx_1 &= x_2 dt \\ dx_2 &= -(2\zeta x_2 + x_1)\, dt - \sigma x_1 \, dB \end{aligned}\right\} \quad (b) \qquad (4.8)$$

For the second order systems, the λ-process is defined on the boundary of the unit circle $\lambda_1^2 + \lambda_2^2 = 1$. Hence, we can study the process on the unit circle via the transformation,

$$\lambda_1 = \cos\varphi, \quad \lambda_2 = \sin\varphi . \qquad (4.9)$$

Thus, we study the one-dimensional φ-process. The generator, as well as the Q function for the φ-process representing (4.8b) is,

$$\mathcal{L} = \frac{\sigma^2}{2}\cos^4\varphi \frac{d^2}{d\varphi^2} - (1+2\zeta \sin\varphi\cos\varphi + \sigma^2\cos^3\varphi \sin\varphi)\frac{d}{d\varphi}$$

$$Q(\lambda(\varphi)) = \frac{\sigma^2}{2}\cos^2\varphi \cos 2\varphi - 2\zeta \sin^2\varphi \qquad (4.10)$$

The corresponding $\mathcal{L}$ and Q for the system (4.8a) is obtained from (4.10) by setting $\zeta = 0$.

For both systems (4.8), we find from $\mathcal{L}$ that the singular points, the points at which the diffusion term vanishes, are the solutions to $\cos^4\varphi = 0$. Hence, $\pm\frac{\pi}{2}$ are the only singular points. The generator of a diffusion process possesses a decomposition into canonical measures m, the speed measure and s, the scale measure. This decomposition can be written as,

$$\mathcal{L} = \frac{d}{dm}\frac{d}{ds} .$$

For the damped system (4.8b), with generator given in (4.10) one easily obtains,

$$m(d\varphi) = \frac{2}{\sigma^2 \cos^2\varphi} \exp\left[-\frac{2}{3\sigma^2} \tan\varphi(3+3\zeta\tan\varphi + \tan^2\varphi)\right] d\varphi$$

(4.11)

$$s(d\varphi) = \frac{1}{\cos^2\varphi} \exp\left[\frac{2}{3\sigma^2} \tan\varphi(3+3\zeta\tan\varphi + \tan^2\varphi)\right] d\varphi \quad .$$

The speed and scale measures for the undamped system (4.8a) are obtained from (4.11) by setting $\zeta = 0$. The nature of singularities of a diffusion process are determined in terms of the canonical speed and scale measures [24].

For an interval $[a, b]$, and any point $c \in (a, b)$, a is:

an exit point if $\int_{a^+}^{c} s(du) \int_{u}^{c} m(dv) < \infty$

(4.12)

an entrance point if $\int_{a^+}^{c} m(du) \int_{u}^{c} s(dv) < \infty$.

The point b is:

an exit point if $\int_{c}^{b^-} s(du) \int_{c}^{u} m(dv) < \infty$

(4.13)

an entrance point if $\int_{c}^{b^-} m(du) \int_{c}^{u} s(dv) < \infty$.

The conditions (4.12), (4.13) applied to study the behavior at the singular points $\pm\pi/2$ for both systems (4.8) show that for the right half circle, $-\pi/2 \leqslant \varphi \leqslant \pi/2$, $\pi/2$ is an entrance and not an exit, and $-\pi/2$ is an exit but not an entrance. Similarly, for the left half circle, $\pi/2 \leqslant \varphi \leqslant 3\pi/2$, $\pi/2$ is an exit point and not an entrance, and $3\pi/2$ is an entrance point and not an exit. It follows, therefore, that the φ-process diffuses around the circle in the clockwise direction and is never trapped at any point, nor held for any time at the singularities $\pm\pi/2$. Thus, the process is ergodic on the the entire circle.

We let R_i be the random time of traversal of the right half circle by the φ-process, and L_i be the random time of traversal of the left half circle during the ith traversal. Therefore, the Markov time T_i for the ith traversal of the complete circle is given by $T_i = R_i + L_i$.

Upon setting $T(n) = T_1 + T_2 + \cdots + T_n$, we have, due to the ergodicity of the φ process, and for any function F with finite expectation,

$$E\left\{F(\varphi_t)\right\} = \lim_{n\to\infty} \frac{1}{T(n)} \int_0^{T(n)} F(\varphi_s)\,ds$$

$$= \lim_{n\to\infty} \frac{\left\{\int_0^{T_1} + \int_{T_1}^{T_1+T_2} + \ldots + \int_{T_{n-1}}^{T(n)}\right\} F(\varphi_s)\,ds}{T_1 + T_2 + \cdots + T_n} \tag{4.14}$$

$$= \lim_{n\to\infty} \frac{\frac{1}{n}\left\{\int_0^{R_1} + \ldots + \int_0^{R_n} F(\varphi_s)ds\right\} + \frac{1}{n}\left\{\int_0^{L_1} + \ldots + \int_0^{L_n} F(\varphi_s)ds\right\}}{\frac{1}{n}\left\{R_1 + \ldots + R_n\right\} + \frac{1}{n}\left\{L_1 + \ldots + L_n\right\}}$$

$$= \frac{E\left\{\int_0^R F(\varphi_s)ds\right\} + E\left\{\int_0^L F(\varphi_s)ds\right\}}{E\{R\} + E\{L\}}$$

with probability one.

The expectations at the bottom of (4.14) can be evaluated through the following relationship from [24, p. 114]. For the internal [a, b], with random time T from entrance at a to exit at b,

$$E\left\{\int_0^T F(\varphi_s)\,ds\right\} = \int_a^b G(\varphi)F(\varphi)m(d\varphi) \ , \tag{4.15}$$

where $m(d\varphi)$ is the speed measure and

$$G(\varphi) = S(b) - S(\varphi) \quad .$$

In our case, G, F and m are known, hence determining the expectations depends upon integration of known functions. Furthermore, since $E\{R\}$

and $E\{L\}$ are positive, only the expectations of the integrals in the numerator of (4.14) need be studied.

We must evaluate,

$$J = J_1 + J_2 = E\left\{\int_0^R Q(\lambda(\varphi_s))ds\right\} + E\left\{\int_0^L Q(\lambda(\varphi_s))\, ds\right\} \tag{4.16}$$

$$= \int_{\pi/2}^{-\pi/2} Q(\lambda(\varphi))\left[s(-\tfrac{\pi}{2}) - s(\varphi)\right] m(d\varphi) + \int_{3\pi/2}^{\pi/2} Q(\lambda(\varphi))\left[s(\tfrac{\pi}{2}) - s(\varphi)\right] m(d\varphi)$$

$$= 2J_1 ,$$

by a simple change in variables.

We finally see that only the algebraic sign of J_1 must be determined in order to secure the stability properties of the systems under investigation.

For the damped system (4.8b), substituting Q from (4.10) and (4.11) into (4.15) yields

$$J_1 = \int_{-\pi/2}^{\pi/2} (\cos 2\varphi - \frac{4\zeta}{\sigma^2}\tan^2\varphi)\, \eta_{\zeta,\sigma^2}(\varphi)d\varphi \quad , \tag{4.17}$$

where

$$\eta_{\zeta,\sigma^2} = \exp\left[-\frac{2}{3\sigma^2}\tan\varphi(3+3\zeta\tan\varphi + \tan^2\varphi\right]$$

$$X \int_{-\pi/2}^{\varphi} \exp\left[\frac{2}{3\sigma^2}\tan\theta(3+3\zeta\tan\theta+\tan^2\varphi)\right] \sec^2\theta\, d\theta \quad .$$

For the undamped system (4.8a), we merely set $\zeta = 0$ in (4.17). In [25], it is found that $J_1 > 0$ for every $\sigma > 0$ for the undamped system (4.8a). For the damped system (4.8b), the complexity of the integral J_1 in (4.17) requires that numerical integration techniques be applied. The results of the

(ζ, σ^2) parameter study shown in Fig. III were determined by numerically solving the equation $J_1(\zeta, \sigma^2) = 0$ which defines the stability boundary. The stability boundary for the system (4.8b). We wish to emphasize the fact that these results are exact results.

They are among the very few exact results that have appeared in the literature related to the question of stability for stochastic systems. As such they add significant insight to the problem in general.

Of very significant importance is the relation between the results obtained in this section concerning Itô differential equations, and the results obtained for "real" noise coefficient systems in Section III. There have been a number of basic papers relating real noise systems and Itô differential equations [see, e.g., [26], [27]]. It is known that one must modify the system equations, by adding correction terms, when the real noise coefficients approach the white noise in some suitable sense. However, for the systems treated in this section the correction terms are identically zero.

Thus, the systems (4.8), with gaussian white noise coefficients, are related to the "real" noise coefficient systems

$$\begin{aligned} &\ddot{x} + (1 + f(t)) x = 0 \qquad &(a) \\ &\ddot{x} + 2\zeta\dot{x} + (1 + f(t)) x = 0 \qquad &(b) \end{aligned} \qquad , \qquad (4.18)$$

of the type treated in Section III.

In many previous simulations studies it has been found that the system (4.18a) is unstable for gaussian coefficients eg, [28] As a result of the analysis in this section, we now suspect that this will always be the case for wide band real noise coefficients. It remains to be seen how the stability of (4.18a) is effected by narrowing the band of the spectrum of the f-process. For the system (4.18b) the curve in Fig. I grows asymptotically like β^{3+}. This is approximately the way the curve grows for the exact stability boundary in Fig. III for the system (4.8b). Although the region in Fig. III will contain the region in Fig. I a comparison of the figures show that the results of Section III are very close to the true stability boundary for the real noise system (3.14).

For further details concerning the computational aspects of the results of this section, the reader is referred to [29] and [30]. A comprehensive study of two dimensional systems following the approach of this section, based upon [30], will appear shortly.

V. Distributed Parameter Systems

In this section we shall introduce some ideas motivated by the development of Section III, relative to the topic of stability for certain distributed parameter systems with non-white coefficients.

One of the first studies of the stability properties of stochastic distributed parameter systems [31] appeared in 1959. In that paper the stability properties of the second moments of the deflection of a simply supported beam which is being subjected to Gaussian white noise axial loads was investigated. The next significant study of this problem [6] was published in 1965, where the almost sure sample stability of the simply supported beam subjected to stationary ergodic end loads was considered. Using Lyapunov techniques developed in the paper for stochastic ordinary differential equations an analysis was made of the beam modes in order to obtain stability criteria. The modal approach was studied, as well, in [32] and [33]. In [34], one finds an application of the properties of semi-groups as well as the Gronwall inequality to study the almost sure stability properties via suitable functional norms defined on the Banach space of sample solutions of the distributed parameter system. Although the results of [34] are applicable to a wide class of systems, the Gronwall inequality is known to yield weak conditions as compared with other techniques. Very recently, Infante and Plaut [35], [36], studied the problem applying Lyapnov functional ideas to beams and plates, making use of Infante's earlier results [11]. Their sufficient conditions are the best that have appeared in the literature so far.

In this section we shall find that significant improvements are obtained from the ideas generated in Section III.

We shall be concerned with specific boundary value problems. We first investigate the stability of the solutions to the boundary value problem,

$$\frac{\partial^2 \omega}{\partial t^2} + 2\beta \frac{\partial \omega}{\partial t} + (f_o + f(t)) \frac{\partial^2 \omega}{\partial x^2} + \frac{\partial^2 \omega}{\partial x^4} = 0 \qquad (5.1)$$

for $x \in [0,1]$, $t \in [0,\infty]$,

where the deflection $\omega(x,t)$ satisfies

$$\omega(o,t) = 0 \qquad \omega(1,t) = 0$$

$$\frac{\partial^2 \omega(0,t)}{\partial x^2} = 0 \qquad \frac{\partial^2 \omega(1,t)}{\partial x^2} = 0 \tag{5.2}$$

The coefficient β may represent damping on the column, f_o a constant axial load and $f(t)$, random excitations on the axial load that comes from a zero mean, stationary ergodic stochastic process whose samples are continuous functions with probability one. We shall study the asymptotic stability of the solutions $\omega(x,t)$ via a Lyapunov functional approach.

That is, we shall define

$$V(t) = \int_0^1 Q\left(\omega, \frac{\partial \omega}{\partial x}, \ldots, \frac{\partial \omega}{\partial t}\right) dx,$$

where Q is a quadratic form in its variables.

The first question is, how to choose the "best" quadratic form Q? What do we mean by best? Clearly, Q should be chosen so that we will obtain the largest region of stability in terms of the parameters to be studied. We study the stability region as a function of the damping, β, and the variance $\sigma^2 = E\{f^2(t)\}$. Hence, Q should be chosen so as to generate the best possible sufficient conditions in terms of (β, σ^2).

We shall use the following approach. Upon expanding $\omega(x,t)$ into its modes, we have

$$\omega(x,t) = \sum_{m=1}^{\infty} \omega_m(x,t) = \sum_{m=1}^{\infty} T_m(t) \sin m\pi x \tag{5.3}$$

Substituting (5.3) into (5.1) yields the modal equations.

$$\ddot{T}_m(t) + 2\beta \dot{T}_m(t) + [(m\pi)^4 - (m\pi)^2 f_o - (m\pi)^2 f(t)]\, T_m(t) = 0, \tag{5.4}$$

$$m = 1, 2, 3, \ldots$$

We have shown that the "best" quadratic Lyapunov function [19] for studying the almost sure stability properties of $T_n(t)$ is given by

$$V_M(t) = (T_M, \dot{T}_M)\begin{pmatrix} 2\beta^2+(m\pi)^2-(m\pi)^2 f_0 & \beta \\ \beta & 1 \end{pmatrix}\begin{pmatrix} T_M \\ \dot{T}_M \end{pmatrix} \quad (5$$

Upon recognizing that

$$\begin{cases} (m\pi)^2\,\omega_M \equiv \dfrac{\partial^2 \omega_M}{\partial x^2} \\ (m\pi)^2\,\omega_M^2 \equiv \left(\dfrac{\partial \omega_M}{\partial x}\right)^2 \end{cases} \quad (5.6)$$

we can apply the orthogonality of the sinusoids sin n πx on [0,1], to obtain the desired Lyapunov functional,

$$V(t) = \int_0^1 \left[\left(\frac{\partial^2 \omega}{\partial x^2}\right)^2 - f_0\left(\frac{\partial \omega}{\partial x}\right)^2 + 2\beta^2\omega^2 + 2\beta\omega\frac{\partial \omega}{\partial t} + \left(\frac{\partial \omega}{\partial t}\right)^2\right] dx \quad (5.$$

V(t) clearly satisfies the desired positive definite properties.

We mention that V(t) is similar to, but not the same as, the Lyapunov functional applied by Infante and Plaut.

Our object now is to obtain bounds on V(t) that will guarantee the desired almost sure stability properties.

Upon differentiating (5.7), we obtain

$$\frac{dV}{dt} = -2\beta V + 2U \quad (5.8)$$

where U is the functional,

$$U(t) = \int_0^1 \left[2\beta^3\omega^2 + 2\beta\omega\frac{\partial \omega}{\partial t} - \beta f(t)\,\omega\frac{\partial^2 \omega}{\partial x^2} - f(t)\frac{\partial \omega}{\partial t}\frac{\partial^2 \omega}{\partial x^2}\right] dx \;. \quad (5.$$

We now attempt to construct a bound,

$$|U(t)| \leq \lambda(t)\, V(t)$$

where $\lambda(t)$ is to be determined.

Due to the orthogonality of $\sin n\pi x$, we can study $U = \sum_{m=1}^{\infty} U_m$, $V = \sum_{m=1}^{\infty} V_m$ and the associated problem,

$$\mu_m(t)\, |U_m(t)| \leq V_m(t) \tag{5.10}$$

In terms of ω_n, (5.10) becomes,

$$\mu_m(t) \left| \int_0^1 \left[2\beta^3 \omega_m^2 + 2\beta^2 \omega_m \frac{\partial \omega_m}{\partial t} - \beta f(t)\, \omega_m \frac{\partial^2 \omega_m}{\partial x^2} - f(t) \frac{\partial \omega_m}{\partial t} \frac{\partial^2 \omega_m}{\partial x^2} \right] dx \right|$$

$$\leq \int_0^1 \left[\left(\frac{\partial^2 \omega_m}{\partial x^2}\right)^2 - f_0 \left(\frac{\partial \omega_m}{\partial x}\right)^2 + 2\beta^2 \omega_m^2 + 2\beta\, \omega_m \frac{\partial \omega_m}{\partial t} + \left(\frac{\partial \omega_m}{\partial t}\right)^2 \right] dx \tag{5.11}$$

Using (5.3), we have from (5.11),

$$\int_0^1 \Big\{ \left[(m\pi)^4 - (m\pi)^2 f_0 + 2\beta^2 \mp 2\beta^3 \mu_m \mp (m\pi)^2 \beta f(t) \mu_m \right] \omega_m^2$$

$$+ \left[2\beta \mp 2\beta^2 \mu_m \mp (m\pi)^2 f(t) \mu_m \right] \omega_m \frac{\partial \omega_m}{\partial t} + \left(\frac{\partial \omega_m}{\partial t}\right)^2 \Big\}\, dx \geq 0 \tag{5.12}$$

The inequality (5.12) will hold iff

$$(m\pi)^4 - (m\pi)^2 f_0 + 2\beta^2 \mp 2\beta^3 \mu_m \mp (m\pi)^2 \beta f(t)\, \mu_m \geq \tag{5.13}$$

$$\geq \left(\beta \mp \beta^2 \mu_m \mp \frac{(m\pi)^2}{2} f(t)\, \mu_m \right)^2$$

After some algebraic manipulations (5.13) yields

$$\mu_m(t) \leq \frac{2\left[(m\pi)^4 - (m\pi)^2 f_0 + \beta^2\right]^{1/2}}{\left|2\beta^2 + (m\pi)^2 f(t)\right|} \tag{5.14}$$

Therefore,

$$U_m \leq \frac{\left|2\beta^2 + (m\pi)^2 f(t)\right|}{2\left[(m\pi)^4 - (m\pi)^2 f_0 + \beta^2\right]^{1/2}} V_m = \frac{1}{2}\lambda_m V_m \tag{5.15}$$

From equation (5. 8), the result (5.15) will obtain,

$$\begin{aligned} \frac{dV}{dt} &= -2\beta V + 2U = -2\beta V + 2\sum_{m=1}^{\infty} U_m \\ &\leq -2\beta V + \sum_{m=1}^{\infty} \lambda_m(t) V_m \\ &\leq -2\beta V + \left(\operatorname*{Max}_m \lambda_m(t)\right) \sum_{m=1}^{\infty} V_m \\ &= \left[-2\beta + \operatorname*{Max}_m \lambda_m(t)\right] V \end{aligned} \tag{5.16}$$

Whereupon,

$$V(t) \leq V_0 \exp\left[-2\beta t + \int_0^t \operatorname*{Max}_m \lambda_m(\tau)\, d\tau\right] \tag{5.17}$$

Thus, since the f-process is ergodic, it immediately follows that the sufficient condition for almost sure stability is

$$E\left\{-2\beta + \operatorname*{Max}_m \lambda_m(t)\right\} < 0. \tag{5.18}$$

"It was assumed in [35] that the maximum occurs for n = 1; this is clearly incorrect. The assumption was corrected in [36], however the maximum was not obtained. Now, $\underset{n}{\mathrm{Max}}\,\lambda_n(t)$ is a function of n, β, f_o and f(t).

It can be determined graphically, and its expectation obtained numerically. To illustrate the way in which this can be done, let us for the moment assume that $f_o = 0$. Therefore, $\lambda_n(t)$ is given by

$$\lambda_n(t) = \frac{|2\beta^2 + (n\pi)^2 f(t)|}{[(n\pi)^4 + \beta^2]^{1/2}}, \quad n = 1, 2, 3, \ldots \tag{5.19}$$

Since, f(t) is a random variable, it can take any value in its range, which we assume to be $(-\infty, \infty)$.

Hence, we can draw the graphs of $\lambda_n(t)$ as a function of f(t) for n = 1, 2, ···, . These lines are shown in Figure . It is quite obvious that $\lambda(t) = \underset{n}{\mathrm{Max}}\,\lambda_n(t)$ is the polygonal envelope of the lines $\lambda_n(t)$.

The polygonal envelope defining λ (t) is

$$\lambda(t) = \begin{cases} |f(t)|, & f(t) \leq f_{01} \\ \lambda_1(t), & f_{01} < f(t) \leq f_{12} \\ \lambda_2(t), & f_{12} < f(t) \leq f_{23} \\ \vdots & \vdots \\ \lambda_n(t), & f_{n-1,n} < f(t) \leq f_{n,n+1} \\ \vdots & \vdots \end{cases} \tag{5.20}$$

where the vertices are at $f_{k-1,k}$, defined as that value of f(t) for which the lines $\lambda_{k-1}(t)$ and $\lambda_k(t)$ intersect. Hence, $f_{k-1,k}$ is that value for which $\lambda_{k-1}(t) = \lambda_k(t)$

After some algebra, we find

$$
\begin{cases}
f_{01} = \dfrac{-2\beta^2}{(\pi^4+\beta^2)^{1/2} + \pi^2} \\[2ex]
f_{k-1,k} = \dfrac{2\beta^2\left(1 - \sqrt{\dfrac{((k-1)\pi)^4+\beta^2}{(k\pi)^4+\beta^2}}\right)}{\pi^2\left(k^2\sqrt{\dfrac{((k-1)\pi)^4+\beta^2}{(k\pi)^4+\beta^2}} - (k-1)^2\right)}
\end{cases}
\tag{5.21}
$$

We can now evaluate the expectation (5.18).

From elementary probability theory, we have,

$$
\begin{cases}
E\{\lambda(t)\} = E\{\operatorname{Max}_m \lambda_m(t)\} \\
\quad = E\{|f| \mid -\infty < f(t) \le f_{01}\} P\{-\infty < f(t) \le f_{01}\} \\
\quad + E\{\lambda_1 \mid f_{01} < f(t) \le f_{12}\} P\{f_{01} < f(t) \le f_1 \\
\quad \vdots \\
\quad + E\{\lambda_m \mid f_{m-1,m} < f(t) \le f_{m,m+1}\} P\{f_{m-1,m} < f(t) \\
\quad \vdots
\end{cases}
\tag{5.22}
$$

If the density function for the f-process is known, as shown for example in Figure IV then we can use numerical integration to evaluate (5.22), for various values of σ^2. Thus, the inequality (5.18) will yield stability regions as a function of (β, σ^2).

The stability regions for the Gaussian as well as for the periodic (sinusoidal) coefficient case is shown in Figure V. The significance of these regions will be discussed below

We mention that adding a continuous non-linear term $g(w)$ to the partial differential equation (5.1), that satisfies

$$\begin{cases} g(w) = -g(-w) \\ w g(w) \geq 0 \\ |g'(w)| \quad \text{monotone increasing,} \end{cases} \tag{5.23}$$

will generate the same stability regions.

For the problem of stability of panels in supersonic flow subjected to random end loads, we consider the infinite panel of unit width, simply supported at x = 0, x = 1. The end loads are assumed to be uniform along the entire boundary $x \equiv 0$, $x \equiv 1$. In that case, the equation (5.1) becomes

$$\frac{\partial^2 w}{\partial t^2} + 2\beta \frac{\partial w}{\partial t} + M \frac{\partial w}{\partial x} + (f_0 + f(t)) \frac{\partial^2 w}{\partial x^2} + \frac{\partial^4 w}{\partial x^4} = 0, \tag{5.24}$$

with the same boundary conditions as (5.1). Here, M represents the mach number of the supersonic flow over the elastic plate. The terms

$M \frac{\partial w}{\partial x} + 2\beta \frac{\partial w}{\partial t}$ represent the aerodynamic forces due to the flow in accordance with the Piston theory approach to supersonic aerodynamics.

We shall apply the Lyapunov functional as defined in (5.7).

Proceeding as above we differentiate (5.7) to obtain the equation (5.8), Where the functional U becomes

$$U = \int_0^1 \Big[2\beta^2 w^2 + 2\beta^2 w \frac{\partial w}{\partial t} - \beta f(t) w \frac{\partial^2 w}{\partial x^2} - f(t)\frac{\partial w}{\partial t}\frac{\partial^2 w}{\partial x^2} - M\frac{\partial w}{\partial x}\frac{\partial w}{\partial t}\Big] dx \tag{5.25}$$

as a result of the aerodynamic terms in (5.24). Again we wish to determine the $\lambda(t)$ that satisfies the inequality,

$$|U(t)| \leq \lambda(t)\, V(t)$$

Because of the term $\frac{\partial w}{\partial x}$, we cannot apply the approach as used for the system (5.1). Instead, we apply the variational calculus to solve the problem,

$$\delta(U - \lambda V) = 0, \tag{5.26}$$

via the associated Euler equations. After extensive but straightforward computations, we find the sequence of λ_n's that satisfy (5.26) to be

$$\lambda_m(t) = \frac{1}{2}\sqrt{\frac{[(m\pi)^2 f(t) + 2\beta^2]^2 + M^2(m\pi)^2}{(m\pi)^4 - (m\pi)^2 f_0 + \beta^2}}, \quad m = 1, 2, 3, \cdots . \tag{5.27}$$

It is obvious that for $M = 0$, (5.27) reduces to (5.14), as it should.

Now, from equation (5.18) we will obtain the desired sufficient conditions as a function of β, σ^2, as well as M.

The Max $\lambda_n(t) = \lambda(t)$ is found in the same fashion as before. The only difference being that (5.27) defines intersecting curves rather than intersecting lines. The stability regions are shown in Figs. VI and VII

The regions shown in Figs. VI and VII were obtained for the Gaussian case as well as the periodic case discussed in Section III.

In order to obtain the regions, we choose discrete values of f and compute the λ_n via equations (5.14) and (5.27). We then choose the largest value corresponding to the given value of f and take the expectations as shown in (5.22). This is accomplished for various values of M by choosing σ^2 and varying β until the inequality (5.18) ceases to hold. In Fig. VIII , a comparison of regions for the case $f_o = 0$, $M = 0, 200, 400$ is made with the region obtained in [36] by Infante and Plaut. The increase in regions is rather significant.

VI. Conclusion

We have attempted to present a few recent ideas and their application to the study of linear stochastic systems. In particular we have shown how they lead to stronger sufficiency conditions, and in the case of the Ito differential equation, to the exact stability boundary for second order systems. For the "real" noise coefficient case, we feel that the results presented are close to the exact stability region for wide band Gaussian processes. Any further extension will require assumptions concerning covariance properties or, equivalently, spectral properties of the coefficient process. Stability conditions in terms of spectral properties of the coefficient process have been treated by a number of authors among them Ariaratnan [32], Gray [37], and Hsu [38]. However, further development appears to be called for on this problem.

Although, we can say significant things concerning second order stochastic systems, the theory does not appear to allow us to study the sample stability for higher order systems in as straightforward a fashion. For the Ito case, the properties of diffusion processes near singularities is not so clear, or so easily classified as they are for one dimensional diffusions. In the non-white case, the eigenvalues of the associated quadratic forms are not so easy to obtain.

However, the rate at which the subject has developed due to the innovation, interest and dedication to the problem of stochastic stability by the many students of the topic leads one to conclude, with a certain sense of optimum, that the outstanding problems cited above will be better understood and in some cases solved in the near future.

References

1. A. Rosenbloom, Analysis of linear systems with randomly time-varying parameters, Proc. Symp. Inf. Nets., Vol. III, Poly. Inst. Brooklyn, p. 145 (1954).

2. R. L. Stratonovich, et al.,"Non-Linear Transformations of Stochastic Processes," Pergamon Press, Oxford (1965).

3. J.E. Bertram and P.E. Sarachik, "Stability of circuits with randomly time-varying parameters," Trans. IRE, PGIT-5, Special Supplement, p. 260 (1959).

4. I.I. Kats and N.N. Krasovskii, "On the stability of systems with random parameters," Prkil. Met. Mek. 24, 809 (1960).

5. J.C. Samuels, " On the mean square stability of random linear systems," Trans. IRE, PGIT-5, Special Supplement, p. 248 (1959).

6. T.K. Caughey and A.H. Gray, Jr., "On the almost sure stability of linear dynamic systems with stochastic coefficients," ASME J. Appl. Mech. 32, 365 (1965).

7. T.K. Caughey, Comments on "On the stability random systems," J. Acoust. Soc. Am. 32, 1356 (1960).

8. H. Kushner, "Stochastic Stability and Control," Academic Press, New York (1967).

9. R.Z. Khas'minskii, "On the stability of the trajectories of Markov processes (in English), Prikl. Mat. Mek. 26, 1554 (1962).

10. N.B. Nevel'son and R.Z. Khas'minskii, "Stability of a linear system with random disturbances of its parameters (in English). Prikl. Mat. Mek. 30, 487 (1966).

11. E.F. Infante, "On the stability of some linear non-autonamous random systems," ASME paper 67-WA/APM-25 (1967).

12. F. Kozin, "On almost sure stability of linear systems with random coefficients," M.I.T. J. Math. Phys. 43, 59 (1963).

13. F. Kozin, "On almost sure asymptotic sample properties of diffusion processes defined by stochastic differential equations," J. Math. Kyoto Univ., 4, 575 (1965).

14. F. Kozin, "A survey of stability of stochastic systems," Automatika, J. IFAC, 5 (1969), pp. 95-112.

15. B.H. Bharucha, "On the stability of randomly varying systems," Ph.D. Thesis Dept. of Elec. Eng., Univ. Calif., Berkeley, July (1961).

16. F. Kozin, "On relations between moment properties and almost sure Lyapunov stability for linear stochastic systems," J. Math. Anal. Appl. 10, 342 (1965).

17. F. Kozin, "Stability of stochastic systems," Paper 3A, 3rd IFAC Congr., London (1966).

18. L. Cesari, "Asymptotic Behavior and Stability Problems in Ordinary Differential Equations," Springer, Berlin (1959).

19. F. Kozin, C.M. Wu, "On the stability of linear stochastic differential equations," (to appear ASME Journ. of App. Mech. (1972)).

20. R.Z. Khaz'minskii, "Necessary and sufficient conditions for the asymptotic stability of linear stochastic systems," Th. Prob. Appls. 1, 144 (1967).

21. K. Ito, "On stochastic differential equations," Mem. Am. Math. Soc. No. 4 (1951).

22. K. Ito, "On a formula concerning stochastic differentials," Nagoya Math. Jr., Japan 3, 55 (1951).

23. R.Z. Khaz'minskii, "Stability of Systems of differential equations with random parametric excitation," Book in Russian (1969).

24. K. Ito and H.P. McKean, Jr., "Diffusion Processes and their sample paths," Springer-Verlag/Academic Press, New York 1965.

25. F. Kozin, S. Prodromou, "Necessary and sufficient conditions for almost sure sample stability of linear Ito equations," SIAM Journ. Appl. Math. Vol. 21, No. 3, pp. 413-424 (1971).

26. E. Wong and M. Zakai, "On the convergence of orginary integrals to stochastic integrals," Ann. Math. Statist., 36 (1965), pp. 1560-1564.

27. R.L. Stratonovich, "A new representation for stochastic integrals," SIAM J. Control, 4 (1966), pp. 362-371.

28. G.F. McDonough, "Stability Problems in the Control of Saturn Launch Vehicles," Dynamic Stability of Structures, Proc. Int. Conf. Pergamon, New York, p. 113 (1967).

29. S. Prodromou, "Necessary and sufficient conditions for stability of stochastic systems," Doctoral thesis, System Sciences, Polytechnic Institute of Brooklyn, New York, 1970.

30. R.R. Mitchell, "Necessary and sufficient conditions for sample stability of second order stochastic differential equations," Doctoral thesis, School of Aero Astro and Engineering Sciences, Purdue University, (1970).

31. J.C. Samuels, A.C. Eringen, "On stochastic linear systems," MIT J. Math. Phys. 38, p. 83 (1959).

32. S. T. Ariaratnam, "Dynamic stability of a column under random loading," Dynamic Stability of Structures, Proc. Int. Conf., Pergamon, New York, p. 267, (1967).

33. J.A. Lepore, and H.C. Shah, "Dynamic stability of axially loaded columns subjected to stochastic excitation," AIAA Journal, vol. 6, No. 8, August 1968, pp. 1515-1521.

34. P.K.C. Wang, "On the almost sure stability of linear stochastic distributed parameter dynamical systems," A.S.M.E. Transactions, Journal of Applied Mechanics, March, 1966, pp. 182-186.

35. E.F. Infante, and R.H. Plaut, "Stability of a column subjected to a time-dependent axial load," AIAA Journal, vol. 7, No. 4, April 1969, pp. 766-768.

36. R.H. Plaut, and E.F. Infante, "On the stability of some continuous systems subjected to random excitation," A.S.M.E. Transactions, Journal of Applied Mechanics, September 1970, pp. 623-627.

37. A.H. Gray, Jr., "Frequency dependent almost sure stability conditions for a parametrically excited random vibrational system," ASME paper 67-APM-9 (1967).

38. C.S. Hsu, T.H. Lee, "A stability study of continuous systems under parametric excitation via Lyapunov's direct method," IUTAM Symposium on Instability of Continuous Systems, July 1969.

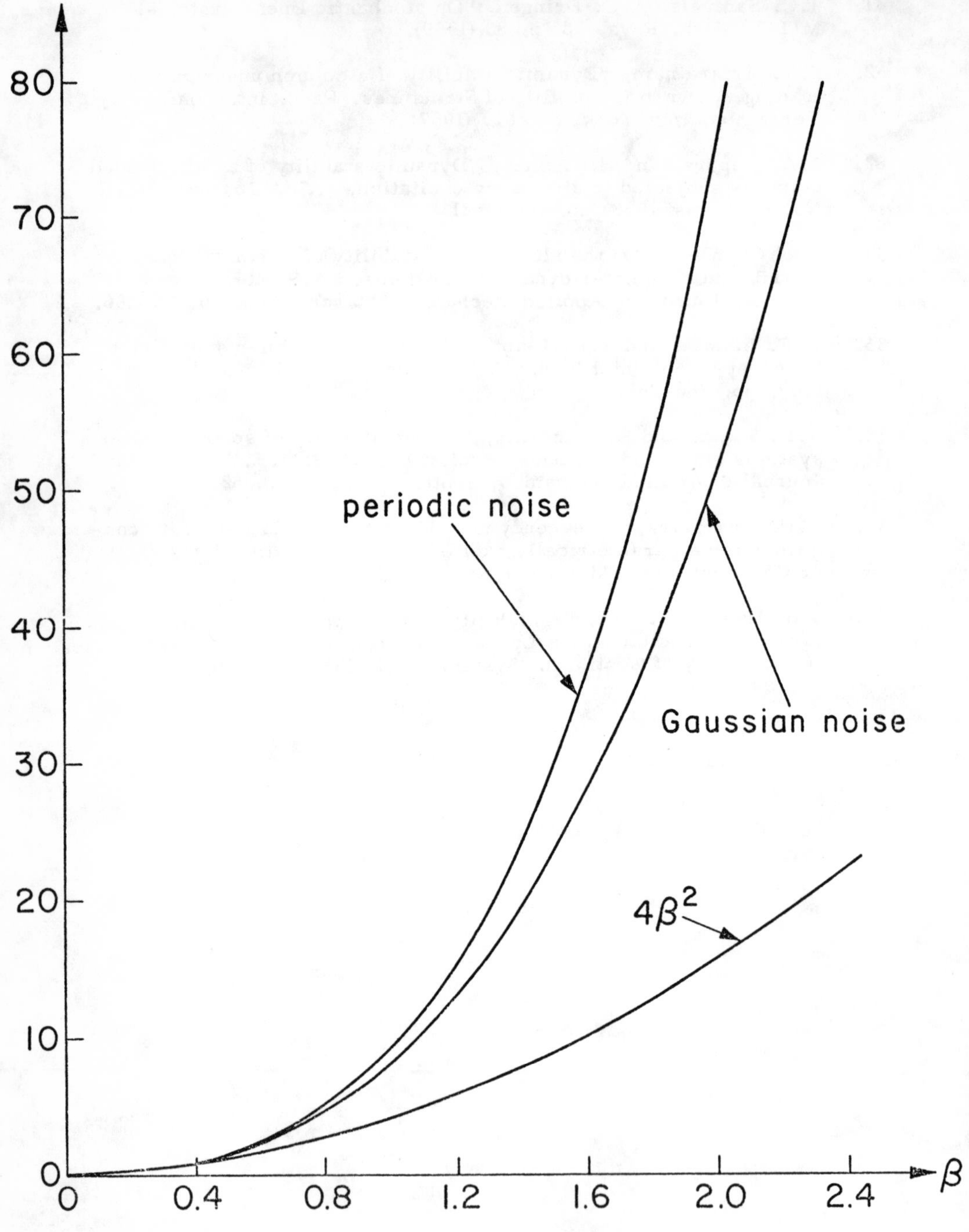
80
70
60
50
40
30
20
10
0
0
0.4
0.8
1.2
1.6
2.0
2.4
β
periodic noise
Gaussian noise
$4\beta^2$

FIGURE 1

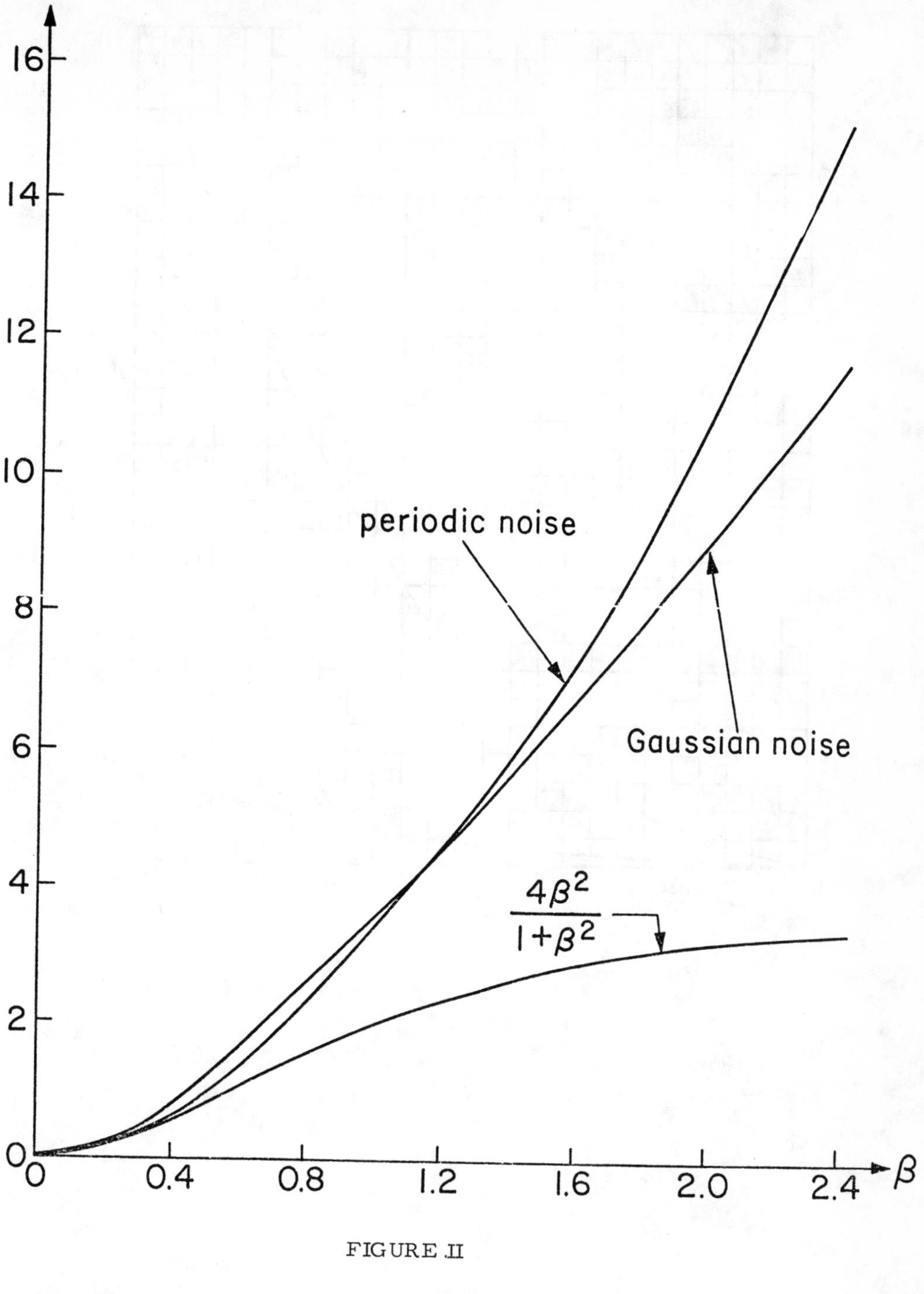
16
14
12
10
8
6
4
2
0
0
0.4
0.8
1.2
1.6
2.0
2.4
β
periodic noise
Gaussian noise
$\frac{4\beta^2}{1+\beta^2}$

FIGURE II

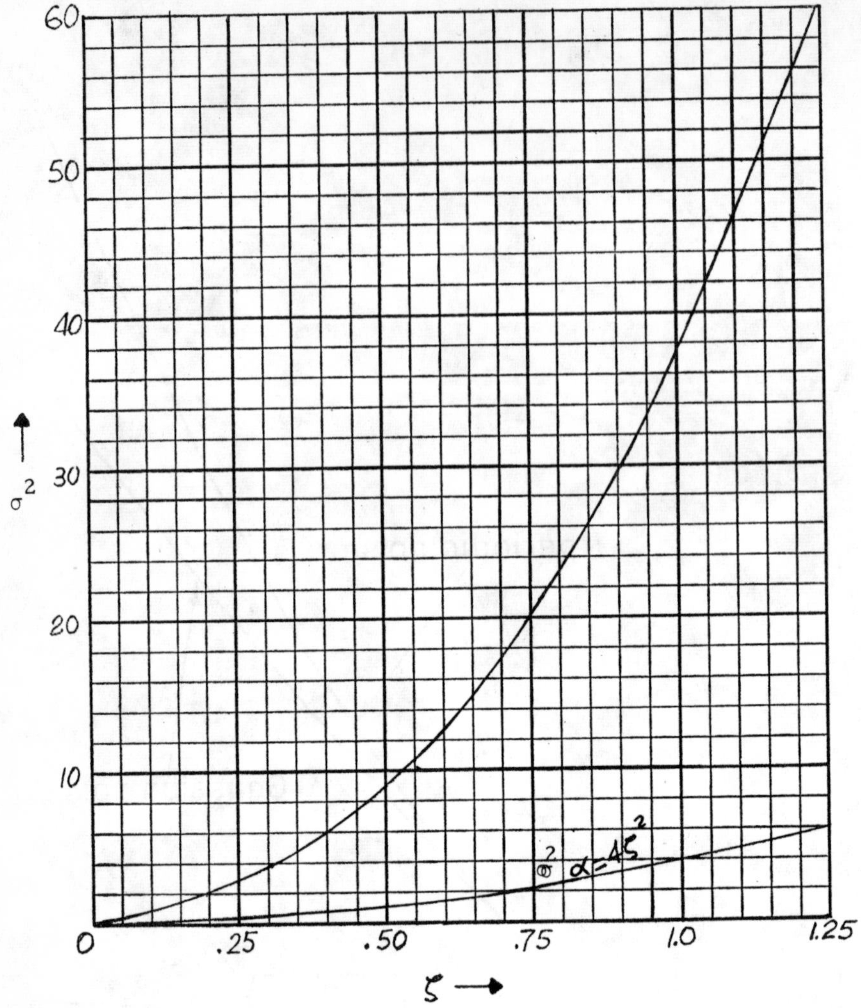
60
50
40
30
20
10
σ^2
0
.25
.50
.75
1.0
1.25
ζ
σ^2 $\alpha = 4\zeta^2$

FIGURE III

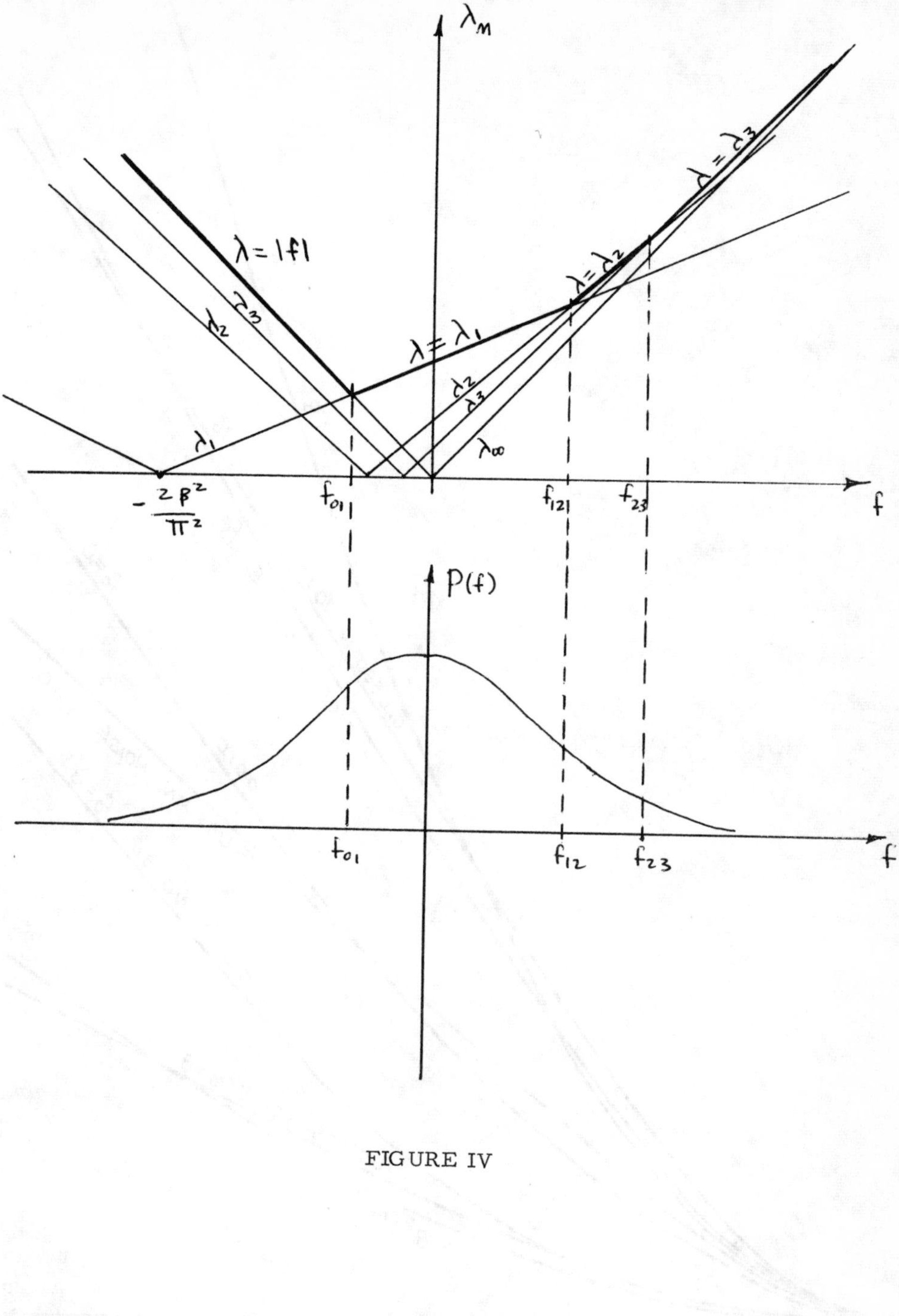
λ_M
$\lambda = |f|$
λ_3
λ_2
$\lambda = \lambda_1$
$\lambda = \lambda_2$
$\lambda = \lambda_3$
λ_2
λ_3
λ_1
λ_∞
$-\frac{2\beta^2}{\pi^2}$
f_{01}
f_{12}
f_{23}
f
P(f)
f_{01}
f_{12}
f_{23}
f

FIGURE IV

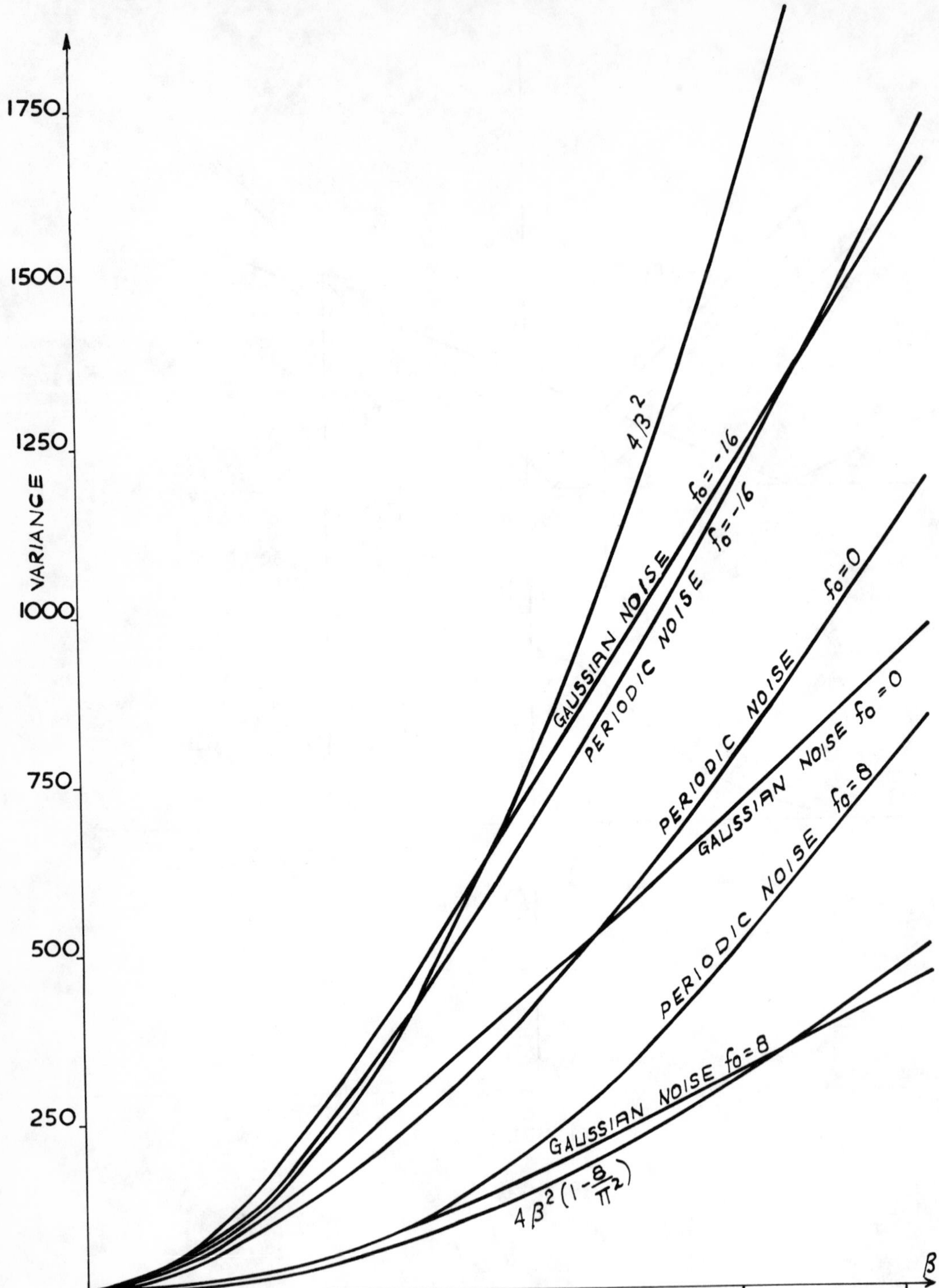

1750
1500
1250
1000
750
500
250
VARIANCE
0
5
10
15
20
25
β
4β²
GAUSSIAN NOISE f0 = -16
PERIODIC NOISE f0 = -16
PERIODIC NOISE f0 = 0
GAUSSIAN NOISE f0 = 0
PERIODIC NOISE f0 = 8
GAUSSIAN NOISE f0 = 8
4β²(1 - 8/π²)

FIG V

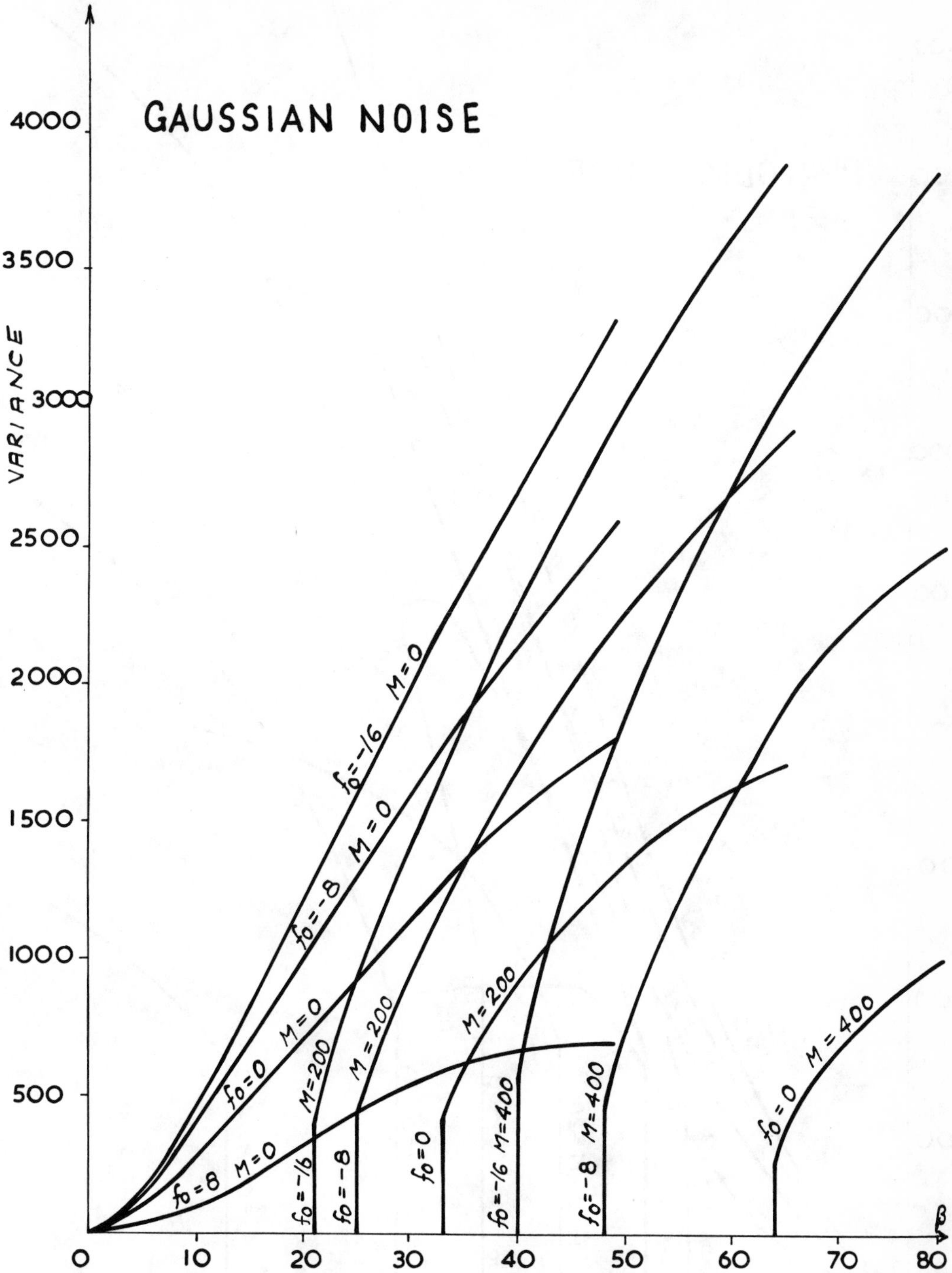
GAUSSIAN NOISE
VARIANCE
4000
3500
3000
2500
2000
1500
1000
500
0
10
20
30
40
50
60
70
80
β
fo=-16 M=0
fo=-8 M=0
fo=0 M=0
fo=8 M=0
M=200
fo=-16
M=200
fo=-8
M=200
fo=0
fo=-16 M=400
fo=-8 M=400
fo=0 M=400

FIG VI

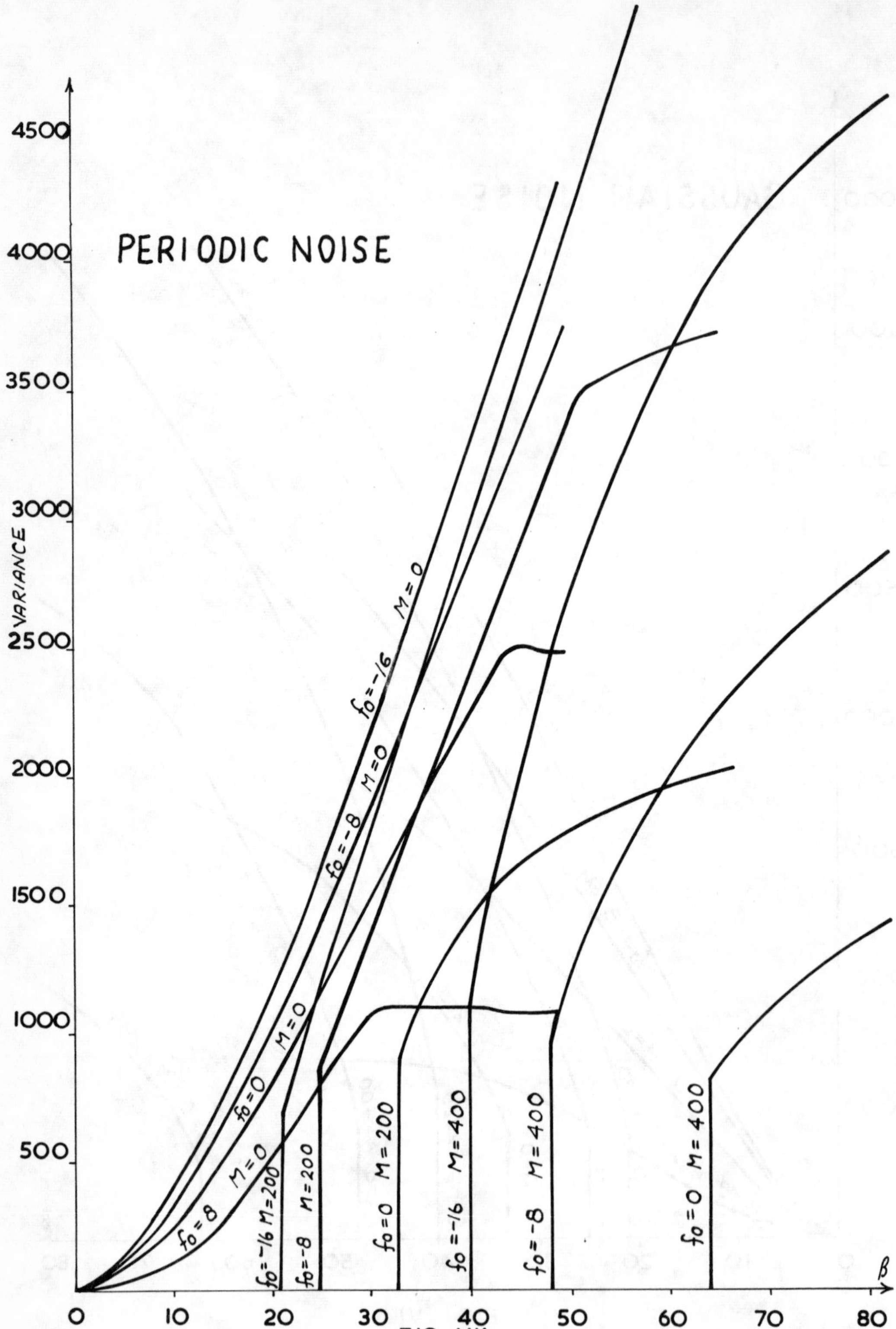
PERIODIC NOISE
VARIANCE
4500
4000
3500
3000
2500
2000
1500
1000
500
0
10
20
30
40
50
60
70
80
β
$f_0=-16$ $M=0$
$f_0=-8$ $M=0$
$f_0=0$ $M=0$
$f_0=8$ $M=0$
$f_0=-16$ $M=200$
$f_0=-8$ $M=200$
$f_0=0$ $M=200$
$f_0=-16$ $M=400$
$f_0=-8$ $M=400$
$f_0=0$ $M=400$

FIG VII

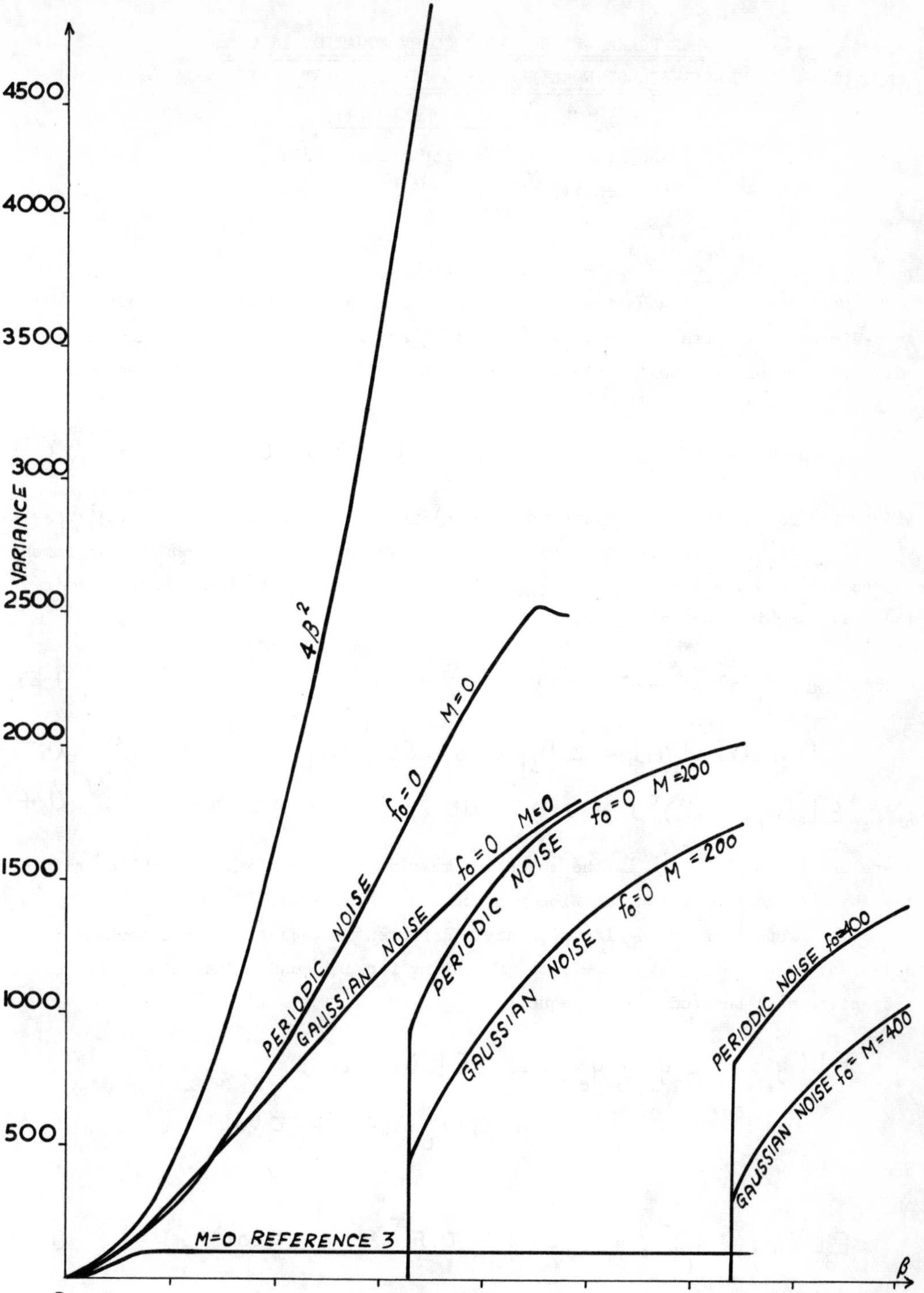
VARIANCE
4500
4000
3500
3000
2500
2000
1500
1000
500
0
10
20
30
40
50
60
70
80
β
4β²
$f_0=0$ M=0
PERIODIC NOISE
GAUSSIAN NOISE $f_0=0$ M=0
PERIODIC NOISE $f_0=0$ M=200
GAUSSIAN NOISE $f_0=0$ M =200
PERIODIC NOISE f_0=400
GAUSSIAN NOISE f_0= M=400
M=0 REFERENCE 3

FIG VIII

THE FOKKER-PLANCK-KOLMOGOROV EQUATION IN THE ANALYSIS OF NONLINEAR FEEDBACK STOCHASTIC SYSTEMS

CHRISTOPHER J. HARRIS

Department of Electronic Engineering,
University of Hull, HU6 7RX,
England.

1. INTRODUCTION

Consider a continuous time, continuous state nonlinear stochastic system with an n-dimensional state vector $y_t = \{y_i(t)\}$. Assuming that y_t is a Markov process, which is the solution to the state dependent noise Itô stochastic differential equation;

$$dy_i(t) = g_i(y_t,t).dt + \sum_j^m f_{ij}(y_t,t).dw_j(t), \quad i=1,\ldots n \tag{1.1}$$

where y(o) or its probability density $p(y_t,o)$ is given, $G(y_t,t) = \{g_i(y_t,t)\}$, $F(y_t,t) = \{f_{ij}(y_t,t)\}$ are the known system dynamics. d. is a stochastic increment in the Itô sense, and $w_t = \{w_i(t)\}$ is an m-dimensional unit Wiener process with the following incremental properties

$$E[dw_i(t)] = 0, \quad i=1,\ldots m \tag{1.2}$$

$$E[dw_i(t).dw_j(t)] = 2D_{ij}(t).\delta_{ij}.dt, \quad (i,j)=1,\ldots m \tag{1.3}$$

or

$$E[dw_t.(dw_t)^T] = 2D(t).dt, \tag{1.4}$$

where $2D(t) = 2\{D_{ij}(t)\}$ is the intensity matrix or the uniform spectral density of the white noise from which the Wiener process w_t is derived.

The properties of the Itô stochastic differential equation are discussed in detail by Doob (1953, Chap. 6,9); the following forward conditional time incremental properties will be used in the sequel:

$$E[(y_{t+\delta t} - y_t)/y_t] = E[\delta y_t./y_t] = G(y_t,t).\delta t + o(\delta t)^{\dagger}, \tag{1.5}$$

and

$$E[\delta y_t.(\delta y_t)^T/y_t] = 2F.D.F^T.\delta t + o(\delta t), \tag{1.6}$$

† where $o(\delta t)$ denotes the limiting property $\lim_{\delta t\to 0} \frac{o(\delta t)}{\delta t} = 0$

2. MOMENT EQUATIONS AND ITÔ'S FUNDAMENTAL LEMMA

Let $\varphi(y_t, t)$ be an arbitary scalar-valued real function, which is continuously differentiable in t the independent variable, and having continuous second mixed partial derivatives with respect to the elements of y_t. Then the stochastic differential $d\varphi$ of φ is by Itô's fundamental differential lemma (Skorokhod, 1965, p.24)

$$d\varphi = \sum_i^n \frac{\partial \varphi}{\partial y_i} \cdot dy_i + \frac{\partial \varphi}{\partial t} \cdot dt + \sum_{i,j}^n \frac{\partial^2 \varphi}{\partial y_i \partial y_j} \cdot (F.D.F^T)_{ij} \cdot dt , \tag{2.1}$$

But since φ is an non-random function of the random variable y_t, then taking the expectation of equation (2.1) over the y_t space, gives the ordinary differential equation,

$$\frac{dE(\varphi)}{dt} = \sum_i^n E\left(\frac{\partial \varphi}{\partial y_i} \cdot g_i\right) + E\left(\frac{\partial \varphi}{\partial t}\right) + \sum_{i,j}^n E\left(\frac{\partial^2 \varphi}{\partial y_i \partial y_j} \cdot (F.D.F^T)_{ij}\right) \tag{2.2}$$

This result, although obtained differently, is similar to that of Skorokhod (1965, p.96) and may be generalised to give an identical result to that of Sancho (1969, p.83) for a stochastic differential equation in the Itô sense with generalised Poisson and Gaussian white noise stochastic parameters.

Similar to Ariaratnam and Graefe (1965, p.239) let $m_N(k_1, k_2, \ldots k_n)$ denote the mixed moment of order N:

$$m_N(k_1, k_2, \ldots k_n) = E[y_1^{k_1} \cdot y_2^{k_2} \cdot \ldots \cdot y_n^{k_n}] \tag{2.3}$$

where non-negative integers k_i satisfy $\sum_i^n k_i = N$

Setting

$$\varphi(y_t, t) = \prod_i^n y_i^{k_i} \tag{2.4}$$

then since φ is not an explicit function of t, $\frac{\partial \varphi}{\partial t} = 0$ and equation (2.2) becomes

$$\frac{dE}{dt}\left[\prod_i^n y_i^{k_i}\right] = \sum_i^n k_i \cdot E[y_i^{k_i - 1} \cdot g_i \cdot \prod_{\substack{j \\ j \neq i}}^n y_j^{k_j}]$$

$$+ \sum_{i,j}^n k_i \cdot k_j \, E[y_i^{k_i - 1} \cdot y_j^{k_j - 1} \cdot (F.D.F^T)_{ij} \cdot \prod_{\substack{k \\ k \neq i,j}}^n y_k^{k_k}], \tag{2.5}$$

for all possible k_i that satisfies $\sum_i^n k_i = N$.

In particular the first two moment equations are

$$\frac{dE[y_i]}{dt} = E[g_i]$$

$$\frac{dE[y_i \cdot y_j]}{dt} = E[y_i \cdot g_j + y_j \cdot g_i] + 2E[(F.D.F^T)_{ij}] , \tag{2.6}$$

for $(i,j) = 1,2,\ldots n$ and given initial conditions $y_t(o)$ or $p(y_t,o)$.

The Itô stochastic differential equations (1.1) are linear if all $g_i(y_t,t)$ and $f_{ij}(y_t,t)$ are linear functions of the state vector y_t, in which case the Nth order moment equation involves only moments of order less than N on the right hand side of equations (2.5), and may be solved explicitly for N. If, in addition each $f_{ij}(y_t,t)$ is independent of y_t, then y_t is Gaussian distributed and only the solution of equations (2.6) is necessary for a complete statistical description of y_t.

3. TRANSFORMATION OF ITÔ'S STOCHASTIC DIFFERENTIAL EQUATIONS

The quantitative analysis of nonlinear systems with stochastic parametric variations can be achieved by the analytic method of transforms, which has been used successfully in the field of deterministic nonlinear systems by Dasarathy and Srinivasan (1968, p.27). These transformations of systems with time dependent variables, which are either deterministic or stochastic, are based upon the concept of equivalent systems in which relationships are evolved between equivalent linear and nonlinear systems. In the following, transformation techniques involving the dependent and independent variables of the governing stochastic state equations, are proposed to convert time dependent stochastic parameter systems into equivalent deterministic systems with stochastic perturbations. The statistical characteristics of the transformed system may be found by the well established techniques of statistical linearisation (Pervozvanskii, 1965) or by the Fokker-Planck-Kolmogorov equation for a diffusion process.

Let the equivalent n-dimensional system with stochastic input perturbations of the same ensemble as the stochastic parametric variations of equations (1.1-6) be,

$$d\varphi_j = h_j(\varphi_t,t).dt + \sum_r^m c_{jr}(t).dw_r(t). , \quad j=1,\ldots n \tag{3.1}$$

where the non-zero deterministic time varying weighting functions $c_{jr}(t)$ are known or specified for all (j,r).

Let the transformation of dependent variables be defined by the scalar real valued functions

$$\varphi_j = \Phi_j(y_t,t) , \quad j=1,\ldots n \tag{3.2}$$

whose second mixed state partial derivatives and derivative with respect to the independent variable t, are jointly continuous and bounded on any finite interval of $\varphi_t = \{\varphi_j\}$ and t. The independent variable t, is retained in the transformation to avoid transformation of the Wiener processes in equations (1.1).

The stochastic differential $d\varphi_j$ of φ_j may be written using Itô's lemma (equation (2.1)) and the system stochastic differential equation (1.1) as

$$d\varphi_j = \sum_i^n \frac{\partial \Phi_j}{\partial y_i} \cdot g_i + \sum_i^n \sum_k^m \frac{\partial \Phi_j}{\partial y_i} \cdot f_{ik} \cdot dw_k$$

$$+ \sum_{i,j}^n \frac{\partial^2 \Phi_j}{\partial y_i \cdot \partial y_j} \cdot (F.D.F^T)_{ij} \cdot dt + \frac{\partial \Phi_j}{\partial t} \cdot dt , \quad j=1,..n \tag{3.3}$$

For the stochastic systems represented by equations (1.1) and (3.1) to be equivalent, then equations (3.1) and (3.3) are identical if and only if

$$c_{jk} = \sum_i^n \frac{\partial \Phi_j}{\partial y_i} \cdot f_{ik} \tag{3.4}$$

and

$$h_j = \sum_i^n \frac{\partial \Phi_j}{\partial y_i} \cdot g_i + \sum_{i,k}^n \frac{\partial^2 \Phi_j}{\partial y_i \partial y_k} \cdot (F.D.F^T)_{ik} + \frac{\partial \Phi_j}{\partial t}$$

$$j = 1,..n \tag{3.5}$$

For m = n the transformation Φ_j and the unknown function h_j may be found exactly, since equation (3.4) can be written in matrix form as:

$$[c_{j1}, \ldots c_{jn}]^T = \left[\frac{\partial \Phi_j}{\partial y_1}, \ldots \frac{\partial \Phi_j}{\partial y_n}\right]^T . F^T , \quad j=1,..n \tag{3.6}$$

Solving the above for φ_j by Cramers rule gives

$$\varphi_j = \Phi_j(y_t,t) = \int \frac{\sum_s^n c_{js} \cdot F_{si}(y_t,t)}{\det . F} . dy_i . \tag{3.7}$$

where $F_{si}(y_t,t)$ is the $(s.i)^{th}$ coefactor of coefficient matrix F. Also

$$h_j = \frac{\partial}{\partial t} \int \sum_k^n c_{jk} \cdot F_{ki} \cdot (\det.F)^{-1} . dy_i$$

$$+ \sum_{r,s}^n \left\{ g_r + (F.D.F^T)_{rs} \cdot \frac{\partial}{\partial y_r} \right\} \sum_k^n c_{jk} \cdot F_{ks} \cdot (\det.F)^{-1} \tag{3.8}$$

For j=1,2,...n equations (3.7) and (3.8) completely define the transformation φ_j and the transformed state equations. These may be studied directly using the Fokker-Planck-Kolmogorov equation (Fuller, 1969) to find the transitional probability density function of the continuous Markov vector process $\{\varphi_j(t)\}$. The results may be retransformed to the y_t space to obtain those statistical properties of interest of the original system via equations (3.7) and the Jacobian determinant method of Pervozvanskii (1965, p.39) for nonlinear lagless transformations of stochastic processes. By considering the transformation of the independent variable (which naturally involves transformations of the stochastic processes), more general forms of systems with stochastic parametric variations reducible to equivalent deterministic systems can be considered.

4. THE FOKKER-PLANCK-KOLMOGOROV EQUATION

If all $h_j(\varphi_t,t)$ of the transformed equations are single valued then the vector process φ_t is Markov since the conditional probability for $\varphi_j(t)$ given $\varphi_i(\tau)$ for $t > t_o \geq \tau$ is dependent only upon $\varphi_j(t_o)$, since the processes

$$\varphi_j(t) = \varphi_j(t_o) + \int_{t_o}^{t} h_j(\varphi_u, u).du + \sum_r^m c_{jr}[w_r(t) - w_r(t_o)]$$

and $\sum_r^m c_{jr}[w_r(t) - w_r(t_o)]$ have independent increments.

If the conditional probability density function $p(\varphi_t, t/\varphi_s, s)$ has continuous second mixed partial derivatives with respect to the elements of φ_t, and if $h_j(\varphi_t,t)$, $\frac{\partial h_j}{\partial \varphi_j}$ are bounded, exist and satisfy the Lipschitz condition within the bounded domain $\Gamma_r^{(n)}$ of state space $\mathcal{E}_o^n$, also if φ_t satisfies the conditions

$$\lim_{\Delta t \to 0} \frac{1}{\Delta t} \int_{|\varphi_s - \varphi_t| \geq \delta} p(\varphi_t, t+\Delta t/\varphi_s, s).d\varphi_s = 0, \quad \text{for any } \delta > 0 \tag{4.1}$$

$$\lim_{\Delta t \to 0} \frac{1}{\Delta t}.E\{\varphi_i(t+\Delta t) - \varphi_i(t)/\varphi_t, t\} = A_i(\varphi_t, t) = h_i(\varphi_t, t), \tag{4.2}$$

$$\lim_{\Delta t \to 0} \frac{1}{\Delta t}.E\{(\varphi_i(t+\Delta t) - \varphi_i(t))(\varphi_j(t+\Delta t) - \varphi_j(t))/\varphi_t, t\}$$

$$= B_{ij}(\varphi_t, t) = \sum_r^n c_{ir}.c_{jr}, \quad (i,j) = 1,..n. \tag{4.3}$$

for A_i, and B_{ij} continuous together with their derivatives and $\{B_{ij}\} \geq 0$; then $p(\varphi_t, t/\varphi_s, s)$ satisfies the Fokker-Planck-Kolmogorov equation:

$$\frac{\partial p}{\partial t} + \sum_i^n \frac{\partial (h_i . p)}{\partial \varphi_i} - \frac{1}{2}\sum_{i,j}^n \frac{\partial^2}{\partial \varphi_i \partial \varphi_j}\left(\sum_r^n c_{ir} . c_{jr} . p\right) = 0, \qquad (4.4)$$

subject to the initial values and boundary conditions for the domain $\Gamma_r^{(n)}$ of solution. The bounded domains $\Gamma_r^{(n)}$ are those regions in $\mathcal{E}^{(n)}$ for which $h_i(\varphi_t, t)$ and $\frac{\partial h_j}{\partial \varphi_j}$ exist and are continuous. The boundaries of each $\Gamma_r^{(n)}$ being defined by the discontinuities of $\{h_j(\varphi_t, t)\}$. Boundary conditions on equation (4.4) require the continuity of the probability density and its first partial derivatives with respect to each φ_j across the surfaces of each domain $\Gamma_r^{(n)}$, where r is the number of discontinuities of $\{h_j(\varphi_t, t)\}$.

Defining an additional state variable $\varphi_{n+1} = t$, the linear differential operator

$$L(p(\varphi_t, t/\varphi_s, s)) = \sum_{i,j}^{n+1}\left\{\frac{1}{2} . \frac{\partial^2 \left(\sum_r^n c_{ir} . c_{jr} . p\right)}{\partial \varphi_i . \partial \varphi_j} - \frac{\partial (h_i . p)}{\partial \varphi_i}\right\}, \qquad (4.5)$$

and its adjoint differential operator

$$K(q(\varphi_t, t/\varphi_s, s)) = \sum_{i,j}^{n+1}\left\{\frac{1}{2} . \sum_r^n c_{ir} . c_{jr} . \frac{\partial^2 q}{\partial \varphi_i . \partial \varphi_j} + h_i . \frac{\partial q}{\partial \varphi_i}\right\}, \qquad (4.6)$$

where $h_{n+1} = 1$, $c_{ir} . c_{jr} = 0$ for $(i, j) = n+1$.
Then

$$q . L(p) - p . K(q) = \sum_i^{n+1} \frac{\partial}{\partial \varphi_i}\left\{\sum_j^{n+1} \frac{q}{2} . \frac{\partial}{\partial \varphi_j}\left(\sum_r^n c_{ir} . c_{jr} . p\right)\right.$$

$$\left. - \frac{p}{2} . \sum_r^n c_{ir} . c_{jr} . \frac{\partial q}{\partial \varphi_j} - h_i . p . q\right\}$$

$$= \sum_i^{n+1} \frac{\partial R_i}{\partial \varphi_i}, \qquad (4.7)$$

where the Riemann function $R_i(\varphi_t; \varphi_s)$ is the fundamental solution of $L(p(\varphi_t, t/\varphi_s, s)) = 0$. R_i and $\frac{\partial R_i}{\partial \varphi_i}$ are continuous throughout $\Gamma_r^{(n+1)}$, then applying Green's formula in (n+1)-dimensions to equation (4.7) gives:

$$\int_{\Gamma_r^{(n+1)}} \cdots \int \sum_i^{n+1} \frac{\partial R_i}{\partial \varphi_i} . d\varphi_t = -\int_{S^{(n)}} \cdots \int \sum_i^{n+1} R_i . \cos(\underline{n+1}, \varphi_i) . dS^{(n)}$$

$$= -\int_{S_0^{(n)}} \cdots \int_{S_0^{(n)}} \sum_{i}^{n+1} R_i \cdot d\varphi_1 \cdot d\varphi_2 \cdots d\varphi_{i-1} \cdot d\varphi_{i+1} \cdots d\varphi_{n+1} , \qquad (4.8)$$

where $S_r^{(n)}$ is the n-dimensional hypersurface which is the projection of the boundary hypersurface $S^{(n)}$ of $\Gamma_r^{(n+1)}$ onto the hyperplane $[\varphi_1, \varphi_2, \ldots \ldots \varphi_{r-1}, \varphi_{r+1}, \ldots \varphi_{n+1}]$.

Solution of integral equations (4.8) for transitional probability density function $p(\varphi_t / \varphi_s)$ over each domain $\Gamma_r^{(n+1)}$ requires that the boundary conditions of $p(\varphi_t / \varphi_s)$ and $\frac{\partial p}{\partial \varphi_j}(\varphi_t / \varphi_s)$ be known; these are given by a similar set of integral equations to equations (4.8).

5. EXAMPLE

The nonlinear stochastic parameter system

$$dy = \left(y^{-2} \cdot \eta\left(\frac{y^3}{2}\right) + \frac{y}{2} - 2y^{-5} \right) . dt + y^{-2} . dw \qquad (5.1)$$

becomes under the transformation $\varphi = \Phi(y,t) = \frac{y^3}{2}$, and equations (3.7), (3.8);

$$d\varphi = (\eta(\varphi) + \varphi) . dt + dw \qquad (5.2)$$

where $\eta(\varphi) = \psi . \operatorname{sgn} \varphi$ and dw is a unit Wiener process. The probability density $p(\varphi, t/\varphi, s)$[†] satisfies the Fokker-Planck-Kolmogorov equation (4.4):

$$\frac{\partial p}{\partial t} - \frac{\partial}{\partial \varphi}((\varphi + \psi) . p) = \frac{1}{2} \cdot \frac{\partial^2 p}{\partial \varphi^2} , \qquad \varphi > 0 \qquad (5.3)$$

$$\frac{\partial p}{\partial t} - \frac{\partial}{\partial \varphi}((\varphi - \psi) . p) = \frac{1}{2} \cdot \frac{\partial^2 p}{\partial \varphi^2} , \qquad \varphi < 0 \qquad (5.4)$$

The transitional probability density boundary conditions at $\varphi = 0$ are

$$[p(\varphi,t)]_{\varphi = 0^-} = [p(\varphi,t)]_{\varphi = 0^+} \qquad (5.5)$$

$$\left[-\psi . p(\varphi,t) - \frac{1}{2} \cdot \frac{\partial}{\partial \varphi} p(\varphi,t)\right]_{\varphi = 0^+} = \left[\psi . p(\varphi,t) - \frac{1}{2} \cdot \frac{\partial}{\partial \varphi} p(\varphi,t)\right]_{\varphi = 0^-} \qquad (5.6)$$

Thence equation (4.7) becomes

---

† The conditional part of $p(\cdot)$ is omitted for simplicity.

$$q(\varphi,t).L(p(\varphi,t)) - p(\varphi,t).K(q(\varphi,t))$$

$$= -\frac{\partial}{\partial t}(q.p) + \frac{\partial}{\partial \varphi}\left\{\frac{1}{2}.q.\frac{\partial p}{\partial \varphi} - \frac{1}{2}.p.\frac{\partial q}{\partial \varphi} + (\varphi + \eta(\varphi)).q.p\right\}, \quad (5.7)$$

Take $q(\varphi,t) = R(\varphi,t)$ as the fundamental solution of $L(p) = 0$, which is the transitional density of the Markov process which satisfies

$$d\varphi_1 = -\varphi_1.dt - \psi.dt + dw \quad (5.8)$$

$$d\varphi_2 = -\varphi_2.dt + \psi.dt + dw \quad (5.9)$$

for all values of (φ_1, φ_2). These equations are linear, then $R(\varphi, t)$ is Gaussian distributed, with mean and dispersion given by equations (2.6) as

$$\frac{dE_1(\varphi_t, t/\varphi_s, s)}{dt} = -\eta(\varphi_t) - E_1 \quad (5.10)$$

$$\frac{dE_2\{(\varphi_t - E_1)^2/\varphi_s, s\}}{dt} = -E_2 + \frac{1}{2} \quad (5.11)$$

for $t > s$ and $E_2(t,t) = 0$. Hence

$$E_1(t) = \varphi_s.\exp\{-(t-s)\} - \eta(\varphi_t).\int_s^t \exp(u-t).du \quad (5.12)$$

$$E_2(t) = \frac{1}{2}(1 - \exp.(-2(t-s))) \quad (5.13)$$

Substituting equation (5.7) in Green's formula equation (4.8) over the domain $t_0 \leq s \leq t - a$, $0 \leq \varphi_s \leq b$, and setting $q = R(\varphi_t, \varphi_s)$ gives on going to the limit $a \to 0$, $b \to \infty$, an integral representation of the desired probability density:

$$p(\varphi_t, t) = \mathrm{sgn}.(\varphi_t).\Big\{\int_0^\infty p(\varphi_0, t_0).[R(\varphi_t ; \varphi_s)]_{s=t_0}.d\varphi_s$$

$$+ \frac{1}{2}\int_{t_0}^t \Big\{(-2\psi.p(\varphi_s, s). - [\frac{\partial p}{\partial \varphi_s}(\varphi_s, s)]).R(\varphi_t ; \varphi_s)$$

$$+ p(\varphi_s, s).[\frac{\partial R}{\partial \varphi_s}]\Big\}_{\varphi_s = 0}.ds\Big\} \quad (5.14)$$

From equations (5.3), (5.4), (5.8), (5.9) and the above the boundary conditions are given by the following linear Volterra integral equations of the second kind:

$$p(0,t) = \int_{-\infty}^{0} p(\varphi_0,t_0).[R(\varphi_2,t;\varphi_s)]_{\substack{\varphi_2=0\\ t=t_0}} . d\varphi_s$$

$$+ \frac{1}{2}. \int_{t_0}^{t} \{(2\gamma . p(\varphi_s,s) + \frac{\partial p(\varphi_s)}{\partial \varphi_s}). R(\varphi_2,t;\varphi_s)$$

$$+ \; p(\varphi_s,s) \; \frac{\partial R(\varphi_2,t;\varphi_s)}{\partial \varphi_s} \}_{\substack{\varphi_s=0\\ \varphi_2=0}} . ds \qquad (5.15)$$

and

$$[\frac{\partial p(\varphi_t)}{\partial \varphi_t}]_{\varphi_t=0} = \int_{0}^{\infty} p(\varphi_0,t_0).[\frac{\partial R(\varphi_1,t;\varphi_s)}{\partial \varphi_1}]_{\substack{\varphi_1=0\\ t=t_0}} . d\varphi_s$$

$$- \frac{1}{2} \int_{t_0}^{t} \{(2.\gamma . p(\varphi_s,s) + \frac{\partial p(\varphi_s)}{\partial \varphi_s}). \frac{\partial R(\varphi_1,t;\varphi_s)}{\partial \varphi_1}$$

$$+ \; p(\varphi_s,s). \frac{\partial^2 R(\varphi_1,t;\varphi_s)}{\partial \varphi_1 \partial \varphi_s} \}_{\substack{\varphi_1=0\\ \varphi_s=0}} . ds \qquad (5.16)$$

The solution of the system of integral equations (5.14-16) can be found by the method of successive approximations.

REFERENCES

Ariaratnam, S.T., and Graefe, P.W.U., Int. J. Control, 1, 239-250 (1965).

Dasarathy, B.V., and Srinivasan, P., J. Sound. Vib. 7,(1), 27-30 (1968).

Doob, J.L., Stochastic Processes, J. Wiley (N.Y). 1953.

Fuller, A.T., Int. J. Control, 9,(6), 603-655 (1969).

Pervozvanskii, A.A., Random Processes in Nonlinear control systems; Academic Press (N.Y) 1965.

Sancho, N.G.F., Int. J. Control, 9,(1), 83-88, (1969).

Skorokhod, A.V., Studies in the theory of random processes, Addison-Wesley (N.Y) 1965.

STABILITY OF LINEAR CYLINDRICAL SHELLS SUBJECTED TO STOCHASTIC EXCITATIONS*

John A. Lepore
University of Pennsylvania
Philadelphia, Pennsylvania

Robert A. Stoltz
Naval Underwater Systems Center
New London, Connecticut

INTRODUCTION

The dynamic stability of cylindrical shells under a deterministic axial loading and external pressure has been investigated by V.V. Bolotin (3) in his comprehensive treatment of the techniques used in the stability analysis of structures. Agamirov and Volmir (1) examined the dynamic buckling of cylindrical shells under single dynamic loadings of axial compression and distributed radial force. Wood and Koval (8) employed linear shell theory to verify experimental data on the buckling of cylinders. The cylinders were loaded by a constant concentrated axial force and a deterministic time-varying radial pressure and the results of these analyses are valid only when the parameters characterizing the system are deterministic. There exists little literature which considers the characteristic system parameters as stochastic. Lepore and Stoltz (6), (7) have investigated the stability of particular elastic systems under stochastic loadings and are applicable only for special excitation-response process behavior.

The intent of this paper is the investigation of the dynamic stability of a thin, linear, circular cylindrical shell subjected to simultaneous, stochastically interdependent axial and radial excitations. The stability bounds developed in this work provide a more general result than that presently found in the literature.

STOCHASTIC STABILITY

In systems which are governed by stochastic differential equations with deterministic coefficients, the concept of stability is still essentially a matter of convergence of solutions of the system. However, it is necessary to deal with convergence in a stochastic sense since one is dealing with limits involving random variables. Of primary concern in this study is the stability which corresponds to mean-square convergence. The null solution $\underline{x}(t) = 0$ of the system $\dot{\underline{x}} = f\{\underline{x}, h(t)\}$ is the equilibrium solution to be investigated. We then can state the following definition:

Definition: An equilibrium state $\underline{x}(t) = 0$ is globally stable in mean-square if, for every $\epsilon > 0$, there exists a real number $\delta(\epsilon) > 0$ such that for any finite initial

---

*This work was supported by the National Science Foundation under Grant No. NSF-GK-4634.

state, $||\underline{x}^2(t_0)|| < \delta(\varepsilon)$ implies the relationship $E\{||\underline{x}^2(t)||\} < \varepsilon$, and further that $\lim_{t\to\infty} E\{||\underline{x}^2(t)||\} \to 0$.

With this definition of stability, the stability of a stochastic system can be determined if a function can be found which satisfies the requirements of the Lyapunov theorem, proofs of which are given, essentially, by Bertram and Sarachik (2), and will not be repeated here:

Theorem: If there exists a Lyapunov function $V(\underline{x},t)$ defined over the entire state space, with the properties: (1) $V(\underline{0},t) = 0$; (2) $V(\underline{x},t)$ is continuous in mean-square in both variables $\underline{x}$ and t, and the first partial derivatives of $V(\underline{x},t)$ in these variables exist; (3) $V(\underline{x},t)$ is a positive-definite function; i.e., $V(\underline{x},t) \geq a\,x^2(t)$ for some $a > 0$; (4) $\lim V(\underline{x},t) \to \infty$ for $||\underline{x}(t)|| \to \infty$, then the equilibrium $\underline{x}(t) = 0$ is globally stable in the mean-square sense if $V(\underline{x},t)$ is a decrescent function whose expected value of its derivative dV/dt is negative-definite over the entire nonzero state space of $\underline{x}(t)$; i.e., $E\{\dot{V}(\underline{x},t)\} < 0$, $\underline{x}(t) \neq 0$, for all $||\underline{x}(t_0)||$.

STABILITY ANALYSIS

We consider a simply-supported, thin-walled, circular cylindrical shell of length L, radius R, and thickness h, subjected to a concentrated axial stochastic load p(t), and a uniformly-distributed stochastic radial loading, q(t).

In addition to those assumptions customarily made in the study of the dynamics of linear, continuous structures, we further assume that the effect of longitudinal and tangential inertial forces are negligible; radial inertia is thus considered to be the principle inertial response. Following the Donnell theory (5), the radial deformation, $w(x,\theta,t)$, of the linear shell is given by the solution to

$$k\nabla^8 w + (1-\nu^2)\frac{\partial^4 w}{\partial x^4} - \nabla^4\{N_x\frac{\partial^2 w}{\partial x^2} + N_\theta\frac{\partial^2 w}{\partial\theta^2} - \xi_1\frac{\partial^2 w}{\partial t^2} - \xi_2\frac{\partial w}{\partial t}\} = 0 \qquad (1)$$

where

ν = Poisson's ratio

x,θ = nondimensionalized coordinates

$k = h^2/12R^2$; $x = y/R$

$\xi_1 = \tilde{\rho}\,(1-\nu^2)R^2/Eh$

$\xi_2 = c(1-\nu^2)R^2/E$

$N_x = N_x^0(1-\nu^2)/Eh$

$N_\theta = N_\theta^0\,(1-\nu^2)/Eh$

and

E = Young's modulus

$\tilde{\rho}$ = shell density per unit area

c = damping

N_x^0, N_θ = initial stress resultants

and

$$\nabla^4 = \{\frac{\partial^2}{\partial x^2} + \frac{\partial^2}{\partial\theta^2}\}^2$$

$$\nabla^8 = \nabla^4\nabla^4$$

For the shell subjected to a concentrated axial load p(t) and a uniformly-distributed radial loading q(t), the initial membrane loads can be determined by assuming that the cylinder remains circular and undergoes a uniform compression circumferentially. Subsequently,

$$N_x^0 = -\{p(t)/2\pi R\} - 1/2\, Rq(t) \tag{2}$$

$$N_\theta^0 = -Rq(t) \tag{3}$$

The simply-supported shell is subject to the following boundary and initial conditions:

$$w(0,\theta,t) = w(L/R,\theta,t) = 0, \quad \frac{\partial^2 w}{\partial x^2}(0,\theta,t) = \frac{\partial^2 w}{\partial x^2}(L/R,\theta,t) = 0 \tag{4}$$

$$w(x,\theta,0) = w_0(x,\theta), \quad \frac{\partial w}{\partial t}(x,\theta,0) = v_0(x,\theta) \tag{5}$$

We seek a solution of Equation (1) in the form

$$w(x,\theta,t) = \sum_{m=1}^{\infty} \sum_{n=1}^{\infty} g_{mn}(t)\, G_{mn}(x,\theta) \tag{6}$$

where $g_{mn}(t)$ are unknown time-dependent functions and $G_{mn}(x,\theta)$ are the spatial-dependent normal modes, forming a complete set and satisfying the boundary conditions given by Equation (4). For the circular cylindrical shell, $G_{mn}(x,\theta) = \sin \alpha_m x$ $\sin n\theta$ where $\alpha_m = m\pi R/L$. Substituting Equations (2), (3), and (6) into (1) and simplifying yields

$$\frac{d^2 g_{mn}}{dt^2} + \frac{c}{\tilde{\rho} h} \frac{dg_{mn}}{dt} + \frac{\alpha_m^2}{\tilde{\rho} hR}\left[\frac{p_{mn}^* - p(t)}{2\pi R^2} - \left(\frac{1}{2} + \frac{n^2}{\alpha_m^2}\right) q(t)\right] g_{mn} = 0 \tag{7}$$

where

$$p_{mn}^* = \frac{2\pi RhE}{\alpha_m^2}\left[\frac{k(\alpha_m^2 + n^2)^2}{(1-\nu^2)} + \frac{\alpha_m^4}{(\alpha_m^2 + n^2)^2}\right]$$

is the critical axial buckling load for the m,n^{th} mode in the absence of the radial loading. Introducing a new independent variable, $\tau = \omega_{mn} t$, Equation (7) becomes

$$g_{mn}'' + 2\Omega_{mn} g'_{mn} + [1 - 2(\beta p)_{mn} p(t) - 2(\beta q)_{mn} q(t)]\, g_{mn} = 0 \tag{8}$$

where

$$\Omega_{mn} = c/w\tilde{\rho} h \omega_{mn}; \qquad (\beta p)_{mn} = 1/2\, p_{mn}^*;$$

$$\omega_{mn}^2 = [\alpha_m^2/2\pi\tilde{\rho} hR^3]\, p_{mn}^*; \qquad (\beta q)_{mn} = [\pi R^2(\alpha_m^2 + n^2)]/2p_{mn}^* = 1/2\, q_{mn}^* \tag{9}$$

The question of the stability of the circular cylindrical shell has been reduced to that of investigating the behavior of the stochastic Hill-type differential equation (8). For convenience, we drop the subscripts m,n from Equation (8), but it is understood that this equation governs each mode.

Reducing this second-order equation to a pair of simultaneous first-order equations, we obtain

$$\underline{Z}' = [\underline{A} + P(t)\ \underline{B} + Q(t)\underline{B}]\ \underline{Z} \tag{10}$$

where

$$A = \begin{bmatrix} 0 & 1 \\ -1 & -2\Omega \end{bmatrix}; \qquad \underline{B} = \begin{bmatrix} 0 & 0 \\ 1 & 0 \end{bmatrix} \tag{11}$$

$$P(t) = 2\beta p\ p(t) = p(t)/p^* \tag{12}$$

$$Q(t) = 2\beta q\ q(t) = q(t)/q^* \tag{13}$$

$$\underline{Z} = \begin{pmatrix} Z_1 \\ Z_2 \end{pmatrix} = \begin{pmatrix} g \\ g' \end{pmatrix} \tag{14}$$

We assume that the stochastic processes $\{p(t), q(t) : t \in (0,\infty)\}$, and consequently $\{P(t), Q(t) : t \in (0,\infty)\}$, are (1) continuous in the interval $0 < t < \infty$ with probability one; (2) are strictly stationary; and (3) possess non-negative mean values.

We select as a possible Lyapunov function the following quadratic form defined over the entire state space:

$$V(\underline{Z},t) = \underline{Z}^t \underline{\underline{S}}\,\underline{Z} \quad . \tag{15}$$

where $\underline{\underline{S}}$ is a real, symmetric, positive-definite matrix, and $\underline{Z}$ and $\underline{Z}^t$ are the state vector and its transpose, respectively. In this form the Lyapunov function approaches zero as the state vector approaches zero, for all time; i.e., it is decrescent. Hence, all the conditions of our theorem are satisfied except that of the behavior of the mean value of the time derivative of $V(\underline{Z},t)$ which is now considered.

Taking the time derivative of Equation (15) and integrating along the trajectory of the system, we have

$$\dot{V}(\underline{Z},t) = \omega\{\underline{Z}^t\,[\underline{\underline{A}}^t\,\underline{\underline{S}} + \underline{\underline{S}}\,\underline{\underline{A}}]\underline{Z} + \underline{Z}^t\,[\underline{\underline{B}}^t\,\underline{\underline{S}} + \underline{\underline{S}}\,\underline{\underline{B}}]\ \underline{Z}P(t)$$

$$+ \underline{Z}^t\,[\underline{\underline{B}}^t\underline{\underline{S}} + \underline{\underline{S}}\,\underline{\underline{B}}]\ \underline{Z}\ Q(t)\} \tag{16}$$

We assume that Equation (15) represents the Lyapunov function of the constant coefficient portion of the system described by Equation (10), i.e.,

$$\underline{Z}' = \underline{\underline{A}}\,\underline{Z} \tag{17}$$

Since all eigenvalues $\lambda_A = -\Omega + (\Omega^2 - 1)^{1/2}$ of the matrix $\underline{\underline{A}}$ have negative real parts, this system is asymptotically stable, and, consequently

$$\dot{V}_D(\underline{Z},t) = \omega \underline{Z}^t [\underline{\underline{A}}^t \underline{\underline{S}} + \underline{\underline{S}}\,\underline{\underline{A}}]\, \underline{Z} < 0 \tag{18}$$

where the subscript D designates the deterministic partial system. The negative-definiteness of $\dot{V}$ will be assured if the matrix sum in the brackets is negative-definite. Hence,

$$\underline{\underline{A}}^t \underline{\underline{S}} + \underline{\underline{S}}\,\underline{\underline{A}} = -\underline{\underline{C}} \tag{19}$$

where $\underline{\underline{C}}$ is a positive-definite matrix, and for this case we assume it to be the unit matrix $\underline{\underline{I}}$. The unknown elements of the $\underline{\underline{S}}$ matrix can now be determined from Equation (19) with $\underline{\underline{C}} = \underline{\underline{I}}$:

$$\underline{\underline{S}} = \begin{bmatrix} \Omega + \frac{1}{2\Omega} & \frac{1}{2} \\ \frac{1}{2} & \frac{1}{2\Omega} \end{bmatrix} \tag{20}$$

Having the V-function completely specified, Equation (16) is written as

$$\dot{V} = \omega\{-\underline{Z}^t \underline{Z} + \underline{Z}^t \underline{\underline{H}}\,\underline{Z}\, P(t) + \underline{Z}^t \underline{\underline{H}}\,\underline{Z}\, Q(t)\} \tag{21}$$

where

$$\underline{\underline{H}} = \underline{\underline{B}}^t \underline{\underline{S}} + \underline{\underline{S}}\,\underline{\underline{B}} = \begin{bmatrix} 1 & \frac{1}{2\Omega} \\ \frac{1}{2\Omega} & 0 \end{bmatrix} \tag{22}$$

Performing the indicated matrix operations, Equation (21) becomes

$$\dot{V} = \omega\{-Z_1^2 + P(t)\, Z_1^2 + Q(t)\, Z_1^2\} + \omega\{-Z_2^2 + \frac{P(t)}{\Omega} Z_1 Z_2 + \frac{Q(t)}{\Omega} Z_1 Z_2\} \tag{23}$$

Since a sufficient condition for stability is being sought, Equation (23) may be simplified, eliminating its explicit dependence on the Z_2 variable, by use of the relation

$$[\frac{P(t)}{2\Omega} Z_1 + \frac{Q(t)}{2\Omega} Z_1 - Z_2]^2 \geq 0 \tag{24}$$

Using this relation in Equation (23) we have

$$\dot{V} \leq \omega\{-Z_1^2 + (P + Q)\, Z_1^2 + \frac{(P + Q)^2}{4\Omega^2} Z_1^2\} \tag{25}$$

The theorem for global mean-square stability will be completely satisfied if the mean value of $\dot{V}$ is negative-definite for nonzero values of the state vector. This requirement is guaranteed if

$$E[(P + Q)^2 Z_1^2] + 4\Omega^2 E[(P + Q) Z_1^2] - 4\Omega^2 E[Z_1^2] < 0 \tag{26}$$

A more convenient form of Equation (26) is accomplished by considering the statistical dependence of the squared amplitude response process, Z_1^2, on the stochastic excitations P(t) and Q(t). In this respect, we introduce the following (reference (4)) multiple linear mean-square regression equation:

$$Z_1^2 = E[Z_1^2] + \rho_1 \frac{\sigma_{Z_1^2}}{\sigma_P}\{P - E[P]\} + \rho_2 \frac{\sigma_{Z_1^2}}{\sigma_Q}\{Q - E[Q]\} \tag{27}$$

where

$$\rho_1 = \frac{\rho_{PZ_1^2} - \rho_{QZ_1^2}\rho_{PQ}}{1 - \rho_{PQ}^2} \qquad \rho_2 = \frac{\rho_{QZ_1^2} - \rho_{PZ_1^2}\rho_{PQ}}{1 - \rho_{PQ}^2} \tag{28}$$

and

$\rho_{PZ_1^2}$ = total correlation coefficient between Z_1^2 and P

$\rho_{QZ_1^2}$ = total correlation coefficient between Z_1^2 and Q

ρ_{PQ} = correlation coefficient between P and Q

We need not explicitly consider the case in which $\rho_{PQ} = 1$, since, in that instance, the P(t) and Q(t) processes are linearly stochastically dependent; and the resulting problem is simplified into the single excitation case. We are concerned with those cases for which

$$0 \le \rho_{PQ} < 1 \tag{29}$$

Substituting Equations (27) into (31), and applying the expectation operator and letting $\upsilon = \sigma_{Z_1^2}/E[Z_1^2]$, we obtain after simplification

$$\begin{aligned} &E^2[P] + E^2[Q] + 2E[P]E[Q] + 4\Omega^2 E[P] + 4\Omega^2 E[Q] + \sigma_P^2 + \sigma_Q^2 + 2\rho_{PQ}\sigma_P\sigma_Q \\ &+ 2\rho_1 \upsilon\{\sigma_P[E[P] + E[Q] + 1/2\, \sigma_P\gamma_P + 2\Omega^2] + \rho_{PQ}\sigma_Q[E[P] + E[Q] + 1/2\, \sigma_Q\gamma_P \\ &+ \sigma_P\gamma_P + 2\Omega^2]\} + 2\rho_2\, \upsilon\{\sigma_Q\, [E[P] + E[Q] + 1/2\, \sigma_Q\gamma_Q + 2\Omega^2] + \rho_{PQ}\sigma_P[E[P] \\ &+ E[Q] + 1/2\, \sigma_P\gamma_P + \sigma_Q\gamma_Q + 2\Omega^2]\} - 4\Omega^2 < 0 \end{aligned} \tag{30}$$

where γ_P and γ_Q are the coefficients of skewness of the P(t) process and the Q(t) process, respectively.

Introducing the partial (or net) correlation coefficients, reference (4), between Z_1^2 and P, and Z_1^2 and Q, we have

$$_{Q}\rho_{PZ_1^2} = \frac{\rho_{PZ_1^2} - \rho_{QZ_1^2}\,\rho_{PQ}}{+\sqrt{(1-\rho_{PQ}^2)(1-\rho_{QZ_1^2}^2)}} \qquad _{P}\rho_{QZ_1^2} = \frac{\rho_{QZ_1^2} - \rho_{PZ_1^2}\,\rho_{PQ}}{+\sqrt{(1-\rho_{PQ}^2)(1-\rho_{PZ_1^2}^2)}} \tag{31}$$

It is quite obvious that the two partial correlation coefficients between excitation and response are positive. Then, comparing Equations (28) and (31), it is clear that ρ_1 and ρ_2 are also positive.

Before Equation (30) can be explicitly applied, the coefficient of variation of the unknown squared amplitude response process, υ, must be investigated. The response amplitude process, Z_1, will, in general, be a nonstationary, nonsymmetric process, attaining both positive and negative values. We consider the case where the probability density function of Z_1 is both continuous and finite at all points in its domain, and that Z_1 is finite, i.e.,

$$-\infty < -\lambda_1 \le Z_1 \le \lambda_2 < +\infty \tag{32}$$

Consider the associated absolute value process, $|Z_1|$. In view of Equation (32), the $|Z_1|$ process is bounded, such that

$$0 \le |Z_1| \le \lambda < \infty \tag{33}$$

Further, the probability density function of $|Z_1|$ is continuous and finite at all points in the domain of $|Z_1|$.

We can now introduce the normalized beta density function, given by

$$f_\beta(\zeta|x_1,x_2) = \frac{1}{B(x_1,x_2-x_1)}\,\zeta^{x_1-1}(1-\zeta)^{x_2-x_1-1} \tag{34}$$

where $0 \le \zeta \le 1$ and $x_2 > x_1 > 0$ and $B(x_1,x_2-x_1)$ is the complete beta function.

A particular subset of the complete ensemble of possible beta distributions will be utilized to represent the potential probability distributions of the $|Z_1|$ process. A subset, rather than the entire family, is employed because of the requirement that the density function of $|Z_1|$ be finite at all points in its domain. From Equation (34), it can be seen that if $x_1 < 1 : f_\beta(\zeta|x_1,x_2) \to \infty$ as $\zeta \to 0$ (35)

and

$$x_2 < x_1 + 1 : f_\beta(\zeta|x_1,x_2) \to \infty \text{ as } \zeta \to 1 \tag{36}$$

Hence, the subset utilized is that group of beta distributions for which

$$x_2 - 1 \geq x_1 \geq 1 \qquad (37)$$

Since it is not known which explicit beta distribution should be used to represent the $|Z_1|$ process, and, indeed, since the probability distribution of the $|Z_1|$ process can be expected to vary as the excitation processes are changed, the final stability condition will be a sufficient condition for all cases in which the $|Z_1|$ process can be represented by some member of the previously discussed subset of beta distributions. Since ρ_1 and ρ_2 are positive, and since the excitation mean values are positive-semidefinite, the upper-bound value of the parameter υ, i.e., $\sqrt{5}$, is used in Equation (30) yielding

$$E^2[P] + E^2[Q] + 2E[P]E[Q] + 4\Omega^2 E[P] + 4\Omega^2 E[Q] + \sigma_P{}^2 + \sigma_Q{}^2 + 2\rho_{PQ}\sigma_P\sigma_Q$$
$$+ 2\sqrt{5}\,\rho_1\{\sigma_P[E[P] + E[Q] + 1/2\,\sigma_P\gamma_P + 2\Omega^2] + \rho_{PQ}\sigma_Q[E[P] = E[Q]$$
$$+ 1/2\,\sigma_Q\gamma_Q + \sigma_P\gamma_P + 2\Omega^2]\} + 2\sqrt{5}\,\rho_2\{\sigma_Q[E[P] + E[Q] + 1/2\,\sigma_Q\gamma_Q + 2\Omega^2]$$
$$+ \rho_{PQ}\sigma_P[E[P] + E[Q] + 1/2\,\sigma_P\gamma_P + \sigma_Q\gamma_Q + 2\Omega^2]\} - 4\Omega^2 < 0 \qquad (38)$$

Equation (38) provides a sufficient condition guaranteeing the global mean-square stability of the circular cylindrical shell excited by combined stochastic excitations. An additional simplification of Equation (38) is available when either or both excitation processes possess symmetric or negatively-skewed probability distributions. In these cases, γ_P and/or γ_Q are taken as zero.

DISCUSSION OF RESULTS

The stability condition for for symmetric, stochastically independent shell excitations is obtained from Equation (38) by allowing the coefficients of skewness and th the excitation cross-correlation, ρ_{PQ}, to go to zero. Figure 1 depicts the regions of global mean-square stability under independent excitations, where both system correlation coefficients have identical values, and one process has a zero mean value. The regions of mean-square stability are those volumes enclosed by the three coordinate axes and a particular set of curves. Because both correlation coefficients are identical, the stability region is symmetric with respect to σ_P and σ_Q. Furthermore, the boundary curves in the $E[P] - \sigma_P$ and $E[P] - \sigma_Q$ planes coincide with those obtained in the single excitation case, reference (5), with the same correlation coefficient. Figure 2 shows the case of unequal correlations. In this instance, the stability

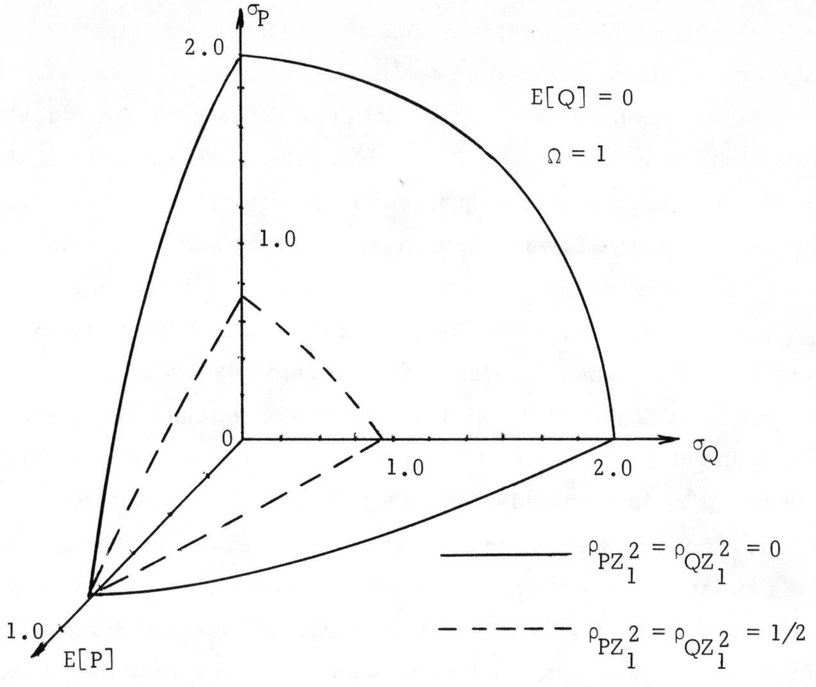

FIGURE 1 - Stability Regions Under Independent Combined Excitations;Identical Correlations

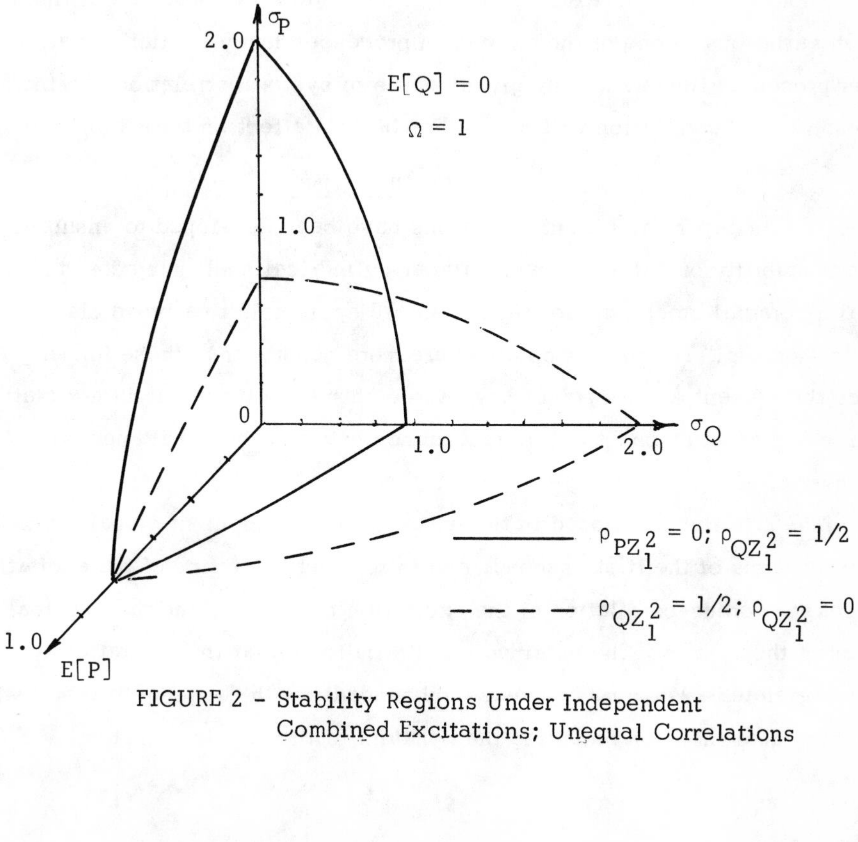

FIGURE 2 - Stability Regions Under Independent Combined Excitations; Unequal Correlations

region is not symmetric with respect to σ_P and σ_Q. Figure 2 also displays the shift in the stability region when the correlation coefficients are directly reversed.

Allowable excitation variances as a function of damping are shown in Figure 3, for excitations with zero mean value. For an uncorrelated system, the boundary curves and the solid region of mean-square stability grow unboundedly with damping. Figure 4 is similar to Figure 3, with the exception that the correlations are unequal.

We now consider the case of stochastically dependent combined excitations; that is, $\rho_{PQ} \neq 0$. Figure 5 shows regions of mean-square stability in terms of the excitation standard deviations, for processes with zero mean values, and for an uncorrelated system. It is seen that with independent excitations, the stability region boundary curve in the σ_P - σ_Q plane is simply the quadrant of a circle with radius 2Ω. This relationship is no longer valid when the excitations themselves are cross-correlated ($\rho_{PQ} > 0$). Figure 5 indicates that the effect of increasing excitation cross-correlation is to decrease the region of mean-square stability, even though the system remains uncorrelated with respect to the excitations. To understand this effect, it is necessary to remember that system uncorrelatedness does not imply that system response is independent of excitation. On the contrary, system correlation is merely an indication of the degree of linear associativity between response and excitation dispersions. Moreover, excitation cross-correlation is a measure of the degree to which variations in one of the excitation processes imply similar variations in the other process. Hence, for any given degree of system correlation, an increase in excitation cross-correlation will have a detrimental effect on the stability region.

CONCLUSIONS

In this paper, sufficient conditions have been developed to ensure the mean-square stability of a thin, linear, circular cylindrical shell subjected to simultaneous axial and radial stochastic loadings, but are applicable to a broad class of linear dynamic systems. The results obtained are more general than those formerly available since the present results permit any degree of system input-output correlation as well as any degree of stochastic interdependency between the simultaneous excitation processes.

The criterion developed in this paper defines regions of global mean-square stability in terms of the first, second, and third order statistics of the excitation processes; the cross-correlation of the excitation processes; and the physical characteristics of the system. The obtained results indicate that the excitation variances of an uncorrelated system may increase unboundedly with damping--a result which satisfies technical intuition regarding the behavior of the physical system.

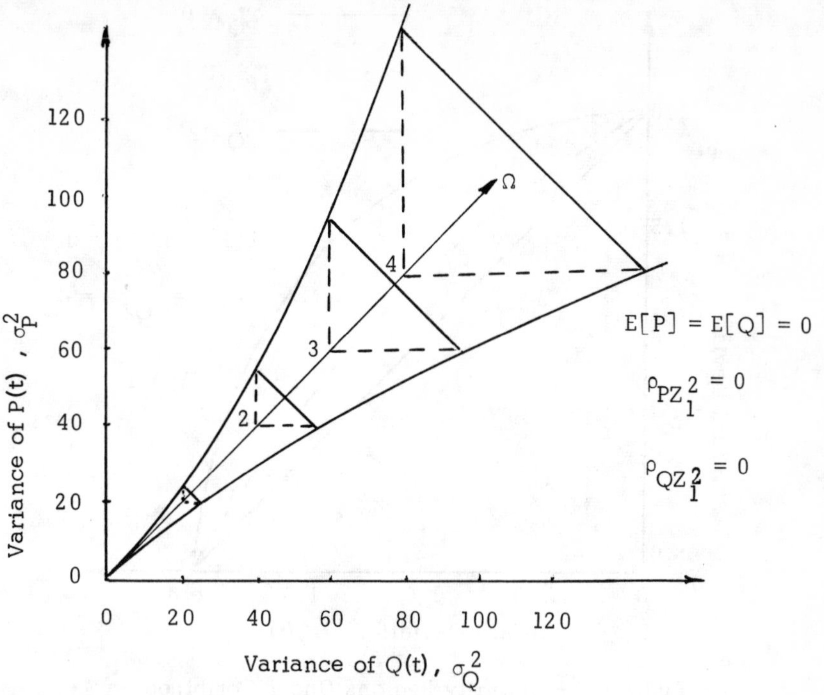

FIGURE 3 - Excitation Variances vs. Damping; Identical Correlations

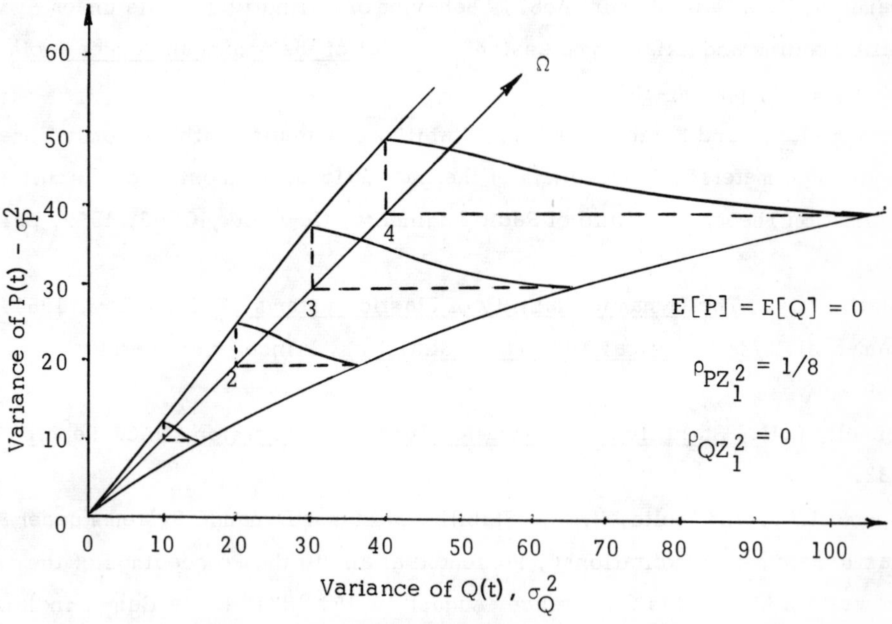

FIGURE 4 - Excitation Variances vs. Damping; Unequal Correlations

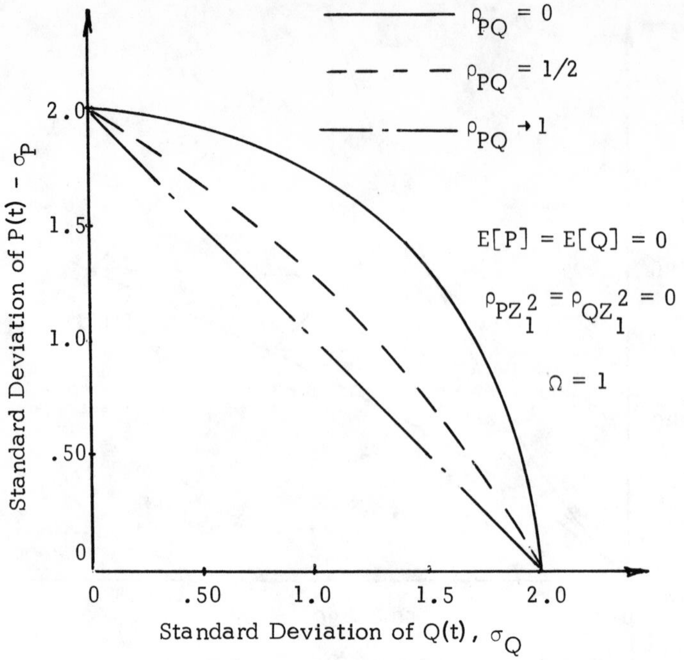

FIGURE 5 - Stability Regions Under Combined Excitations with Zero Mean

REFERENCES

1. Agamirov, V.L., and Volmir, A.S., "Behavior of Cylindrical Shells under Hydrostatic Loading and Axial Compression", Journal of the American Rocket Society, January 1961, pp. 98-101.

2. Bertram, J.E., and Sarachik, P.E., "Stability of Circuits with Randomly Time Varying Parameters", Proceedings of the International Symposium on Circuit and Information Theory, Institute of Radio Engineers, New York, CT-6, 1959, pp. 260-270.

3. Bolotin, V.V., The Dynamic Stability of Elastic Systems, Holden-Day, 1964.

4. Cramer, H., Mathematical Methods of Statistics, Princeton University Press, 1958.

5. Donnell, L.H., Stability of Thin-Walled Tubes under Torsion, NACA Report 479, 1933.

6. Lepore, J.A., and Stoltz, R.A., "Stability of Linear Dynamic Systems under Stochastic Parametric Excitations", Presented at and in the Proceedings of the 12th Midwestern Mechanics Conference, August 16-18, 1971, Notre Dame, Indiana.

7. Lepore, J.A., and Stoltz, R.A., "Dynamic Stability of Cylindrical Shells under Combined Axial and Radial Stochastic Excitations", Proceedings of the AIAA/ASME

12th Structures, Structural Dynamics and Materials Conference, April 19-21, 1971, Anaheim, California.

8. Wood, J.D., and Koval, L.R., "Buckling of Cylindrical Shells under Dynamic Loads", AIAA Journal, November 1963, pp. 2576-2582.

AVERAGE VALUE CRITERIA FOR STOCHASTIC STABILITY

Roger W. Brockett*
Division of Engineering
and Applied Physics
Harvard University
Cambridge, Mass. 02138, U.S.A.

Jan C. Willems**
Department of Electrical Engineering
Massachusetts Institute of Technology
Cambridge, Mass. 02139, U.S.A.

INTRODUCTION

Many problems in control and other areas of applied mathematics lead to stability questions for dynamical systems which are described by mathematical models involving time-varying parameters. Frequently one may assume that these time-varying parameters are stochastic processes with known statistics. Typical examples of interesting applications which lead to such stochastic stability questions are the stability analysis of numerical computations in the face of round-off error, systems involving the human operator, sampled data systems with jitter in the sampling rate, mechanical systems subject to random vibrations, and economic systems which model some of the uncertainties as variable lags.

Essentially all of the above examples lead to mathematical models in which the stochastic processes enter the model in a multiplicative way. It is for this class of systems that the stochastic stability question becomes interesting and challenging. In contrast, when the stochastic processes enter the model in an additive way as, for example, in the linear quadratic theory, then the stochastic stability question usually reduces to the stability of the deterministic system obtained by putting the stochastic processes equal to zero.

In this paper we will analyze a class of stochastic systems and obtain various explicit stability criteria. Before we describe the model let us introduce the following notation: R denotes the real number system, R^n denotes n-dimensional real Euclidean space, R^{mxp} denotes the real mxp matrices, prime denotes transpose,

---

*Supported in part by the U.S. Office of Naval Research under the Joint Services Electronics Program by Contract N00014-67-A-0298-0006 and by the National Aeronautics and Space Administration under Grant NGR 22-007-172.

**Supported in part by the National Aeronautics and Space Administration, Ames Research Center, under Grant NGL 22-009-124 and by the National Science Foundation under Grant No. GK-25781.

≥ 0 (> 0) means that a symmetric matrix is nonnegative (positive) definite, $\lambda[\cdot]$ denotes an arbitrary eigenvalue of a matrix, whereas $\lambda_{max}[\cdot]$ ($\lambda_{min}[\cdot]$) denotes the maximum (minimum) eigenvalue of a matrix with real eigenvalues, Re denotes the real part of a complex number, max $[\cdot,\cdot]$ (min $[\cdot,\cdot]$) denotes the maximum (minimum) of two real numbers, and $\mathscr{E}\{\cdot\}$ denotes the expected value of a random variable.

We will study the stability of the linear system Σ described by the differential equation:

$$\Sigma : \dot{x} = Ax - BK(t)Cx ,$$

where $x \in R^n$ and $A \in R^{nxn}$, $B \in R^{nxm}$, and $C \in R^{pxn}$ are constant matrices and $K(t)$ is a time-varying function taking values in R^{mxp}. The differential equation Σ will be viewed as describing the closed loop dynamics of the feedback interconnection of the stationary linear system

$$\Sigma_1 : \dot{x}_1 = Ax_1 + Bu_1 \quad ; \quad y_1 = Cx_1$$

in the forward loop, and the memoryless time-varying linear system

$$\Sigma_2 : y_2 = K(t)u_2$$

in the feedback loop. The feedback interconnection equations are given by: $u_1 = -y_2$, $u_2 = y_1$. It is easily verified that we indeed have $\Sigma = \Sigma_1 x \Sigma_2|$ feedback. This feedback system is shown in Figure 1.

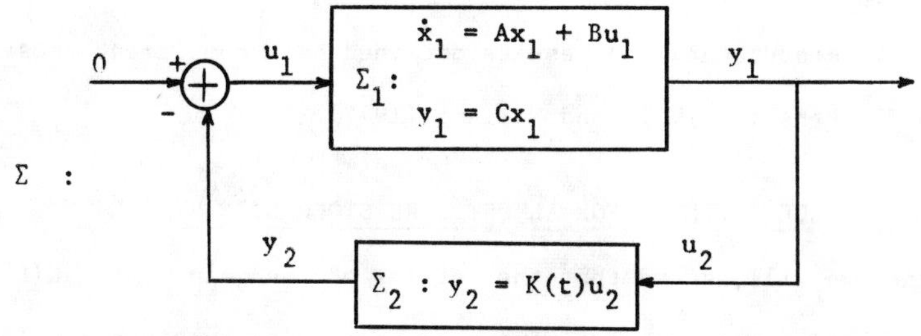

Figure 1: Σ *viewed as* $\Sigma_1 x \Sigma_2|$ feedback.

We will assume throughout, for simplicity, that $\Sigma_1 = \{A,B,C\}$ is minimal (i.e., (A,B) is controllable and (A,C) is observable). The transfer function of Σ_1 is given by $G(s) = C(Is-A)^{-1}B$. The gain matrix $K(t)$ is assumed to be a stochastic process whose properties will be described in more detail later. We seek conditions on the statistics of K(t) which guarantee the stability of Σ (to be defined later).

If we consider the equation for Σ from a state space point of view then it is apparent that the case where K(t) is a colored process is quite distinct from the case that K(t) is white. If K(t) is white noise then the system behaves pretty much like a linear one and we may use most of the theory on stochastic differential equations directly as for example the Lyapunov techniques for stochastic systems (see e.g., Kushner [1967], Chapter 2). If on the other hand K(t) is a colored process then we should model Σ as something like:

$$\dot{z} = Fz + Gw \ ; \quad K = Hz$$
$$\dot{x} = Ax - BKCx$$

with w white noise. This case is thus inherently nonlinear. The results obtained in this paper fall into two categories. In the first class we consider the colored case and show how one may use what are essentially linear techniques to obtain conditions for almost sure asymptotic stability of Σ. The method of proof uses Wazewski's inequality previously exploited in this context by Infante [1968]. These criteria are thus independent of the autocorrelation function of K(t).

The second class of results considers the white noise case and shows how one may use the frequency-domain stability criteria for linear systems in order to obtain criteria for mean square stability of Σ. This question has been studied extensively in the literature and the results obtained here complement those obtained by Willems and Blankenship [1971] and Willems [1972].

1. AVERAGE VALUE CRITERIA FOR ALMOST SURE STOCHASTIC STABILITY

In this section we will assume that the entries of the gain matrix $K(t) \in R^{mxp}$ are stationary stochastic processes satisfying an ergodicity hypothesis which ensures the almost sure equality of time averages and ensemble averages. Thus if $F : R^{mxp} \to R$ is integrable then we assume that almost surely:

$$\mathscr{E}\{F(K(t))\} = \mathscr{E}\{F(K(0))\} = \lim_{T\to\infty} \frac{1}{T}\int_{t_o}^{t_o+T} F(K(\tau))d\tau \ .$$

We will consider almost sure asymptotic stability. This is defined as:

<u>Definition 1</u>: Σ is said to be *almost surely asymptotically stable* if the equality

$\lim_{t\to\infty} x(t) = 0$ holds with probability one for all given initial conditions $x(t_0)$.

1.1 A Stability Criterion for Completely Symmetric Systems

Consider the system Σ_1. There are various ways of describing its response function from the inputs to the outputs. The most commonly used input/output descriptions of Σ give either its *transfer function* $G(s) \triangleq C(Is-A)^{-1}B$ or its *impulse response* $W(t) \triangleq Ce^{At}B$ $(t \geq 0)$. There is however an alternative input/output description which, although it has roots going back at least as far in time as do the concepts of transfer function and impulse response, has become particularly prevelant in the last half decade. This description gives the so-called *Hankel matrix* of Σ_1 defined by:

$$H \triangleq \left[\begin{array}{c|c|c|c|c} CB & CAB & \cdots & CA^NB & \cdots \\ \hline CAB & CA^2B & \cdots & CA^{N+1}B & \cdots \\ \hline \vdots & \vdots & \ddots & \vdots & \ddots \\ \hline CA^NB & CA^{N+1}B & \cdots & CA^{2N-1}B & \cdots \\ \hline \vdots & \vdots & \ddots & \vdots & \ddots \end{array}\right] = [W^{(i+j-2)}(0)] .$$

It turns out that many qualitative input/output properties of Σ_1 are most easily described in terms of H.

It is well-known that there exist many minimal realizations $\{A,B,C\}$ of a given $G(s)$, $W(t)$, or H, but that they all may be recovered from one of them by the transformation group $\{A,B,C\} \overset{S}{\to} \{SAS^{-1},SB,CS^{-1}\}$ with S an arbitrary invertible element of R^{nxn}. The dimension of a minimal realization of a given transfer function is called the *McMillan degree*.

We will consider the following class of systems Σ_1:

Definition 2: Σ_1 is said to be *completely symmetric* if m=p and* $H = H' \geq 0$.

---

*The infinite matrix H is said to be *nonnegative definite* (denoted by ≥ 0) if all its finite truncations are nonnegative definite, i.e. if $\sum_{i,j=0}^{N} z_i' CA^{i+j}Bz_j \geq 0$ for all N and for all sequences $\{z_i\}_0^N$.

The following lemma gives a very useful alternative characterization of completely symmetric systems. Its proof, which is not germane to our purposes, is an immediate consequence of some known facts in realization theory and is left to the reader.*

Lemma 1: *Σ is completely symmetric if and only if its transfer function* $G(s) = C(Is-A)^{-1}B$ *admits a realization* $\{A_1, B_1, C_1\}$ *with* $A_1 = A_1'$ *and* $B_1 = C_1'$.

Thus Σ_1 is completely symmetric if and only if there exists a nonsingular (nxn) matrix S such that $SAS^{-1} = (SAS^{-1})'$ and $SB = (CS^{-1})'$. Completely symmetric systems have the property that the eigenvalues of A are all real. This is in fact also the case after applying symmetric feedback and it may be shown that Σ_1 is completely symmetric if $G(s) = G'(s)$ and if A-BKC has real eigenvalues for all $K = K'$. Note also that Σ is completely symmetric if and only if its transfer function admits the partial fraction expansion $G(s) = \sum_{i=1}^{k} \frac{R_i}{s+\lambda_i}$ with $R_i = R_i' \geq 0$. If m=p=1 then Σ_1 is completely symmetric if and only if the poles and the zeros of the transfer function G(s) are real and interlace, i.e. if $\lambda_1, \lambda_2, \ldots, \lambda_n$ are the poles and if $z_1, z_2, \ldots, z_r$ are the zeros of G(s), then $r = n-1$, λ_i and z_i are real, and $\lambda_1 > z_1 > \lambda_2 > \ldots > z_{n-1} > \lambda_n$. This pole-zero pattern is illustrated in Figure 2.

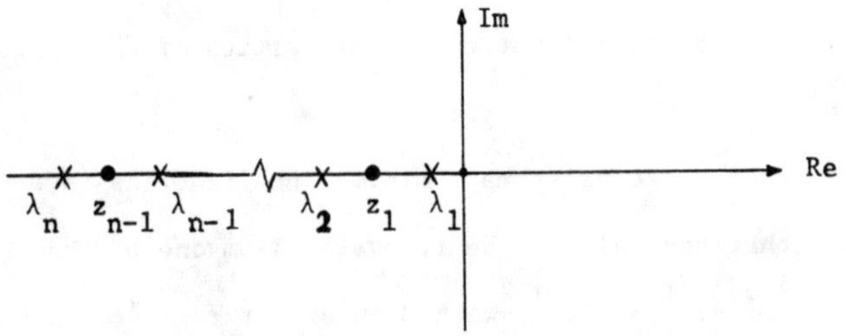

Figure 2: *Typical pole/zero pattern of a completely symmetric system.*

Completely symmetric systems are a natural generalization of *relaxation systems* (see Willems [1972]) which are completely symmetric systems which satisfy the additional stability requirement Re $\lambda[A] \leq 0$. Thus Σ_1 is a relaxation system if and only if its transfer function admits a realization $\{A_1, B_1, C_1\}$ with $A_1 = A_1' \leq 0$ and $B_1 = C_1$. There are various other ways of defining a relaxation system. It may be

---

*The background material of realization theory used here may be found in Brockett [1970], Chapter 2, or Kalman [1969], Section 10.11.

shown that Σ_1 defines a relaxation system if and only if $H = H' \geq 0$ *and* $\sigma H = \sigma H' \leq 0$, where σH denotes the *shifted* Hankel matrix of Σ_1, i.e., H with the first block row (or column) deleted. Alternatively, Σ_1 defines a relaxation system if and only if its impulse response $W(t) = Ce^{At}B$ is a *completely monotonic function* on $[0,\infty)$, i.e. $W(t) = W'(t)$ and $(-1)^k \frac{d^k}{dt^k} W(t) \geq 0$ for all $t \geq 0$ and $k = 0,1,2,\ldots$. Relaxation systems play an important role in physics. They describe the response of various classes of systems such as R-C and R-L electrical networks, viscoelastic materials thermal systems, and chemical reactions.

We now state the main result of this section.

Theorem 1: *Assume that Σ_1 is completely symmetric and that $K = K'$ almost surely. Let $\overline{\lambda_{max}} \triangleq \mathscr{E}\{\lambda_{max}[A-BKC]\}$. Then Σ is almost surely asymptotically stable if* $\overline{\lambda_{max}} < 0$.

Proof: The proof of Theorem 1 follows an argument due to Wazewski adapted to the case under consideration (as in Brockett [1970], Section 32, Exercise 6).

Since Σ_1 is completely symmetric, there exists a nonsingular matrix S such that $A_1 = SAS^{-1} = (SAS^{-1})' = A_1'$ and $B_1 = SB = (CS^{-1})' = C_1'$. Let $x_1 = Sx$. Then x_1 satisfies the equation:

$$\dot{x}_1 = (A_1 - B_1K(t)B_1')x .$$

Let $V(x_1) = x_1'x_1$. Then along solutions of the above equation we have:

$$\dot{V}(x_1) = 2x_1'(A_1 - B_1K(t)B_1')x_1 ,$$

which, since $A_1 - B_1K(t)B_1'$ is symmetric, shows that:

$$\dot{V}(x_1) \leq 2\lambda_{max}[A_1 - B_1K(t)B_1']V(x_1) .$$

Since $A_1 - B_1K(t)B_1'$ and $A-BK(t)C = S^{-1}(A_1 - B_1K(t)B_1')S$ are similar matrices, this yields:

$$\dot{V}(x) \leq 2\lambda_{max}[A-BK(t)C]V(x_1) .$$

Thus

$$V(x_1(t)) \leq V(x_1(t_o))\exp(2\int_{t_o}^{t} \lambda_{max}[A-BK(\tau)C]d\tau) .$$

Finally by the ergodic hypothesis

$$\lim_{T\to\infty} \frac{1}{T}\int_{t_o}^{t_o+T} \lambda_{max}[A-BK(\tau)C]d\tau = \overline{\lambda_{max}}$$

is almost surely negative, which shows that $\lim_{t\to\infty} V(x_1(t)) = 0$ almost surely. Thus $\lim_{t\to\infty} x_1(t) = S \lim_{t\to\infty} x(t) = 0$ almost surely, which proves the theorem. ■

Note: 1. Theorem 1 predicts stability if $\text{Re}\,\lambda(A) < 0$ and $K = K' \geq \varepsilon I > 0$ almost surely. It then reduces to a special case of the multivariable circle criterion.

The major difficulty in applying Theorem 1 is that as a rule $\overline{\lambda_{max}}$ will be difficult to compute from the distribution of K since $\lambda_{max}[A-BKC]$ is a very nonlinear function K which does not even admit a general analytic expression. This difficulty may however be overcome in the important special case that there is only one stochastic gain in Σ:

Theorem 2: *Assume that* $m=p=1$ *and that* Σ_1 *is completely symmetric with transfer function* $g(s) = C(Is-A)^{-1}B$. *Let* z_1 *be the largest zero of* $g(s)$ *and assume that* $K(t)$ *possesses the density function* $p(K)$. *Then* Σ *is almost surely asymptotically stable if:*

$$\int_{z_1}^{\infty} \sigma p\left(-\frac{1}{g(\sigma)}\right) \frac{\partial g(\sigma)/\partial\sigma}{g^2(\sigma)}\, d\sigma > 0 .$$

Proof: By Theorem 1 it suffices to prove that the integral in the theorem statement equals $-\overline{\lambda_{max}}$. Consider therefore $\lambda_{max}[A-BKC]$. Since the eigenvalues of A-BKC are the poles of the system obtained after putting the constant feedback gain K around Σ_1, it follows that these eigenvalues are the zeros of $1+Kg(s)$. Since the poles and zeros of $g(s)$ are real and interlacing it follows from a simple root-locus consideration that the maximum zero of $1+Kg(s)$ is a monotone decreasing function of K which varies from z_1 for $K = \infty$ to $+\infty$ for $K = -\infty$. The gain K and $\lambda_{max}[A-BKC]$ are in fact related by $g(\lambda_{max}[A-BKC]) = -\frac{1}{K}$. Thus, by a standard formula from probability theory we have that:

$$\mathscr{E}\{\lambda_{max}[A-BKC]\} = \int_{z_1}^{\infty} \sigma p\left(-\frac{1}{g(\sigma)}\right) \left|\frac{\partial g(\sigma)/\partial\sigma}{g^2(\sigma)}\right| d\sigma$$

which yields the desired result. ■

Notes: 2. Figure 3 shows the behavior of the functions $g(\sigma)$, $-1/g(\sigma)$, and $\lambda_{max}(A-BKC)$. The qualitative behavior of these functions is very well understood as a result of exhaustive analysis of R-C and R-L electrical networks (see, e.g., Guillemin [1957], Chapter 4).

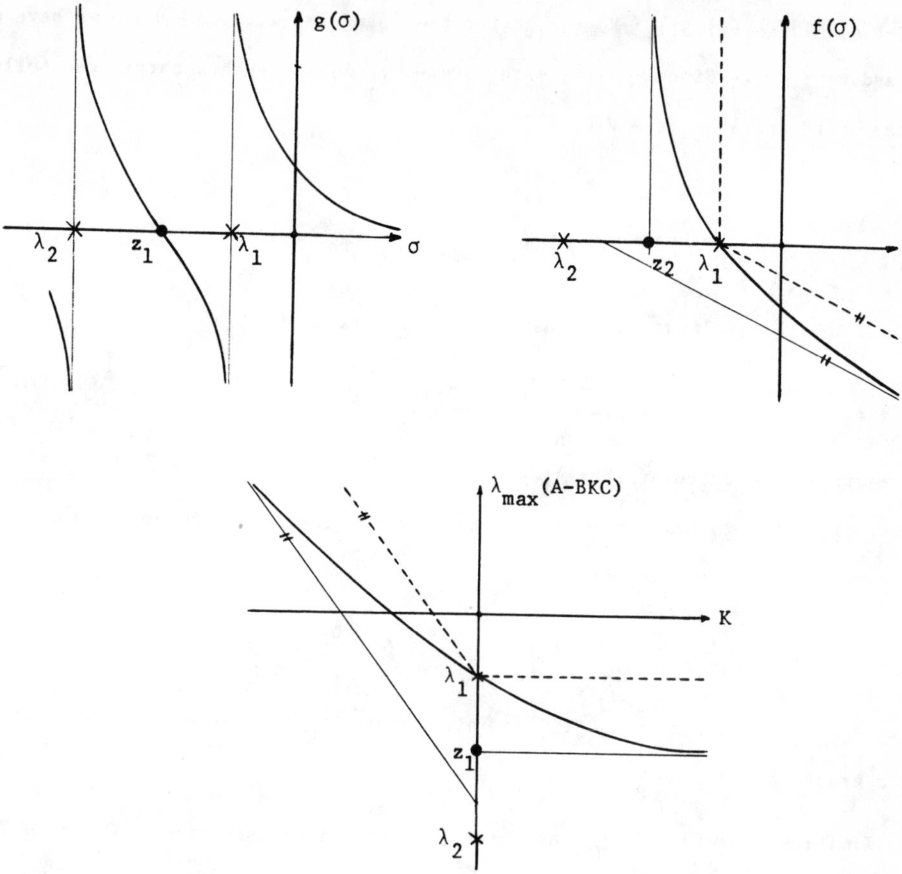

<u>Figure 3</u>: *Sketch of* $g(\sigma)$, $f(\sigma) = -\frac{1}{g(\sigma)}$, *and* $\lambda_{max}(A\text{-}BKC)$

3. Theorem 2 indicates the destabilizing effect of the stochastic gain. To see this, let us assume (essentially without loss of generality) that $\mathscr{E}(K) = 0$. It may be shown that $\lambda_{max}(A\text{-}BKC)$ is a strictly convex function of K which by Jensen's inequality (see Feller [1966], p. 151) implies that $\overline{\lambda_{max}} \geq \lambda_1$ with equality holding if and only if K = 0 almost surely. Note also that Theorem 2 is easily extended to the case where K does not possess a density function.

4. Let $g(s) = \dfrac{q_{n-1}s^{n-1}+\ldots+q_o}{s^n+p_{n-1}s^{n-1}+\ldots+p_o}$ and let $\lambda_1,\ldots,\lambda_n$ and $z_1,\ldots,z_{n-1}$ denote the poles and the zeros of g(s). Thus $\lambda_1 > z_1 > \lambda_2 > \ldots > z_{n-1} > \lambda_n$. Let $\lambda_{max}(K) = \lambda_1(K) > \ldots > \lambda_n(K)$ denote the zeros of $p(s)+Kq(s)$. From root-locus considerations it is easily seen

that $z_i < \lambda_i(K) < \lambda_i(0) < \lambda_i(-K) < z_{i-1}$ for $K > 0$ and $i=1,2,\ldots,n$ (where we have put $z_o \triangleq \infty$ and $z_n \triangleq -\infty$). Since $-\sum_{i=1}^{n} \lambda_i + Kg_{n-1} = -\sum_{i=1}^{n} \lambda_i(K)$ we thus obtain the following upper bound for λ_{max} (see Figure 3):

$$\lambda_{max} \leq \begin{cases} \lambda_1 & \text{for } K \geq 0 \\ \lambda_1 - Kg_{n-1} & \text{for } K \leq 0 \end{cases} .$$

This shows that Σ is almost surely asymptotically stable if:

$$\mathscr{E}\{\min[-\frac{\lambda_1}{q_{n-1}}, \frac{K-\lambda_1}{q_{n-1}}]\} > 0$$

which requires in particular that $\lambda_1 < 0$.

<u>Examples</u>: 1. If K is uniformly distributed between the limits K_- and K_+ then Σ is almost surely asymptotically stable if:

$$\frac{z_+}{g(z_+)} - \frac{z_-}{g(z_-)} + \int_{z_+}^{z_-} \frac{d\sigma}{g(\sigma)} > 0$$

where $z_+ \triangleq \lambda_{max}(K_+)$ and $z_- \triangleq \lambda_{max}(K_-)$. This inequality is easily verified directly from the graph of $f(\sigma) = -\frac{1}{g(\sigma)}$.

2. The limiting behavior of λ_{max} as $K \to \pm\infty$ is given by (see Figure 3):

$$\lambda_{max} \to \begin{cases} z_1 & \text{for } K \to \infty \\ -Kq_{n-1}+\alpha & \text{for } K \to -\infty \end{cases}$$

where $\alpha = \sum_{i=1}^{n} \lambda_i - \sum_{i=1}^{n-1} z_i = \frac{q_{n-2}}{q_{n-1}} - p_{n-1}$. Thus as K becomes more and more distributed at large absolute values we see that almost sure asymptotic stability results if:

$$z_1 P_+ + (\sum_{i=1}^{n} \lambda_i - \sum_{i=1}^{n-1} z_i)P_- - q_{n-1} \int_{-\infty}^{0} K\, p(K) dK < 0$$

where $P_+ \triangleq P(K > 0)$ and $P_- \triangleq P(K < 0)$. For the uniformly distributed case studied in Example 1 with $K_+ > 0$ and $K_- < 0$ this condition requires

$$z_1 K_+ + (\sum_{i=1}^{n} \lambda_i - \sum_{i=1}^{n-1} z_i)K_- + q_{n-1} \frac{K_-^2}{2} < 0 .$$

3. Consider the equation studied by Infante [1968], p. 11:

$$\dot{n} = \frac{f(t)-\beta}{\ell} n + \lambda c \quad ; \quad \dot{c} = \frac{B}{\ell} n - \lambda c$$

where β, ℓ, $\lambda > 0$. This equation describes the kinetics of a simple nuclear reactor problem. It is easily seen that Theorem 2 applied to this case with

$$g(s) = \frac{1}{\ell}\ \frac{s+\lambda}{s(s+\frac{\beta}{\ell}+\lambda)} \quad \text{and } k(t) = -f(t)\,.$$

Thus almost sure asymptotic stability results if:

$$\int_0^\infty (\sigma-\lambda)p(\frac{(\sigma-\lambda)(\ell\sigma+\beta)}{\sigma})(\ell + \lambda\beta\frac{1}{\sigma^2})d\sigma < 0$$

where $p(\cdot)$ denotes the density function of f.

1.2 A Frequency-Domain Stability Criterion

In this section we will derive another criterion for almost sure asymptotic stability of the system Σ. We first recall the definition of a positive real function:

Definition 3: Let $H(s)$ be a matrix of real rational functions of the complex variable s. It is said to be *positive real* if $H(s) + H'(\bar{s}) \geq 0$ for all Re $s \geq 0$, $s \neq$ poles of $H(s)$.

There exist various equivalent conditions for positive realness. Such conditions may be found in most books on electrical network synthesis (see, for example, Guillemin [1957], Chapter 1, or Newcomb [1966]). Positive real functions play a fundamental role in the theory of passive systems, particularly in the analysis and synthesis of electrical networks. They have recently also shown to be an essential tool for obtaining frequency-domain stability criteria for feedback systems. A time-domain condition for positive realness is given in the following lemma, the celebrated *Kalman-Yacubovich-Popov lemma:*

Lemma 2: *Consider the minimal system:*

$$\dot{z} = Fz + Gv \ ; \ w = Hz,$$

and let σ *be a real number. Then* $H(I(s-\sigma)-F)^{-1}G$ *is positive real if and only if there exists a solution* $Q = Q' > 0$ *to the relations:*

$$F'Q + QF \leq -2\sigma Q \ ;$$

$$QG = H' \ .$$

For a proof of Lemma 2 we refer the reader to Willems [1972].

The value of the above lemma in stability analysis lies in the fact that the quadratic form induced by the matrix Q yields a very suitable candidate for a Lyapunov function. It plays a crucial role in the following theorem which is the main result of this section:

Theorem 3: *Let* $m = p$. Then Σ *is almost surely asymptotically stable if there exists a constant (mxm) matrix* Λ *and a real number* σ *such that:*

(i) $\Lambda + \Lambda' > 0$;

(ii) $F(s-\sigma) \triangleq G(s-\sigma)(I-\Lambda(s-\sigma)G(s-\sigma))^{-1}$ *is positive real;*

and (iii) $\mathscr{E}\{\min[\sigma,\lambda_{\min}[(K+K')(\Lambda+\Lambda')^{-1}]]\} > 0$.

Proof: We will assume that $(I-\Lambda CB)$ is invertible and that the McMillan degree of $F(s)$ is n. The general case may be resolved by a subsequent limiting argument which is left to the reader.

It is easily seen that $F(s)$ is the transfer function of the system:

$$\dot{z} = Az + B(v+\Lambda\dot{w}) \quad ; \quad w = Cz ,$$

or

$$\dot{z} = (A+B(I-\Lambda CB)^{-1}\Lambda CA)z+B(I-\Lambda CB)^{-1}v \quad ; \quad w = Cz .$$

This system is minimal since the McMillan degree of $F(s)$ is assumed to be n. Thus by condition (ii) and Lemma 2 there exists a matrix $Q = Q' > 0$ such that

$$[A+B(I-\Lambda CB)^{-1}\Lambda CA]'Q+Q[A+B(I-\Lambda CB)^{-1}\Lambda CA] \leq -2\sigma Q$$

and

$$QB(I-\Lambda CB)^{-1} = C'$$

Let S be an invertible (nxn) matrix such that $S'S = Q$ and let $x_1 = Sx$. The equation for x_1 is given by:

$$\dot{x}_1 = (A_1-B_1K(t)C_1)x_1$$

where $A_1 = SAS^{-1}$, $B_1 = SB$, and $C_1 = CS^{-1}$. Moreover, $B_1 = C_1'(I-\Lambda C_1B_1)$ and $(A_1+C_1'\Lambda C_1A_1)' + (A_1+C_1'\Lambda C_1A_1) \leq -2\sigma I$. Consider now the derivative of $V(x_1) = x_1'x_1+y_1'\Lambda y_1$, where $y_1 = C_1x_1$, along solutions of the above differential equation. A simple calculation using the above relations shows that:

$$\dot{V}(x_1) \leq -2\sigma x_1'x_1-2y_1'K(t)y_1 = -2\sigma V(x_1)+2y_1'(\sigma\Lambda-K(t))y_1 .$$

Let $\lambda(t) = \lambda_{min}[(K(t)+K'(t))(\Lambda+\Lambda')^{-1}]$ and let P be a nonsingular matrix such that $P'P = \Lambda + \Lambda'$. Since $\lambda(t) = \lambda_{min}[P^{-1}(K(t)+K'(t))(P')^{-1}]$ it thus follows that $y_1'K(t)y_1 \geq \lambda(t)y_1'\Lambda y_1$ for all y_1. Hence

$$\dot{V}(x_1) \leq -2\sigma V(x_1) + 2(\sigma-\lambda(t))y_1'\Lambda y_1 \ .$$

We now distinguish two cases:

(i) $\lambda(t) \geq \sigma$ which implies $\dot{V}(x_1) \leq -2\sigma V(x_1)$;

and (ii) $\lambda(t) \leq \sigma$ which, since $V(x_1) \geq y_1' \Lambda y_1$, implies:

$$\dot{V}(x_1) \leq -2\sigma V(x_1) + 2(\sigma-\lambda(t))V(x_1) = -2\lambda(t)V(x_1) \ .$$

Hence

$$\dot{V}(x_1) \leq -2\min[\sigma,\lambda(t)]V(x_1)$$

and

$$V(x_1(t)) \leq V(x_1(t_o))\exp(-2\int_{t_o}^{t} \min[\lambda,\sigma(t)]dt) \ .$$

By the ergodic hypothesis and condition (iii) this indeed implies that $\lim_{t\to\infty} V(x_1(t))=0$ almost surely. Thus $\lim_{t\to\infty} x_1(t) = S \lim_{t\to\infty} x(t) = 0$ almost surely, which proves the theorem. ■

Notes: 5. If $K + K' \geq \varepsilon I > 0$ almost surely and if G(s) is positive real then Theorem 2 predicts almost sure asymptotic stability by considering the limit $\sigma \to 0$ and $\Lambda \to 0$. In this sense Theorem 2 is thus a generalization of the circle criterion. The advantage of the theorem is that it allows the gain K(t) to become negative provided however this is compensated by K(t) being sufficiently positive at some other time.

One of the disadvantages of Theorem 3 is the inherent difficulty in verifying the average value condition from the distribution of K since $\lambda_{min}\{[(K+K')(\Lambda+\Lambda')^{-1}]\}$ is a very nonlinear function of K. In the scalar case however one may resolve the various conditions in Theorem 3 much further. Thus we arrive at the following more explicit criterion for systems with a single stochastic parameter:

Theorem 4: *Assume that* $m = p = 1$ *and let* $g(s) = C(Is-A)^{-1}B = \dfrac{q_{n-1}s^{n-1}+\ldots+q_o}{s^n+p_{n-1}s^{n-1}+\ldots+p_o}$ *denote the transfer function of* Σ_1. *Then* Σ *is almost surely asymptotically stable if there exists a real constant* β *such that*

(i) $\mathscr{E}\{\min[\beta,K]\} > 0$;

(ii) *the poles of* $G(s)$ *lie in* $\operatorname{Re} s < -q_{n-1}\beta$;

and (iii) *the locus of* $G(j\omega - q_{n-1}\beta)$, $-\infty < \omega < \infty$, *does not encircle or intersect the closed disc centered on the negative real axis of the complex plane and passing through the origin and the point* $-\frac{1}{\beta}$.

Proof: By Theorem 3 it suffices to show that there exists a constant $\lambda > 0$ such that $F(s-\sigma) = g(s-\sigma)(1-\lambda(s-\sigma)g(s-\sigma))^{-1}$ is positive real and $\mathscr{E}\{\min[\sigma\lambda, k]\} > 0$. Note that this implies $\sigma > 0$. Now $F(s-\sigma)$ is positive real if and only if $F^{-1}(s-\sigma) = \frac{1}{g(s-\sigma)} - \lambda(s-\sigma)$ is positive real. Since $F^{-1}(s-\sigma) = (\frac{1}{q_{n-1}} - \lambda)(s-\sigma) + \frac{r(s-\sigma)}{q(s-\sigma)}$ with $r(s)$ a polynomial of degree at most $(n-1)$ it follows that $\lambda \leq \frac{1}{q_{n-1}}$ and that $F^{-1}(s-\sigma)$ will be positive real for some λ if and only if it is positive real for $\lambda = \frac{1}{q_{n-1}}$, which is thus the optimal value of λ to consider. The condition $q_{n-1} > 0$ follows from the frequency domain condition (iii) as a result of the behavior of $g(j\omega-\sigma)$ for $\omega \to \infty$. Pick now $\sigma = \beta q_{n-1}$.

In order to complete the proof of the theorem it suffices to show that $F^{-1}(s-\sigma) = \frac{1}{g(s-q_{n-1}\beta)} - \frac{1}{q_{n-1}} s + \beta$ is positive real. By one of the test of positive realness this can be achieved by proving that $\operatorname{Re} F^{-1}(s-\sigma)]_{s=j\omega} \geq 0$ and (since $F^{-1}(s-\sigma)$ has no more zeros than poles) that the roots of $q(s-\sigma)$ lie in $\operatorname{Re} s < \sigma$. The real part condition comes down to asking $g(s-\sigma)]_{s=j\omega}$ to have the non-intersection property stated in condition (iii). By the non-encirclement condition the roots of $p(s-\sigma) + kq(s-\sigma)$ lie in $\operatorname{Re} s < \sigma$ for $k > \beta$. By letting $k \to \infty$ this implies that the roots of $q(s-\sigma)$ lie indeed in $\operatorname{Re} s \leq \sigma$. By the non-intersection property $g(j\omega-\sigma) \neq 0$ for $-\infty < \omega < \infty$ and we conclude that the roots of $q(s-\sigma)$ indeed lie in $\operatorname{Re} s < \sigma$ as desired.

Notes: 6. It may be shown that conditions (ii) and (iii) of Theorem 4 will be verified for $\beta \leq \beta_1$ if they are verified for β_1. Thus the optimal β to consider is the smallest number which satisfies condition (i) of the theorem.

7. If K has density function $p(K)$ then condition (i) of Theorem 4 requires that:

$$h(\beta) \triangleq \beta \int_{\beta}^{\infty} p(K)\,dK + \int_{-\infty}^{\beta} Kp(K)\,dK > 0$$

Now $\frac{dh(\beta)}{d\beta} \geq 0$, $h(0) \geq 0$ and $h(\infty) = \mathscr{E}\{K\}$. Thus there exists a β such that $h(\beta) > 0$ if and only if $\mathscr{E}\{K\} > 0$, and if so, then there exists a β^* such that $h(\beta) > 0$ for $\beta > \beta^*$. Thus Theorem 4 will predict almost sure asymptotic stability of Σ

if $\mathscr{E}[K] > 0$, if the poles of g(s) lie in Re s < $q_{n-1}\beta^*$ and if $g(j\omega - q_{n-1}\beta^*)$ satisfies the frequency domain condition of Theorem 4. This procedure lends itself very nicely to the graphical analysis illustrated in Figure 4.

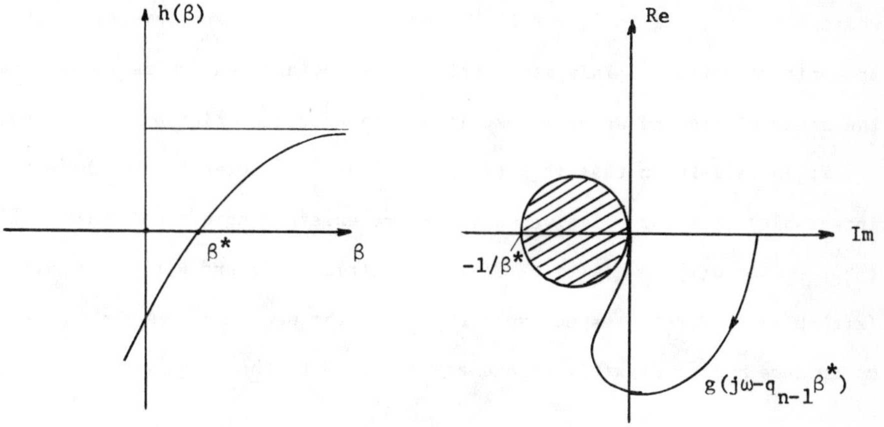

Figure 4: *Illustrating the application of Theorem 4.*

Examples: 4. Assume that K is uniformly distributed between K_- and K_+ with $K_- \leq 0$ and $K_+ + K_- \geq 0$. Then $\beta^* = K_+ - \sqrt{K_+^2 - K_-^2}$. Expressed in terms of the spread $\Delta K = K_+ - K_-$ and the mean $M = \frac{K_+ - K_-}{2}$ this yields $\beta^* = (\sqrt{\frac{\Delta K}{2}} - \sqrt{M})^2$ which in the range of interest $\frac{\Delta K}{2} \geq M \geq 0$ shows that β^* increases with ΔK for fixed M. This again indicates the destabilizing effect due to the uncertainty in K.

5. Let Σ_1 be a completely symmetric system as defined in Section 1.1. Then conditions (ii) and (iii) of Theorem 4 will be satisfied as long as $q_{n-1}\beta < -\lambda_1$ with λ_1 the largest pole of g(s). The stability condition then becomes $\mathscr{E}\{\min[-\frac{\lambda_1}{q_{n-1}}, K]\} > 0$ which is similar to, but more conservative than, the condition obtained in Note 4. Thus Theorem 2 which only applies to completely symmetric systems gives a sharper stability estimate than Theorem 4 which applies to general systems.

2. ANALYSIS OF THE MEAN AND THE COVARIANCE EQUATIONS

This last section of the paper is concerned with the stability analysis of the mean and the covariance of the state of Σ where K(t) is assumed to be a white stochastic process. For simplicity we will consider only the case in which the process K(t) is scalar valued, but we will treat the non-stationary case. If we denote the mean of K(t) by $\bar{k}(t)$ and the variance by $\bar{q}^2(t)$ then Σ is described by the stochastic differential equation:

$$\Sigma' : dx = (A-\bar{k}(t)bc)x\ dt + \bar{q}(t)bcx\ d\beta ,$$

where $A \in R^{nxn}$, $b \in R^{nx1}$, $c \in R^{1xn}$, and β denotes a Wiener process with zero mean and unit covariance. This stochastic differential equation is to be interpreted in the sense of Ito and we will take it as the starting point of our analysis.

It is well-known that if $\bar{k}(t)$ and $\bar{q}(t)$ are sufficiently smooth (e.g., locally integrable) then for all given $x(t_o)$ there exists a unique solution to Σ' for $t \geq t_o$. Let $\mu(t) \triangleq \mathscr{E}\{x(t)\}$, $\Gamma(t) \triangleq \mathscr{E}\{x(t)x'(t)\}$, and $R(t) \triangleq \mathscr{E}\{(x(t)-\mu(t))(x(t)-\mu(t))'\}$ denote respectively the *mean*, the *second moment matrix*, and the *covariance matrix* of x(t). These are governed by the equations:

$$\dot{\mu} = (A-\bar{k}(t)bc)\mu ;$$

$$\dot{\Gamma} = (A-\bar{k}(t)bc)\Gamma+\Gamma(A-\bar{k}(t)bc)'+q^2(t)bc\Gamma c'b' \quad ;$$

and
$$R(t) = \Gamma(t)-\mu(t)\mu'(t),$$

with initial conditions $\mu(t_o) = x(t_o)$ and $\Gamma(t_o) = x(t_o)x'(t_o)$.

We will be concerned with the asymptotic properties of these variables. The relevant stochastic stability concepts are now defined:

Definition 4: Σ' is said to be *asymptotically stable in the mean, in the mean square,* or *in the covariance* if, respectively, $\lim_{t\to\infty} \mu(t) = 0$, $\lim_{t\to\infty} \Gamma(t) = 0$, or $\lim_{t\to\infty} R(t) = 0$ for all given initial conditions $x(t_o)$.

It is easily seen from the relations $\Gamma(t) = R(t)+\mu(t)\mu'(t)$ and $R(t) = R'(t) \geq 0$ that mean square asymptotic stability implies stability in the mean and in the covariance. The stability of the mean is a standard deterministic stability problem for which many criteria have been derived. These criteria involve the transfer function $g(s) = c(Is-A)^{-1}b$ and properties of k(t) as, for example, its bounds (e.g. in the circle criterion: see Brockett [1970], Section 35), bounds on its derivative, or its periodicity. The stability of the differential equation which expresses the evolution of the second moment matrix $\Gamma(t)$ is much more intricate to analyze and we will show how criteria like the multivariable circle criterion may be used. If $\overline{q}^2(t) = 0$ then its stability is equivalent to the stability of the mean equation, whereas if $\overline{q}^2(t) \neq 0$ then more stringent conditions will have to be imposed.

2.1 Multilinear System Theory

It is easy to see that if x_1 and x_2 are vectors which satisfy the linear equations:

$$\dot{x}_1 = A_1(t)x_1 \quad ; \quad x_1 \in R^{n_1} ,$$

and

$$\dot{x}_2 = A_2(t)x_2 \quad ; \quad x_2 \in R^{n_2} ,$$

then the product x_1x_2' satisfies also a linear equation, namely:

$$\frac{d}{dt} x_1x_2' = A_1(t)x_1x_2' + x_1x_2'A_2'(t) .$$

By taking $x_1 = x_2$ we see that if x satisfies a linear equation, then so does xx'.

This idea generalizes from quadratic forms to homogeneous p-th degree forms. These facts have been known at least since Lyapunov's thesis, but they have to the present time been used very little in system theory. They may for example be exploited in the minimization of homogeneous performance measures of degree $p > 2$ for linear dynamical systems.

The above ideas may be used in setting up transfer functions for a class of bilinear systems. We will make some use of the Kronecker product denoted here by $\otimes$. Thus the Kronecker product of $M \in R^{nxm}$ and $R \in R^{pxq}$ is the element $M \otimes R \in R^{npxmq}$ defined by:

$$M \otimes R \triangleq \left[\begin{array}{c|c|c|c} m_{11}R & m_{12}R & \cdots & m_{1q}R \\ \hline m_{21}R & m_{22}R & \cdots & m_{2q}R \\ \hline \vdots & \vdots & \vdots\vdots\vdots & \vdots \\ \hline m_{p1}R & m_{p2}R & \cdots & m_{pq}R \end{array}\right]$$

The main use of this notation is that if an (nxn) matrix Q is written in lexographic notation as the n^2-vector

$$Q_v = \mathrm{col}(q_{11}, q_{12}, \ldots, q_{1n}, \ldots, q_{n1}, q_{n2}, \ldots, q_{nn})$$

then $(MQ)_v = (I \otimes M)Q_v$.

Consider now the following lemma:

Lemma 3: *Let* $\{A,b,c\}$ *be a minimal realization of the transfer function* $g(s) = c(Is-A)^{-1}b$. *Then the differential equation:*

$$\dot{Q} = AQ + QA + bv' + vb' \quad ; \quad w = cQ ,$$

defines a minimal realization on the $\frac{n(n+1)}{2}$ *dimensional space of symmetric* (nxn) *matrices of the transfer function:*

$$g^{[2]}(s) = (c \otimes I+I \otimes c)(Is-I \otimes A-A \otimes I)^{-1}(b \otimes I+I \otimes b) .$$

We will not give a detailed proof of this lemma. The proof exploits the fact that the above matrix equation describes the bilinear system

$$\frac{d}{dt} xx' = Axx' + xx'A' + bux' + xu'b' ;$$

$$yx' = cxx'$$

where $\dot{x} = Ax + bu$; $y = cx$.

The dynamical system identified in the statement of Lemma 3 plays an important role in the analysis of the covariance equation under consideration. We know from this lemma that controllability and observability will be preserved. The poles of $g^{[2]}(s)$ are given by $\{\lambda_i(A)+\lambda_j(A)\}$, i,j=1,...,n. There appears to be no convenient general formula for deriving $g^{[2]}(s)$ from g(s). In a specific case however, it is a relatively straightforward matter to calculate $g^{[2]}(s)$.

Example: 6. Let [A,b,c] be the standard controllable representation (see Brockett [1970], p. 106) of $G(s) = \frac{1}{s^2+as+b}$. Then

$$g^{[2]}(s) = -\frac{1}{s^3+3as^2+(2a^2+4b)s+4ab}\begin{bmatrix} 2(s+2a) & 4 \\ s(s+2a) & 2s \end{bmatrix} .$$

2.2 The Circle Criterion for the Covariance Equation

We now return to the covariance equation:

$$\dot{\Gamma} = (A-\bar{k}(t)bc)\Gamma + \Gamma(A-k(t)bc)' + \bar{q}^2(t)bc\Gamma c'b'$$

which we model as the feedback system:

$$\Sigma_1' : \dot{Q} = AQ + QA'+bv'+vb'+bwb' \quad ; \quad y = cQ, \; z = cQc' ,$$

$$\Sigma_2' : v = -\bar{k}(t)y, \quad w = \bar{q}^2(t)z \quad .$$

It follows from Lemma 3 that Σ_1' is completely controllable and completely observable. Let

$$\tilde{G}(s) \triangleq \left[\begin{array}{c|c} G_{11}(s) & G_{12}(s) \\ \hline G_{21}(s) & G_{22}(s) \end{array}\right] ,$$

where $\qquad y(s) = G_{11}(s)v(s) + G_{12}(s)w(s)$

and $\qquad z(s) = G_{21}(s)v(s) + G_{22}(s)w(s)$,

denote the transfer function of Σ_1'. It is easily calculated that $\tilde{G}(s)$ is given by:

$$\tilde{G}(s) = \left[\begin{array}{c} c \otimes I + I \otimes c \\ \hline c \otimes c \end{array}\right] (Is-A \otimes I-I \otimes A)^{-1} \; [b \otimes I+I \otimes b \mid b \otimes b]$$

Thus the stability of the covariance equation is equivalent to the stability of a deterministic feedback system with (n+1) feedback loops, with transfer function $\tilde{G}(s)$ in the forward loop and gain matrix

$$F(t) = \left[\begin{array}{c|c} \bar{k}(t)I & 0 \\ \hline 0 & -\bar{q}^2(t) \end{array}\right]$$

in the feedback loop.

The multivariable circle criterion and its various generalizations is thus immediately applicable to this situation. We will illustrate this only in the simplest case. Let $||\cdot||$ denote some norm on R^{n+1} and let matrix norms be induced norms. The small loop gain theorem due to Zames [1966] thus leads to:

Theorem 5: *Assume that* Re $\lambda[A] < 0$. *Then* Σ' *is asymptotically stable in the mean square if:*

$$\left(\sup_{-\infty<\omega<\infty} ||\tilde{G}(j\omega)|| \right) \left(\sup_{-\infty<t<\infty} ||F(t)|| \right) < 1 \; .$$

Unfortunately it does not appear to be an easy matter to express the above criterion as direct conditions on the original transfer function g(s) and the functions $\bar{k}(t)$ and $\bar{q}^2(t)$. In the case that $\bar{k}(t)$ or $\bar{q}^2(t)$ are time-invariant however it is possible to obtain a criterion which is a great deal more specific:

Corollary 1: *Assume that* $\overline{k}(t) = k$ *is constant. Then* Σ' *is asymptotically stable in the mean square if:*

$$\Big(\sup_{-\infty<t<\infty} \overline{q}^2(t)\Big)\Big(\int_0^\infty (ce^{(A-kbc)t}b)^2 dt\Big) < 1 \quad .$$

Proof: The equation for Γ may be modelled as the feedback system:

$$\dot{Q} = (A-kbc)Q+Q(A-kbc)' + bwb' \quad ; \quad z = cQc' \ ,$$

$$w = q^2(t)z \quad .$$

The first system has $(ce^{(A-kbc)t}b)^2$ as impulse response. Since this is always nonnegative it follows that its Fourier transform attains it maximum for $\omega = 0$. Since this maximum is given by $\int_0^\infty (ce^{(A-kbc)t}b)^2 dt$ we obtain the corollary by applying the circle criterion in the scalar case. ∎

Corollary 2: *Assume that* $\overline{q}^2(t) = q^2$ *is constant and let*

$$\hat{G}(s) \triangleq (c \otimes I+I \otimes c)(Is-A \otimes I-I \otimes A - q^2(b \otimes b)(c \otimes c))^{-1}(b \otimes I+I \otimes b) \ .$$

then Σ' *is asymptotically stable in the mean square if:*

$$\text{(i)} \quad \overline{q}^2 \int_0^\infty (ce^{At}b)^2 dt < 1 \quad ;$$

and $\text{(ii)} \Big(\sup_{-\infty<t<\infty} \overline{k}(t)\Big)\Big(\sup_{-\infty<\omega<\infty} ||\hat{G}(j\omega)||\Big) < 1$

Proof: The equation for Γ may be modelled as the feedback system:

$$\dot{Q} = AQ + QA' + bcQc'b' + bv' + vb' \quad ; \quad y = CQ \ ,$$

$$v = -k(t)y \quad .$$

The first system has $\hat{G}(s)$ as transfer function and is stable if condition (i) is satisfied. The corollary thus follows from the multivariable circle criterion (see Brockett [1970], Section 33). ∎

Notes: 8. The conditions of Corollary 1 may be expressed in terms of frequency-domain data. They then lead to conditions very similar to the deterministic circle criterion (see Willems and Blankenship [1971]).

9. J.L. Willems [1972] has obtained a number of criteria for systems as the one studied here. His criteria which are in the vein of Corollary 1 are sharper and

more explicit than those studied here.

10. It is well-known that the circle criterion gives the best conditions which may be proven by means of a quadratic Lyapunov function. However in the case under consideration one can obtain results by using "linear" Lyapunov functions. Indeed, one may view the equation describing Γ as a differential equation on the space P of nonnegative definite symmetric (nxn) matrices. Restricting our attention to this subset of the vector space S of symmetric (nxn) matrices does not buy us anything as far as stability is concerned (i.e. stability on P is equivalent to stability on S). However it enhances the likelihood that a particular function will be definite and thus greatly enlarges the class of Lyapunov functions. For example the function Trace $[P\Gamma]$ with $P = P' > 0$ is positive definite on P but not on S. It hence defines a suitable Lyapunov function for studying the mean square stability question. This method is exploited in Willems [1972].

CONCLUSIONS

We have presented here a number of results on the stability of linear systems with stochastic coefficients. Two average value criteria for almost sure stability were derived and we showed how one may use deterministic stability results like the multivariable circle criterion in order to obtain mean square stability criteria in the case the stochastic parameters are white noise processes.

REFERENCES

Brockett, R.W., *Finite Dimensional Linear Systems*, New York: Wiley (1970).

Feller, W., *An Introduction to Probability Theory and its Applications, Vol. II*, New York: Wiley (1966).

Guillemin, E.A., *Synthesis of Passive Networks*, New York: Wiley (1957).

Infante, E.F., *On the Stability of Some Linear Nonautonomous Random Systems*, J. of Applied Mechanics, 35, 7-12, (1968).

Kalman, R.E., Falb, P.L., and Arbib, M.A., *Topics in Mathematical System Theory*, New York: McGraw-Hill (1969).

Kushner, H.J., *Stochastic Stability and Control*, New York, Academic Press (1967).

Newcomb, R.W., *Linear Multiport Synthesis*, New York: McGraw-Hill (1966).

Willems, J.C. and Blankenship, G.L., *Frequency Domain Stability Criteria for Stochastic Systems*, IEEE Trans. on Automatic Control, AC-16, 292-299 (1971).

Willems, J.C., *Dissipative Dynamical Systems, Part I: General Theory; Part II: Linear Systems with Quadratic Supply Rates*, Archive for Rational Mechanics and Analysis, 45, 321-393 (1972).

Willems, J.L., *Lyapunov Functions and Global Frequency Domain Stability Criteria for a Class of Stochastic Feedback Systems*, presented at the IUTAM Symposium on Stability of Stochastic Dynamical Systems, Univ. of Warwick, Coventry, England (1972).

Zames, G., *On the Input-Output Stability of Time-Varying Nonlinear Feedback Systems. Part I: Conditions Derived Using Concepts of Loop Gain, Conicity, and Positivity; Part II: Conditions Involving Circles in the Frequency Plane and Sector Nonlinearities*, IEEE Trans. on Automatic Control, AC-11, 228-238, 465-476 (1966).

ULTIMATE BEHAVIOUR OF A CLASS OF STOCHASTIC DIFFERENTIAL SYSTEMS DEPENDENT ON A PARAMETER *

P. R. SETHNA

Department of Aerospace Engineering

and Mechanics

University of Minnesota

I. Introduction

We will discuss systems of the form:

$$\dot{x} = \varepsilon f(H(t),x) \qquad (1.1)$$

where $x \varepsilon R^n$, $\varepsilon \varepsilon R$, $0 \leq \varepsilon \leq \varepsilon_o$ and f is a function into R^n which depends on x and on an nxN matrix H with elements $h_{ij}(t)$ that are statistically stationary processes with the property of ergodicity of the mean and $h_{ij}(t)$ are assumed to be bounded and continuous for $-\infty<t<\infty$ with probability one. Our object is to study the behaviour of the process x as $t\to\infty$.

The vector function f in (1.1) will be assumed to have a special form so that if f_i is the ith component of f then:

$$f_i(H(t),x) = \sum_{j=1}^{N} g_{ij}(x)h_{ij}(t) \qquad (1.2)$$

where N is some positive integer.

Let $<\ >$ denote the time average, then taking time averages of both side of (1.2) and using the property of ergodicity of the mean we have :

$$< f_i(H(t),x) > = \sum_{j=1}^{N} g_{ij}(x)\ E\{h_{ij}\} \qquad (1.3)$$

---

* This work has been supported under a grant AF-AFOSR-2284-72

The time average of f then is a function of the ensemble average of the matrix H(defined as the matrix of the ensemble average of its elements). We thus use the notation :

$$<f(H(t), x> \equiv f^{o} (E\{H\},x)$$

Our results for the stochastic differential system (1.1) are related to the solutions of the nonstochastic, autonomous differential system :

$$\dot{\xi} = \varepsilon f^{o}(E\{H\},\xi) \qquad (1.4)$$

where $\xi \varepsilon R^{n}$.

We base our results on certain results for classes of nonstochastic, nonautonomous differential equations that are related to the "method of averaging". There are available in the literature other results that relate results associated with the "method of averaging" to stochastic differential systems, most notably the works of Kolomiets and Mitropolskiy [1]*and Khas'minskii [2]. The former work is concerned with systems with random white noise and the latter with finite time results. Our results are in a different category since they are concerned with classes of ergodic processes and the results are valid for all positive t.

II. Some Results For Nonstochastic Differential Systems

Our main result for system (1.1) is based on a nonlocal theorem for a special class of nonlinear nonautonomous ordinary differential equations. In this section we give the outline of a proof for this theorem. This theorem is a generalization of a theorem by Sethna and Moran [3]. The theorem in [3] is restricted to systems that are almost periodic in time. In the generalization to be discussed here we give results for systems that are merely bounded in time.

Our proof is based on a recent result of Hale [4], a lemma of Diliberto [5] and a well known theorem of Bagoliuboff [6]. The proofs of the theorems of Hale and Bagoliuboff are easily available in the literature. The lemma of Diliberto, in the form that we need, is less accessible and so we will outline its proof.

---

* Numbers in square brackets refer to references at the end of the paper

Lemma (Diliberto)

Consider a scalar linear differential system :

$$\dot{\omega} = \varepsilon\omega + h(t) \qquad (2.1)$$

where $\omega\varepsilon R$, $\varepsilon\varepsilon R$, $\varepsilon>0$, and h is a continuous mapping $R\to R$. We assume that there exist two positive numbers M_1 and M_2, independent of t and T such that

$$|h(t)|<M_1 \quad \text{and} \quad \left| \int_t^{t+T} h(s)ds \right| < M_2 \qquad (2.2)$$

for all $t\varepsilon R$, $T\varepsilon R$.

Then, there exists a unique bounded solution of (1):

$$\phi(t,\varepsilon) = -\int_0^\infty e^{-\varepsilon\sigma} h(t+\sigma)d\sigma \quad \text{with} \quad \left|\phi(t,\varepsilon)\right| \leq M_2 \qquad (2.3)$$

for all $t\varepsilon R$, $\varepsilon>0$ and the bound M_2 is also independent of ε.

Proof:

For any solution $\psi(t,\omega_o,t_o)$ with $\psi(t_o,\omega_o,t_o) = \omega_o$

$$\psi(t,\omega_o,t_o) = \omega_o e^{\varepsilon t} + \int_{to}^{t} e^{\varepsilon(t-s)}h(s)ds$$

The particular solution passing through $\omega_o = \omega_o^\dagger$ at $t=t_o$ with

$$\omega_o^\dagger = -\int_{t_o}^\infty e^{-\varepsilon s}h(s)ds$$

is the solution (2.3) and clearly both $\omega_o^\dagger$ and $\phi(t,\varepsilon)$ exist because of hypothesis (2.2) and furthermore ϕ is unique because all solutions of (2.1) are unique. We now prove the bound on $\phi(t,\varepsilon)$. We observe that ϕ can be written as :

$$\phi(t,\varepsilon) = -\int_0^\infty e^{-\varepsilon\sigma} \left[\frac{d}{d\sigma} \int_0^\sigma h(t+\theta)d\theta\right] d\sigma$$

then on integration by parts and using bounds (2.2) it is easy to show that $|\phi(t,\varepsilon)| \leq M_2$.

We now give a statement of the theorem of Hale [4] for bounded systems in a form specialized to our needs.

Theorem(Hale)

Consider the system:

$$\dot{x} = Ax + g(t,x,\varepsilon) \tag{2.4}$$

where $X\varepsilon R^n$, and A is an $n \times n$ constant matrix with characteristic roots with non-zero real parts. In (2.4), g is a continuous function into R^n from $\{t: t\varepsilon R, -\infty < t < \infty\} \times \{x: x\varepsilon R^n, |X| \leq K\} \times \{\varepsilon: \varepsilon\varepsilon R, 0 \leq \varepsilon \leq \varepsilon_o\}$. It is assumed that g has continuous partial derivatives with respect to the components of x and is bounded uniformly for all t and $|X| \leq K$ for each fixed ε. If g_x is the Jacobian matrix of g with respect to x then it is assumed that:

$$\lim_{\varepsilon \to 0, \mathbf{x} \to 0} g_x(t,x,\varepsilon) = 0, \qquad \text{uniformly in } t,$$

and

$$\lim_{\varepsilon \to 0} g(t,0,\varepsilon) = 0, \qquad \text{uniformly in } t,$$

Under these conditions, there exist positive constants σ and ε_1, $0<\sigma\leq K$, $0<\varepsilon_1\leq\varepsilon_o$ such that for each ε, $0<\varepsilon\leq\varepsilon_1$, there exists a unique solution $x^*(t,\varepsilon)$ of (2.4) which is bounded for all t, $|x^*(t,\varepsilon)|<\sigma$ and x^* is continuous in ε and $\lim_{\varepsilon\to 0} x^*(t,\varepsilon) = 0$ uniformly in t.

If A has characteristic roots with negative real parts x^* is stable. If at least one characteristic root of A has positive real part then x^* is unstable.

We now state and prove a local theorem for a nonstochastic differential system of the general structure of (1.1).

Theorem 1

In the vector differential system

$$\dot{x} = \varepsilon f(t,x) \tag{2.5}$$

$x\varepsilon R^n$, $0\leq\varepsilon\leq\varepsilon_o$ and the n vector function f has its ith component of the form:

$$f_i = \sum_{j=1}^{N} g_{ij}(x)\, h_{ij}(t) \qquad i = 1,2,\ldots.n$$

where N is some positive integer and g_{ij} are continuous functions of x with continuous derivatives for $|x| \leq K$. With regard to the functions $h_{ij}(t)$, i=1,2...n, j=1,2..N. it will be assumed that they are continuous and bounded for all t and that their time average

$$h^{o}_{ij} = \lim_{T\to\infty} \frac{1}{T} \int_{o}^{T} h_{ij}(t)dt$$

exists and that each function satisfies conditions analogous to (2.2) i.e:

$$|h_{ij}(t)| \leq M_1 \quad \text{and} \quad \left| \int_{t}^{t+T} |h_{ij}(\sigma) - h^{o}_{ij}| d\sigma \right| \leq M_2 \qquad (2.6)$$

for all $t\varepsilon R, T\varepsilon R$, where M_1 and M_2 are some numbers independent of t and T.

Consider the autonomous differential system

$$\dot{\xi} = \varepsilon f^{o}(\xi) \qquad (2.7)$$

where $\xi \varepsilon R^n$ and the function f^o is a mapping into R^n with its ith component of the form :

$$f^{o}_{i}(\xi) \equiv \sum_{j=1}^{N} g_{ij}(\xi) h^{o}_{ij} \qquad i = 1,2,\ldots n.$$

Let the system (2.7) have a constant solution ξ_o such that the matrix $\frac{\partial f^{o}(\xi_o)}{\partial \xi}$ has characteristic roots with nonzero real parts.

Under these conditions, there exist positive constants σ and ε_1, $0<\sigma\leq K$, $0<\varepsilon_1\leq\varepsilon_o$, such that for each ε, $0<\varepsilon\leq\varepsilon_1$ there exists a unique bounded solution $x^*(t,\varepsilon)$ of (2.5) bounded for all t and $|x^*(t,\varepsilon) - \xi_o|<\sigma$ for $-\infty<t<\infty$, x^* is continuous in ε and $\lim_{\varepsilon\to 0} |x^*(t,\varepsilon) - \xi_o| = 0$

The solution x^* is stable if the matrix $\frac{\partial f^{o}(\xi_o)}{\partial \xi}$ has characteristic roots with negative real parts and x^* is unstable if this matrix has at least one characteristic root with positive real part.

<u>Proof</u>

The method of proof is to use a transformation, based on the lemma of Diliberto, to reduce (2.5) to the form (2.4). We then have the result immediately from the Theorem of Hale.

Let $x = \xi + \varepsilon\,\beta\,(t,\xi)$ (2.8)

where the n vector function β of t, parameterized by ξ, is a unique bounded solution of :

$$\frac{\partial \omega_i}{\partial t} = \varepsilon \omega_i + \sum_{j=1}^{n} g_{ij}(\xi)\,\left[h_{ij}(t) - h_{ij}^{o}\,\right] \qquad (2.9)$$

We note that equation (2.9) is analogous to (2.1) and because of conditions (2.6) satisfies all the conditions of the lemma of Diliberto. Thus β exists and is bounded for all $|\xi| \leq K$.

Substituting (2.8) in (2.5)and noting that $\beta(t,\xi)$ satisfies (2.9) we have :

$$\dot{\xi} = \varepsilon f^{o}(\xi) + \varepsilon^2 \theta(t,\xi,\varepsilon) \qquad (2.10)$$

where

$$\theta \equiv \phi + \frac{\partial f(\xi_i t)}{\partial \xi}\,\beta + B(t,\xi,\varepsilon)\,f^{o}(\xi) + O(\varepsilon^3)$$

where ξ_i are values of the components of ξ chosen to apply the mean value theorem for vector valued functions and the B is defined as follows:

$$\left[I + \varepsilon\,\frac{\partial\beta}{\partial\xi}(t,\xi)\right]^{-1} = I + \varepsilon\,B(t,\xi,\varepsilon)$$

and B is a quantity of order zero in ε.

Now let $\xi = \xi_0 + \eta$

then equation (2.10) can be written in the form

$$\frac{d\eta}{dt} = A\eta + \varepsilon\gamma\,(\tau,\eta,\varepsilon) \qquad (2.11)$$

where $A = \dfrac{\partial f^{o}(\xi_0)}{\partial \xi}$, $\varepsilon t = \tau$ and

$$\gamma\,(\tau,\eta,\varepsilon) = f^{o}(\xi_0 + \eta) - \frac{\partial f^{o}(\xi_0)}{\partial \xi}\,\eta + \varepsilon^2 \theta\,(\tfrac{\tau}{\varepsilon},\xi_0 + \eta,\varepsilon)$$

Now (2.11) satisfies all the condition of system (2.4) and from the Theorem of Hale our result follows immediately.

The desired general nonlocal theorem is proved by using Theorem 1 and the theorem of Bogolinboff [6]. We will state this theorem without proof since the proof is similar to that in [3] for the corresponding theorem for almost periodic systems.

<u>Theorem 2</u>

Let the system (2.5) satisfy all the conditions on the function f of Theorem 1. Let ξ_0 be a constant solution of (2.7) which with its δ neighborhood lies in the ball $B(K) = \{\mathbf{x}: \mathbf{x}\varepsilon R^n, |\mathbf{x}| \leq K\}$ and let the characteristic values of the matrix $\frac{\partial f^o(\zeta_o)}{\partial \xi}$ have negative real parts.

Given any $\rho > 0$, let $\phi(t)$ be a solution of (2.7) with $\phi(0)\varepsilon B(K)$ which, together with its ρ neighborhood, remains in B(K) for all $t \geq 0$ and $\lim_{t\to\infty} \phi(t) = \xi_o$.

Then, given any $\eta > 0$, there exists an ε^*, $0 < \varepsilon^* \leq \varepsilon_o$ such that for each $\varepsilon, 0 < \varepsilon \leq \varepsilon^*$, the solution $\psi(t,\varepsilon)$ of (2.5), with $\psi(0,\varepsilon) = \phi(0)$, satisfies $|\psi(t,\varepsilon) - \phi(t)| < \eta$ for all $t \geq 0$. Furthermore, there exists an n vector function $\mathbf{x}^*(t,\varepsilon)$, continuous in t and ε, bounded in t for all $-\infty < \quad < \infty$ which is a solution of (2.5) and such that

$$\lim_{\varepsilon\to 0} |\mathbf{x}^*(t,\varepsilon) - \xi_o| = 0 \text{ and } \lim_{t\to\infty} |\psi(t,\varepsilon) - \mathbf{x}^*(t,\varepsilon)| = 0$$

III. The Main Result

Theorem

Consider the stochastic system (1.1) where f has the special form (1.2). The functions $g_{ij}(\mathbf{x})$ are assumed to be continuous and to have continuous derivatives with respect to the components of $\mathbf{x}$ for $|\mathbf{x}| \leq K$. The n x N matrix H has for each of its elements $h_{ij}(t)$ statistically stationary processes with the property of ergodicity of the mean. The processes $h_{ij}(t)$ are assumed to be bounded and continuous for $-\infty < t < \infty$ with probability one and to satisfy conditions (2.6) with probability one.

Suppose that all solutions of the derived autonomous system (1.4) approach a constant solution $\xi_o(E\{H\})$ and the matrix $\frac{\partial f^o(\xi_o)}{\partial \xi}$ has characteristic roots with negative real parts. Then, there exists an $0 < \varepsilon^* \leq \varepsilon_o$ such that for each $\varepsilon, 0 < \varepsilon \leq \varepsilon^*$ we have the following:

For each sample matrix H there exists with probability one an n vector function $\mathbf{x}^*(t,\varepsilon)$ continuous in t and ε bounded for $-\infty < t < \infty$ which is a solution of (1.1) and $\lim_{\varepsilon\to 0} |\mathbf{x}^*(t,\varepsilon) - \xi_o| = 0$
Furthermore there exist with probability one n vector functions

$\psi(t,\varepsilon)$, $\psi(0,\varepsilon)$ R^n which are solutions of (1.1) and such that $\lim_{t\to\infty} |\psi(t,\varepsilon) - \mathbf{x}^*(t,\varepsilon)| = 0$

If $E\{\mathbf{x}^*(t,\varepsilon)\}$ is the expectation of $\mathbf{x}^*(t,\varepsilon)$ then

$$E\{\mathbf{x}^*(t,\varepsilon)\} = \xi_o\{E\{H\}\} + O(\varepsilon) \tag{3.1}$$

Proof

The results for each sample matrix H follow immediately from Theorem 2. We merely note that here we assume that all solutions of (1.4) tend to ξ_0 and consequently we get a global result.

In order to prove (3.1), we have for each sample matrix H(t) from (2.8)

$$x = \xi + \varepsilon\beta(t,\xi,\varepsilon)$$

If $\psi(t,\varepsilon)$ is a solution (1.1) and $\phi(t)$ is the corresponding solution of (1.4) (as in Theorem 2) then :

$$\psi(t,\varepsilon) = \phi(t) + \varepsilon\,\beta\,(t,\phi,\varepsilon)$$

Taking limits as $t\to\infty$ of this expression and noting that β is bounded for all t we have :

$$x^*(t,\varepsilon) = \xi_0 + O(\varepsilon) \tag{3.2}$$

Taking the expected value of (3.2) we have the result (3.1) and this completes the proof.

Remark 1

We will now show that among the processes that satisfy our conditions are a class of periodic processes. Let h(t) be a continuous periodic, statistically stationary process with a periodic autocorrelation function $R(\tau)$ of period T. Then $R(\tau)$ can be represented in a Fourier series :

$$R(\tau) = \sum_{-\infty}^{+\infty} \alpha_n e^{i\,n\,\omega_0 \tau} \quad , \quad \omega_0 = \frac{2\pi}{T}$$

Furthermore, the process h(t) itself has a representation [7] in the mean :

$$h(t) = \sum_{n=-\infty}^{+\infty} A_n\, e^{in\omega_0 t} \tag{3.3}$$

where A_n , n = 0, 1, ··· are uncorrelated and orthogonal random variables with

$$A_n = \frac{1}{T} \int_0^{+} h(u)\, e^{-in\omega_0 u}\, du$$

with

$$E\{A_n\} = E\,\{h(t)\} \quad n = 0$$
$$= 0 \quad n \neq 0$$

and

$$E\{A_n A_m^*\} = \alpha_n \quad n = m$$
$$= 0 \quad n \neq m$$

Although the representation (3.3) is in the mean, by taking an appropriate subsequence of the sequence of partial sums we also have a representation with probability one.

The necessary and sufficient condition for the process to have ergodicity of the mean [8] is :

$$\lim_{T\to\infty} \int_0^{T_1} R(\tau)d\tau = E^2 \{h(t)\}$$

Thus the ergodicity condition in our case is :

$$E \{ |A_o{}^2| \} = E^2 \{A_o\} \qquad (3.4)$$

Furthermore, the process h(t) with the representation (3.3), obviously satisfies the integrability conditions (2.6).

Thus any periodic process satisfies our conditions provided A_o satisfies (3.4) i.e. if A_o is a constant with probability one.

Remark 2

If the system (1.4) has more than one stable constant solution ξ_o, we have a result for the expectation for $\mathbf{x}^*$ that depends on initial values of the solutions $\psi(t,\varepsilon)$.

Conjecture

It is conjectured that all statistically stationary processes with the property of ergodicity of the mean will satisfy our hypothesis provided that the spectral density $S(\omega)$ of the process is zero for ω in the neighborhood of zero.

REFERENCES

1. Y.A.Mitropolskiy and V.G.Kolomiets, "Application of the averaging principle to the investigation of the influence of random effects on oscillatory systems" (in Russian). Mathematical Physics (Edited by Y.A.Mitropolskiy), Kiev, p.146,(1967)

2. R.Z. Khas'minskiy, "On stochastic processes defined by differential equations with a small parameter"(in English). Th.Prob.Appls. 11,211(1966)

3. P.R.Sethna and T.J.Moran, "Some nonlocal results for weakly nonlinear dynamical systems". Quarterly of Applied Mathematics Vol. XXvI,No.2. p.175-185 (1968)

4. J.K.Hale, "Ordinary Differential Equations" Wiley-Interscience (1969). Chap.IV.

5. S. Diliberto, "New results on periodic surfaces and the averaging principle". U.S.Japanese Conference on Integral and Differential Equations,1966.

6. N.M.Bogoliuboff and Yu.A.Mitropolskiy, " Asymptotic methods in the theory of nonlinear oscillations". Gordon and Breach, New York, 1962, Chapter 6.

7. A.Papoulis, "Probability, Random Variables and Stochastic Processes". McGraw Hill (1965),Chapter 10.

8. Reference 7. Chapter 13.

STABLE PERIODIC SOLUTIONS OF WEAKLY NONLINEAR STOCHASTIC DIFFERENTIAL EQUATIONS

HELGA BUNKE

Central Institute of Mathematics and Mechanics
German Academy of Sciences, Berlin, GDR

Using results on the existence of periodic and stationary solutions of stochastic linear differential equations, then the existence of periodic and stationary solutions of stochastic differential equations with small nonlinear terms can be proved by approximation methods. The following result is a generalization of a theorem of A.Ja.Dorogevcev. If Z_t, $t \in R^1$, is a stochastic process defined over a probability space $(\Omega, \mathfrak{A}, P)$, we denote the sample functions by $Z_t(\omega)$, $\omega \in \Omega$. Let a stochastic process Z_t, $t \in R^1$, be called Θ-periodic if all finite dimensional distribution functions $F(z_1,\dots,z_k;\ t_1+\tau,\dots,t_k+\tau)$ are Θ-periodic functions in τ. A stochastic process Z_t, $t \in R^1$ is called stationary, if all finite dimensional distribution functions $F(z_1,\dots,z_k;\ t_1+\tau,\dots,t_k+\tau)$ are independent of τ. If the processes $Z_t^{[k]}$, $k = 1,2,\dots$ are Θ-periodic and if Z_t is a stochastic process with $\lim_{k\to\infty} Z_t^{[k]} = Z_t$ a.s., $t \in R^1$, then Z_t is also Θ-periodic.

Lemma 1: Let be Z_t, t R^1, a Θ-periodic n-dimensional stochastic Process and $y(t,z)$ a function $y: R^1 \times R^n \to R^m$, which is Θ-periodic in t for fixed $z \in R^n$ and $\mathfrak{B}^n$-measurable for fixed $t \in T^1$. Then $Y_t = y(t,Z_t)$ is a Θ-periodic stochastic process.

The solutions of an n-dimensional linear nonstochastic differential equation system with a continuous matrix function $A(t)$:

$$\frac{d\,x(t)}{dt} = A(t)\,x(t) \qquad (1)$$

form an n-dimensional linear vector space [Coddington, E.A. and Levinson N.]. An $n \times n$-matrix $Q(t)$ with linear independent solutions of (1) as columns is called fundamental matrix of (1). If $A(t)$ is θ-periodic, then $Q(t) = P(t)\, e^{tR}$ holds, where $P(t)$ is a 2θ-periodic matrix function, nonsingular for all t and R is a constant matrix. The characteristic numbers of R are called characteristic exponents of $A(t)$. Furthermore the matrix function $Q(t)\, Q^{-1}(t - \alpha)$ is θ-periodic for all $\alpha \in R^1$.

Now we consider the linear stochastic differential equation system

$$\frac{dX_t}{dt} = A(t)\, X_t + Z_t \tag{2}$$

A stochastic process X_t is called a solution of a stochastic differential equation [Bunke, H.] if almost all sample functions are solutions of the corresponding "sample differential equations" (with sample functions in place of Z_t).

Lemma 2: Let be

1. $A(t)$ is a continuous θ-periodic matrix function on R^1.
2. Z_t, $t \in R^1$, is a θ-periodic stochastic process with continuous sample functions.
3. $\int_{-\infty}^{0} E\left\{\| e^{-\tau R} P^{-1}(\tau)\, Z_\tau \|\right\} d\tau < \infty$.

Then $W_t^o = (Y_t^o, Z_t)$ (and in consequence of this also Y_t^o) is a θ-periodic process, where Y_t^o is defined by

$$Y_t^o = P(t) \int_{-\infty}^{t} e^{(t-\tau)R}\, P^{-1}(\tau)\, Z_\tau\, d\tau .$$

If moreover $A(t)$ is a constant matrix ($A(t) \equiv A$) and Z_t is stationary then W_t^o is a stationary process.

Remark: If 1. and 2. of Lemma 2 hold then the following assumption is sufficient for 3.:

All characteristic exponents λ_k of the matrix function A(t) have negative real parts ($\max_k \operatorname{Re} \lambda_k < -\varrho < 0$) and $\int_0^\Theta E\{\|Z_\tau\|\} d\tau < \infty$ holds.

After this preliminaries we consider the following differential equation

$$\frac{dX_t}{dt} = f(X_t,t,Z_t) + g(X_t,t,Z_t) . \tag{3}$$

We prove the following result:

<u>Theorem 1</u>: Assume that:

1. f(x,t,z) and g(x,t,z) are n-dimensional continuous vector functions on $R^n \times R^1 \times R^m$, which are Θ-periodic on R^1 for any $(x,z) \in R^n \times R^m$ and with $f(o,t,z) = 0$, $(t,z) \in R^1 \times R^m$.
2. Z_t, $t \in R^1$ is a Θ-periodic m-dimensional stochastic process with continuous sample functions.
3. There holds a.s.

$$\|g(x,t,Z_t)\| \leq \vartheta(t,Z_t) \; ; \; (x,t) \in R^n \times R^1$$

   with $\int_0^\Theta E\{\vartheta(\tau,Z_\tau)\} d\tau = K < \infty$

   and

$$\|g(x_1,t,Z_t) - g(x_2,t,Z_t)\| \leq \beta \|x_1 - x_2\|$$
$$(x_1,x_2,t) \in R^n \times R^n \times R^1 ,$$

   where β is sufficiently small.
4. There is a continuous Θ-periodic matrix function $A_o(t)$ with characteristic exponents, which have negative real parts ($\max_k \lambda_k < -\varrho < 0$) and for which

$$\|f(x_1,t,Z_t) - f(x_2,t,Z_t) - A_o(t)(x_1 - x_2)\|$$
$$\leq \varkappa \|x_1 - x_2\| , \; (t,x_1,x_2) \in R^1 \times R^n \times R^n$$

holds a.s. with sufficiently small $\varkappa$.

Then there exists a solution X_t^o of (3), where (X_t^o, Z_t) is a Θ-periodic stochastic process and any solution X_t of (3), tends a.s., exponentially to X_t^o (there is an $\varepsilon > 0$ with:

$$\lim_{t\to\infty} \| X_t - X_t^o \| e^{\varepsilon t} = 0 \text{ a.s.})$$

<u>Proof</u>: Let be $Q_o(t)$ the fundamental matrix of $\frac{dx}{dt} = A_o(t)\, x$ with $Q_o(0) = I$. It holds $Q_o(t) = P_o(t)\, e^{tR_o}$ with a $2\,\Theta$-periodic matrix $P_o(t)$ and with a constant matrix R_o. The real parts of the characteristic numbers of R_o are smaller than $-\varrho$. Let X_{1t}^o denote the Θ-periodic solution of the differential equation

$$\frac{dX_{1t}^o}{dt} = A_o(t)\, X_{1t}^o + g(0, t, Z_t) \tag{4}$$

The existence of this solution follows from lemma 1 and 2. Furthermore holds

$$\int_0^\theta E\{\| X_{1\tau}^o \|\}\, d\tau \le \int_0^\theta \| P_o(\tau)\| \int_0^\infty E\{\| e^{\alpha R_o}\, P^{-1}(\tau-\alpha)\, g(0,\tau-\alpha, Z_{\tau-\alpha})\, d\alpha\, d\tau$$

$$\le K\, c \int_0^\theta \| P_o(\tau)\| e^{\varrho\tau}\, d\tau \sum_{k=o}^{\infty} (e^{-2\varrho\theta})^k < \infty\,, \tag{5}$$

and (X_{1t}^o, Z_t) is θ-periodic. Now we consider the differential equation

$$\begin{aligned} \frac{dX_{nt}^o}{dt} &= A_o(t)\, X_{nt}^o + f(X_{n-1t}^o, t, Z_t) \\ &\quad - A_o(t)\, X_{n-1t}^o + g(X_{n-1t}^o, t, Z_t)\,. \end{aligned} \tag{6}$$

Under the assumption that (X_{n-1}^o, Z_t) is a Θ-periodic process with $\int_0^\theta E\{\| X_{n-1\,\tau}^o \|\}\, d\tau < \infty$ it follows from the remark to lemma 2 that also (6) has a solution with Θ-periodic (X_{nt}^o, Z_t) in consequence of the Θ -periodicity of

$$V_t = f(X^o_{n-1t}, t, Z_t) - A_o(t)\, X^o_{n-1t} + g(X^o_{n-1t}, t, Z_t)$$

(Lemma 1) and of

$$\int_0^\Theta E\{\|V_\tau\|\}\, d\tau \leq \varkappa \int_0^\Theta E\{\|X^o_{n-1\tau}\|\}\, d\tau + \int_0^\Theta E\{\varphi(\tau, Z_\tau)\}\, d\tau < \infty .$$

The inequality $\int_0^\theta E\{\|X^o_{n\tau}\|\, d\tau < \infty$ can be proved in the same way as in the case n = 1.

By induction the (X^o_{nt}, Z_t) n = 1,2,... form a sequence of θ -periodic processes. From (6) we obtain in consequence of the assumptions 3 and 4 a.s.

$$\begin{aligned}\|X^o_{nt} - X^o_{n-1t}\| &= \Big\| \int_{-\infty}^t Q_o(t)\, Q_o^{-1}(\tau)\, f(X^o_{n-1\tau}, Z(\omega)) - f(X^o_{n-2\tau}, \tau, Z(\omega)) \\ &\qquad - A_o(\tau)[X^o_{n-1\tau} - X^o_{n-2\tau}] \\ &\qquad + g(X^o_{n-1\tau}, \tau, Z_\tau(\omega)) - g(X^o_{n-2\tau}, \tau, Z_\tau(\omega))\, d\tau \Big\| \\ &\leq (\varkappa + \beta) \int_{-\infty}^t \|Q_o(t)\, Q_o^{-1}(\tau)\|\, \|X^o_{n-1\tau} - X^o_{n-2\tau}\|\, d\tau \\ &\leq \gamma(\varkappa + \beta) \int_{-\infty}^t e^{-\rho(t-\tau)} \|X^o_{n-1\tau} - X^o_{n-2\tau}\|\, d\tau, \quad t \in R^1,\end{aligned}$$

where the constant γ is chosen such that

$$\|Q_o(t)\, Q_o^{-1}(\tau)\| \leq \gamma\, e^{-\rho(t-\tau)}, \quad t, \tau \in R^1, \ \tau \leq t$$

holds. At almost all $\omega \in \Omega$ and for any finite interval $T \subset R^1$ the following inequality holds:

$$\|X^o_{nt}(\omega) - X^o_{n-1t}(\omega)\| \leq \sup_{t \in T} \|X^o_{1t}(\omega)\|\, ((\varkappa + \beta)\, \gamma\, \rho^{-1})^{n-1} < \infty \qquad (7)$$

and for $\varkappa + \beta < \rho\, \gamma^{-1}$ the sequence of $X^o_{nt}(\omega)$ tends for $n \to \infty$ uniformly in T to $X^o_t(\omega)$:

$$X_t^o(\omega) = \lim_{n\to\infty} \{ X_{no}^o(\omega) + \int_o^t [f(X_{n-1\tau}^o(\omega), \tau, Z_\tau(\omega) + $$

$$+ A_o(\tau)[X_{n\tau}^o(\omega) - X_{n-1\tau}^o(\omega)] + g(X_{n-1\tau}^o, \tau, Z_\tau)] d\tau$$

$$= X_o(\omega) + \int_{t_o}^t \{ \lim_{n\to\infty} [f(X_{n-1\tau}^o(\omega), \tau, Z_\tau(\omega)) +$$

$$+ A_o(\tau) [X_{n\tau}^o(\omega) - X_{n-1\tau}^o(\omega)]$$

$$+ g(X_{n-1\tau}^o, \tau, Z_\tau(\omega))]\} d\tau$$

$$= X_o(\omega) + \int_o^t [f(X_\tau^o(\omega), \tau, Z_\tau(\omega)) + g(X_\tau^o(\omega), \tau, Z_\tau(\omega))] d\tau,$$

$$t \in T.$$

Thus, for almost all $\omega \in \Omega$ there is a function $X_t^o(\omega) = \lim_{n\to\infty} X_{nt}^o(\omega)$, defined on R^1, which is a solution of the differential equation

$$\dot{x} = f(x, t, Z_t(\omega)) + g(x, t, Z_t(\omega)).$$

It follows,that there is a stochastic process X_t^o with $\lim_{n\to\infty} X_{nt}^o = X_t^o$ $(t \in R^1)$ a.s., which is a solution of (3). In consequence of the Θ-periodicity of (X_{nt}^o, Z_t) $n = 1,2,\ldots$ also $(X_t^o, Z_t) = \lim_{n\to\infty} (X_{nt}^o, Z_t)$ $(t \in R^1)$ a.s. is a Θ-periodic process. Let be X_t an arbitrary solution of (3). Then we have:

$$\frac{d[X_t - X_t^o]}{dt} = A_o(t)[X_t - X_t^o] +$$

$$+ f(X_t, t, Z_t) - f(X_t^o, t, Z_t) - A_o(t)[X_t - X_t^o] \qquad (8)$$

$$+ g(X_t, t, Z_t) - g(X_t^o, t, Z_t), \quad (t \in R^1) \text{ a.s.}$$

From (8) we obtain

$$\| X_t - X_t^o \| \le \| X_o - X_o^o \| \gamma e^{-\varrho t} + \gamma(\varkappa + \beta) \int_o^t e^{-\varrho(t-\tau)} \| X_\tau - X_\tau^o \| d\tau,$$

$$(t \in R^1) \text{ a.s.} \qquad (9)$$

and from this

$$\|X_t - X_t^o\| \leq \|X_o - X_o^o\| \gamma \exp[(\varkappa+\beta)\gamma - \varrho)t \quad , (t \in R^1) \text{ a.s.} \tag{10}$$

This implies the exponential convergence

$$\lim_{t\to\infty} \| X_t - X_t^o \| e^{\varepsilon t} = 0, \text{ a.s., for } \varkappa + \beta < \gamma^{-1} (\varrho - \varepsilon).$$

Now let us consider the differential equation

$$\frac{dX_t}{dt} = f(X_t, Z_t) + g(X_t, Z_t). \tag{11}$$

The following theorem is a simple corollary to theorem 1.

Theorem 2: Let be

1. $f(x,z)$ and $g(x,z)$ are n-dimensional continuous functions defined on $R^n \times R^m$ with $f(o,z) = o$, $z \in R^m$.
2. Z_t, $t \in R^1$ is a stationary m-dimensional stochastic process with continuous sample functions.
3. It holds $\|g(x,Z_t)\| \leq \varphi(Z_t)$ $(t \in R^1)$ a.s. with $E\{\varphi(Z_t)\} < \infty$ and $\|g(x_1,Z_t) - g(x_2,Z_t)\| \leq \beta \|x_1 - x_2\|$ $(x_1, x_2, t) \in R^n \times R^n \times R^1$ a.s., where β is sufficiently small.
4. There is a real matrix A_o, with characteristic numbers which have negative real parts $(\max_k \mathrm{Re}\,\lambda_k < -\varrho \leq 0)$ such that holds
$$\| f(x_1,Z_t) - f(x_2,Z_t) - A_o(x_1 - x_2) \| \leq \varkappa \| x_1 - x_2 \| \quad (x_1, x_2, t) \in R^n \times R^n \times R^1 \text{ a.s.}$$
with sufficiently small $\varkappa$.

Then a solution X_t^o of (11) exists on R^1 such that (X_t^o, Z_t) is stationary and any solution of (11) tends a.s. exponentially to X_t^o.

Example: We consider the differential equation

$$\ddot{Y}_t + (a + Z_{1t}) \dot{Y}_t + (b + Z_{2t}) Y_t = \sigma(Y_t, \dot{Y}_t) Z_{3t} + Z_{4t} \tag{12}$$

$Z_t = (Z_{1t},\ldots,Z_{4t})$ is a stationary process, $a > 0$, $4b - a^2 > 0$ and

$E\{|Z_{4t}|\} < \infty$. σ is a real function continuous and bounded on R^2 with

$$|\sigma(x_1, y_1) - \sigma(x_2, y_2)| \leq \mu[|x_1 - x_2|^2 + |y_1 - y_2|^2]^{\frac{1}{2}} .$$

From (12) written as a system of differential equation for

$$X_t^T = (Y_t, \dot{Y}_t)$$

it follows

$$\| Q_0(t)\, Q_0^{-1}(\tau) \| \leq \gamma\, \exp^{\frac{1}{2}}[a(t - \tau)]$$

with $\gamma = (\alpha + \frac{2}{\lambda})\,(1 + \frac{1}{2}(a + \lambda))$,

$$\lambda = \sqrt{4b - a^2} \text{ and } \alpha = |\sin \operatorname{arc\,tg} \lambda a^{-1}|^{-1} .$$

From theorem 2 and from the proof of theorem 1 we get the following result:

If

$$|Z_{1t}| + |Z_{2t}| + \mu|Z_{3t}| < \gamma^{-1} (\tfrac{a}{2} - \varepsilon) \; (t \in R^1) \text{ a.s.}$$

holds, then there is a solution Y_t^o of (12) such that (Y_t, Y_t^o, Z_t) is stationary, and for any solution Y_t of (12) and some $\varepsilon > 0$ holds:

$$\lim_{t \to \infty} e^{\varepsilon t}[|Y_t - Y_t^o| + |\dot{Y}_t - \dot{Y}_t^o|] = 0, \quad a.s.$$

References

Bunke, H. Ordinary differential equations with random parameters (in German) Akademie-Verlag, Berlin 1972

Coddington, E.A., Levinson N. Theory of Ordinary Differential Equations. McGraw-Hill, New York 1955

Dorogevcev, A.Ja. Some remarks about differential equations, which are perturbated by a periodic stochastic process (in Russian), Ukrain. Mat. Z., 14, 119 - 128 (1962).

STABILITY OF MECHANICAL SYSTEMS UNDER STOCHASTIC PARAMETRIC EXCITATION

S.T. Ariaratnam

Solid Mechanics Division, University of Waterloo, Waterloo, Ontario, Canada

INTRODUCTION

This paper presents a survey of some results in the stochastic stability of a class of mechanical systems which, under static loads, lose their stability by a 'bifurcation' of their equilibrium configurations. The distinguishing feature of such systems is that the external loads appear in the governing equations of motion in the form of a coefficient or parameter. Thus, when the loads are dynamic, the instability of these systems is referred to as 'parametric instability'.

Parametric instability under deterministic periodic excitation has been extensively investigated both theoretically and experimentally and several important instability phenomena have been established. A corresponding investigation when the excitation is stochastic has recently become necessary. This is due to the fact that there now exist several situations where the exciting forces in mechanical systems cannot be satisfactorily described in deterministic form. Some examples of such excitations are forces generated by jet and rocket devices in modern aircraft and missile structures, and excitations due to wind gusts and earthquakes. These forces fluctuate in a random manner over a wide band of frequencies and have to be regarded as stochastic processes. Furthermore, even when the excitation is principally deterministic, it may be more realistic to investigate the stability of the system by subjecting it to a further random perturbation.

In this survey an attempt is made to present in a systematic manner some results in the stability of coupled linear oscillatory systems under stochastic parametric excitation. A comparison is made with corresponding results of deterministic theory when the excitation is a harmonic time function. To aid in this comparison, a brief review of the principal results of deterministic theory is first presented.

STABILITY UNDER HARMONIC EXCITATION

We first consider systems that are described by equations of motion of the form

$$\ddot{q}_i + \omega_i^2 q_i + \varepsilon \cos \omega t \sum_{j=1}^{n} k_{ij} q_j = 0, \quad i = 1,2,\ldots,n, \tag{1}$$

where the q_i are the generalised normal coordinates and the ω_i denote the natural frequencies of the system. These equations describe exactly the parametrically excited motion of a class of discrete mechanical systems with n degrees of freedom about the equilibrium configuration $q_i = 0$. They also describe approximately the motion of certain continuous systems since the solution of the governing partial differential equations can often be reduced to that of a finite number of ordinary differential equations by suitable discretisation techniques.

The stability of the trivial solution $q_i = 0$ has been extensively investigated. Some results together with application to the stability of several elastic systems are found in the text by Bolotin (1964). A survey of some of the more recent results have been given by Mettler (1965, 1967). For a given system, the instability conditions define certain regions in the ε-ω plane which correspond to instability of the equilibrium configuration q = 0, Fig. 1.

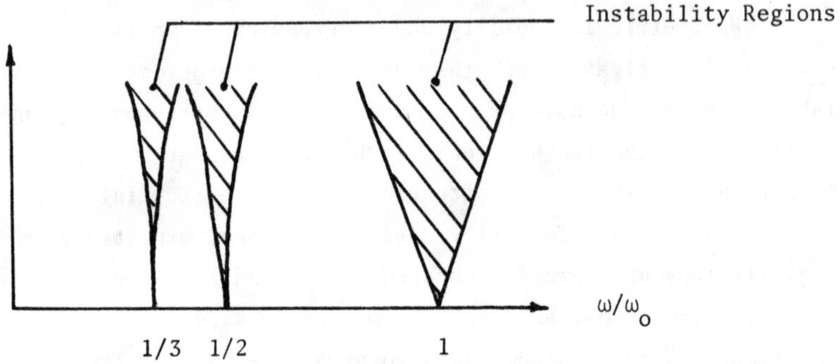

Fig. 1

These regions have peaks on the ω-axis at the discrete points $\omega = \omega_o/p$ where p is a positive integer determining the <u>order</u> of the instability and ω_o depends on the nature of the coupling coefficients k_{ij}. For $\omega_o = 2\omega_i$, the regions are referred to as instability regions of the <u>first kind</u>. Instability regions of the <u>second kind</u> are obtained for

$$\omega_o = \omega_i + \omega_j \quad (i \neq j) \quad \text{if} \quad k_{ij}k_{ji} > 0$$

$$= |\omega_i - \omega_j| \quad (i \neq j) \quad \text{if} \quad k_{ij}k_{ji} < 0$$

The instabilities corresponding $\omega_o = |\omega_i \pm \omega_j|$ are generally referred to as <u>combination resonances</u>. Common examples of mechanical systems exhibiting instability of the first kind are a simple pendulum whose support is given a vertical sinusoidal oscillation, and an elastic column subjected to a harmonically varying axial thrust. The thin elastic strip in lateral bending and torsional vibrations under a transverse harmonic load having constant direction, Fig.2, is an example of a two degree of freedom system that can show combination resonance corresponding to $\omega_o = \omega_1+\omega_2$, while the same load acting in a 'follower' fashion as in

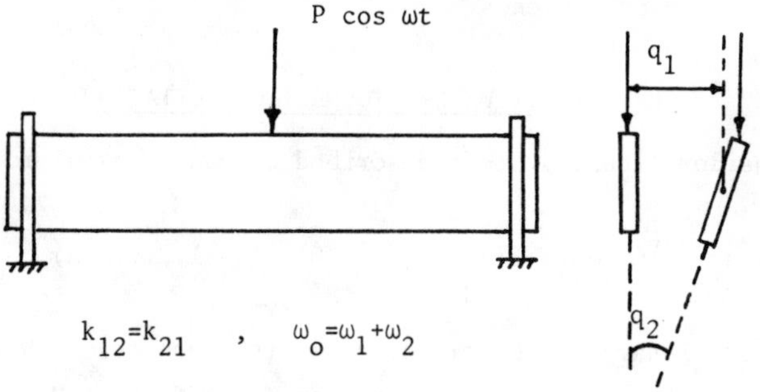

Fig. 2

Fig.3 can cause combination resonance for $\omega_o = \omega_1-\omega_2$.

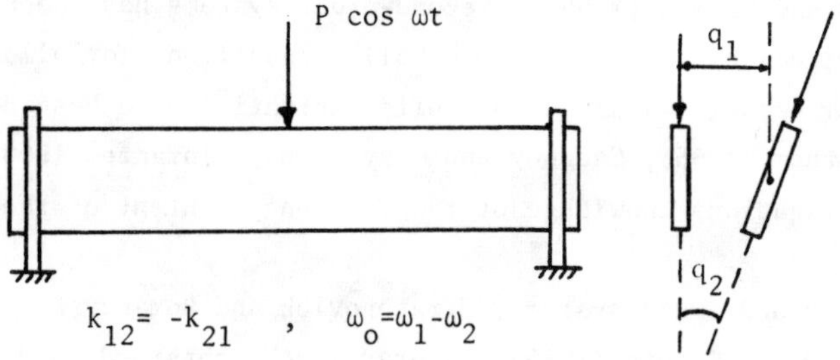

Fig. 3

The presence of a small amount of dissipative damping does not alter the general nature of these results except that the peaks in Fig.1 are raised slightly above the ω-axis, and, depending on the ratios between the damping coefficients, the instability regions of the second kind (combination resonance) may be made wider or narrower than in the undamped case.

In gyroscopic systems the results are different. The equations of motion then take the form

$$\ddot{q}_i + \omega_i^2 q_i + \sum_{j=1}^{n} c_{ij}\dot{q}_j + \varepsilon \cos \omega t \sum_{j=1}^{n} k_{ij} q_j = 0, \quad i = 1,2,\ldots,n, \tag{2}$$

where the matrix (c_{ij}) is antisymmetric, i.e. $c_{ij} = -c_{ji}$. If $\bar{\omega}_i$ denote the eigenvalues of the system (2) with the time-dependent terms left out, instabilities of the second kind occur now for both $\omega_o = \bar{\omega}_i+\bar{\omega}_j$, and $\omega_o = |\bar{\omega}_i-\bar{\omega}_j|$ even if the matrix (k_{ij}) is symmetrical. A typical example is provided by the tranverse vibration of a rotating shaft of unsymmetrical section subjected to a pulsating axial thrust or couple at its ends.

STABILITY UNDER STOCHASTIC EXCITATION

We consider linear systems described by equations of motion of the form

$$\ddot{q}_i + \omega^2 q_i + \sum_{j=1}^{n} d_{ij}\dot{q}_j + f(t)\sum_{j=1}^{n} k_{ij}q_j = 0, \quad i = 1,2,\ldots,n, \tag{3}$$

where f(t) is a stationary stochastic process with zero mean. When f(t) is a Gaussian white noise process, the phase vector $(q,\dot{q})$ forms a Markov process. The moment stability of the solutions can be investigated either through the associated Fokker-Planck equation or by the use of Itô's differential rule to set up differential equations governing the response moments. Such studies have been carried out for particular systems by Bogdanoff and Kozin (1962), Caughey and Dienes (1962), and Ariaratnam and Graefe (1965). Second-order systems have been investigated using other methods by Samuels (1960, 1961). Conditions for almost sure asymptotic stability under ergodic Gaussian non-white excitation have been derived by Kozin (1963), Ariaratnam (1965), Caughey and Gray (1965), Infante (1968) and Man (1970). In these investigations knowledge of the frequency content of the excitation has not been utilized.

For second-order systems, Stratonovich and Romanovskii (1958), Weidenhammer (1964), Graefe (1966) and Gray (1967) obtained conditions which showed that stability depended on the excitation spectral density at twice the system natural frequency, a result analogous to that for the deterministic Mathieu equation. Corresponding results for oscillatory multi-degree of freedom systems without gyroscopic terms were reported by Ariaratnam (1967, 1969). Here we shall briefly outline the method of obtaining such results for both non-gyroscopic and gyroscopic systems.

Non-Gyroscopic Systems

The systems considered are given by the equations of motion

$$\ddot{q}_i + \omega_i^2 q_i + \epsilon\sum_{j=1}^{n} d_{ij}\dot{q}_j + f(t)\sum_{j=1}^{n} k_{ij}q_j = 0, \quad i = 1,2,\ldots,n. \tag{4}$$

The damping is assumed to be small and dissipative, i.e. $\varepsilon << 1$, $d_{ii} > 0$ for every value of i. The systems described by Eqs. (4) are then oscillatory and it is meaningful to speak of the amplitude and phase of the response. To investigate the stability, Eqs. (4) are first converted to standard form by means of the transformation

$$q_i = a_i \cos \Phi_i, \; \dot{q}_i = -a_i\omega_i \sin \Phi_i, \; \Phi_i = \omega_i t + \phi_i, \; i = 1,2,\ldots,n.$$

This procedure yields the set of 2n first order equations:

$$\dot{a}_i\omega_i = -\varepsilon \sum_{j=1}^{n} d_{ij}a_j\omega_j \sin \Phi_i \sin \Phi_j + f(t) \sum_{j=1}^{n} k_{ij}a_j \sin \Phi_i \cos \Phi_j \qquad 5(a)$$

$$a_i\omega_i\dot{\phi}_i = -\varepsilon \sum_{j=1}^{n} d_{ij}a_j\omega_j \cos \Phi_j \sin \Phi_j + f(t) \sum_{j=1}^{n} k_{ij}a_j \cos \Phi_i \cos \Phi_j \qquad 5(b)$$

It is now assumed that the excitation f(t) is a stationary random process with zero mean and spectral density $\varepsilon S(\omega)$. For small values of ε, asymptotic solutions of Eqs. (5) may be obtained by employing the method of averaging. A special form of this method applicable to stochastic differential equations has been developed by Stratonovich (1963). In this method, the terms not involving the random excitation f(t) are first averaged according to the usual rule, namely

$$\{F(a,\phi,t)\}_{av} = \lim_{T\to\infty} \frac{1}{T} \int_0^T F(a,\phi,t)dt,$$

where the integration is performed over explicit t. The terms containing f(t) are next replaced by the sum of the mean value and the fluctuation about the mean. The procedure for this is found in Stratonovich (1963). After these steps are performed and retaining only terms of the first-order in ε, Eq. 5(a) becomes

$$\dot{u}_i = -\varepsilon d_{ii}u_i + m_i(u) + f_i(t) \qquad (6)$$

where $\quad u_i = a_i^2, \quad m_i = \sum m_{ij}, \quad f_i = \sum f_{ij},$

$$m_{ij} = \frac{k_{ij}}{\omega_i} E[a_i a_j f(t) \sin \Phi_i \sin \Phi_j]$$

$$= \frac{\varepsilon}{4\omega_i} \{\frac{k_{ij}^2}{\omega_i} u_j S_{ij}^{+} + \frac{k_{ij}k_{ji}}{\omega_j} u_i S_{ij}^{-}\},$$

$$f_{ij} = \frac{k_{ij}}{\omega_i} a_i a_j f(t) \sin \Phi_i \cos \Phi_i - m_{ij},$$

$$S_{ij}^{\pm} = S(\omega_i+\omega_j) \pm S(\omega_i-\omega_j).$$

The processes $f_i(t)$ have zero mean values. Now, the relaxation time of the $u_i(t)$ processes is $O(1/\varepsilon)$. Hence, if the correlation time τ_c of the f(t) process is much smaller than this relaxation time, to a first approximation, the processes $f_i(t)$ may be replaced by equivalent delta-correlated processes $f_i^*(t)$ whose correlation functions are given by

$$E[f_i^*(t)f_j^*(t+\tau)] = \delta(\tau)\int_{-\infty}^{\infty} E[f_i(t)f_j(t+\tau)]d\tau.$$

Then, Eq.(6) may be replaced by the Itô equation

$$\dot{u}_i = -\varepsilon \sum_{j=1}^{n} A_{ij}u_j + f_i^*(t) \tag{7}$$

where

$$\begin{aligned} A_{ii} &= d_{ii} - \frac{k_{ii}^2}{2\omega_i^2} S(2\omega_i) - \sum_{\substack{j=1 \\ j\neq i}}^{n} \frac{k_{ij}k_{ji}}{4\omega_i\omega_j} S_{ij}^- \\ A_{ij} &= -\frac{k_{ij}^2}{4\omega_i^2} S_{ij}^+, \quad i \neq j. \end{aligned} \tag{8}$$

The equation governing the mean square amplitudes $\bar{u}_i = E[a_i^2]$ may be set up by taking the expectation of both sides of Eq.(7):

$$\dot{\bar{u}} + \varepsilon A\bar{u} = 0 \tag{9}$$

where $u = \{u_1, u_2, \ldots, u_n\}$ and A is the $n \times n$ matrix whose elements are given by (8). The mean square amplitudes are bounded as $t \to \infty$ if and only if the eigenvalues of A have all positive real parts. The conditions for this may be obtained using the Routh-Hurwitz criterion. It is worth noting that only those values of the excitation spectrum at the frequencies $2\omega_i$, $\omega_i \pm \omega_j$ will appear in the stability conditions. Also, the coupling of the damping terms does not influence the stability in the first approximation. Since $q_i = a_i \cos \Phi_i$, $|q_i^2| \leq a_i^2$ and these conditions therefore guarantee stability in the mean of the response.

For a single degree of freedom system, the condition for mean square stability reduces to

$$d_{ii} > \frac{k_{ii}}{2\omega_1^2} S(2\omega_1), \tag{10}$$

a result obtained independently by Stratonovich and Romanovskii (1958) and Graefe (1966). For particular forms of $S(\omega)$, similar results were also obtained by Chelpanov (1962) and Weidenhammer (1964).

Suppose $S(\omega)$ vanishes outside the bandwidth $\omega_o - \frac{1}{2}\Delta\omega_o < \omega < \omega_o + \frac{1}{2}\Delta\omega_o$. The correlation time of such a process is $O(1/\Delta\omega_o)$ so that if $\Delta\omega_o >> \varepsilon$, the Markov approximation made previously would still be valid. Taking $\omega_o = \omega_r + \omega_s$ $(r \neq s)$, the elements of the matrix A reduce to

$$A_{ii} = d_{ii} \qquad i \neq r,s$$

$$A_{ij} = 0 \qquad i,j \neq r,s$$

$$A_{rr} = d_{rr} - \frac{k_{rs}k_{sr}}{4\omega_r\omega_s} S(\omega_r+\omega_s)$$

$$A_{ss} = d_{ss} - \frac{k_{rs}k_{sr}}{4\omega_r\omega_s} S(\omega_r+\omega_s)$$

$$A_{rs} = - \frac{k_{rs}^2}{4\omega_r^2} S(\omega_r+\omega_s)$$

$$A_{sr} = - \frac{k_{sr}^2}{4\omega_s^2} S(\omega_r+\omega_s)$$

from which the condition for mean square stability is obtained as

$$\frac{d_{rr}d_{ss}}{d_{rr}+d_{ss}} > \frac{k_{rs}k_{sr}}{4\omega_r\omega_s} S(\omega_r+\omega_s) \tag{11}$$

Thus no instability occurs if $k_{rs}k_{sr} < 0$.

Similarly, if $\omega_o = |\omega_r-\omega_s|$, the corresponding condition is

$$\frac{d_{rr}d_{ss}}{d_{rr}+d_{ss}} > - \frac{k_{rs}k_{sr}}{4\omega_r\omega_s} S(\omega_r-\omega_s). \tag{12}$$

Again, there is no instability if $k_{rs}k_{sr} > 0$. The qualitative similarity between these conditions and those for harmonic excitation is worth noting.

Stability with probability one may be investigated by the Liapunov method as discussed in Kushner (1967). For the system defined by Eq. (7), the scalar function

$$V(a) = \sum_1^n a_i^2$$

satisfies the requirements of a Liapunov function. According to the theorem on stochastic Liapunov functions, if $\mathcal{L}\{V(a)\} \leq 0$ then $a \to [a:\mathcal{L}\{V(a)\} = 0]$ w.p.1., where $\mathcal{L}$ is the differential generator of the process $a(t)$. After performing the necessary differentiation, we obtain

$$\mathcal{L}[V(a)] = -\varepsilon \sum_{i=1}^{n} \sum_{j=1}^{n} A_{ij} a_j^2$$

where the coefficients A_{ij} are as defined by (8). Hence, sufficient conditions for asymptotic stability w.p.1 are

$$\sum_i A_{ij} > 0, \quad j = 1,2,\ldots,n,$$

that is,

$$d_{ii} - \frac{k_{ii}^2}{2\omega_i^2} S(2\omega_i) - \sum_{\substack{j=1 \\ j \neq i}}^{n} \frac{k_{ij}}{4\omega_i} \left(\frac{k_{ij}}{\omega_i} S_{ij}^+ + \frac{k_{ji}}{\omega_j} S_{ij}^- \right) > 0, \quad i = 1,2,\ldots,n. \tag{13}$$

For a single degree of freedom system, the result reduces to the condition (10) for stability in the mean. For $\omega_o = \omega_r \pm \omega_s$, (13) leads to the pair of conditions

$$\begin{aligned} d_{rr} - \frac{k_{rs}}{4\omega_r} \left(\frac{k_{rs}}{\omega_r} \pm \frac{k_{sr}}{\omega_s} \right) S(\omega_r \pm \omega_s) > 0 \\ d_{ss} - \frac{k_{sr}}{4\omega_s} \left(\frac{k_{sr}}{\omega_s} \pm \frac{k_{rs}}{\omega_r} \right) S(\omega_r \pm \omega_s) > 0 \end{aligned} \tag{14}$$

Gyroscopic Systems

The class of systems considered has equations of motion of the form

$$\begin{aligned} \ddot{q}_1 + \varepsilon\beta\dot{q}_1 - 2\Omega\dot{q}_2 + (\omega_1^2 - \Omega^2) q_1 + f(t) q_1 = 0 \\ \ddot{q}_2 + \varepsilon\beta\dot{q}_2 + 2\Omega\dot{q}_1 + (\omega_2^2 - \Omega^2) q_2 + f(t) q_2 = 0 \end{aligned} \tag{15}$$

where $f(t)$ is a stationary stochastic process with zero mean. These equations represent the transverse flexural motion of a light elastic rotating shaft of unequal flexural rigidities carrying a single rigid mass at its mid-length and subjected to a randomly varying axial thrust, see for e.g. Dimentberg (1961). The coordinates q_1, q_2 denote the transverse displacement of the mass measured with respect to rotating principal axes. Ω is the angular velocity of the shaft and ω_1, ω_2, $(\omega_1 < \omega_2)$ are the natural frequencies of transverse vibration; β represents the coefficient of internal damping, Fig.4.

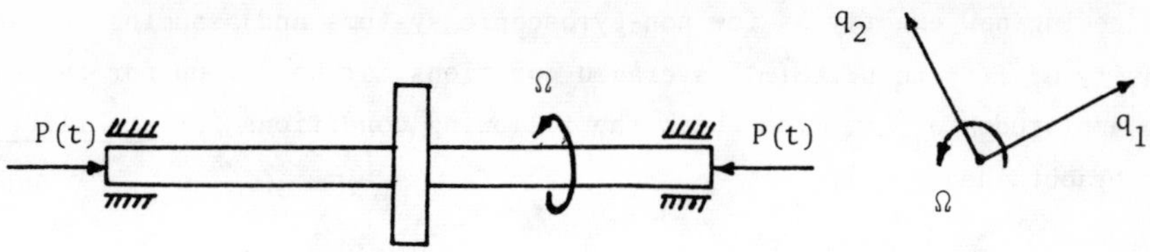

Fig. 4

The eigenfrequencies $\overline{\omega}_1$, $\overline{\omega}_2$ of the system (15) for $\varepsilon = 0$, $f(t) = 0$ are given by the roots of the equation

$$\overline{\omega}^4 - (\omega_1^2+\omega_2^2+2\Omega^2)\overline{\omega}^2 + (\omega_1^2-\Omega^2)(\omega_2^2-\Omega^2) = 0.$$

For real $\overline{\omega}$, Ω should be outside the range

$$\omega_1 \leq \Omega \leq \omega_2$$

Considering only these values of Ω (for which the unloaded shaft is stable), Eqs. (15) may be converted to standard form by the transformation

$$q_1 = a_1 \sin \Phi_1 + a_2 \sin \Phi_2$$

$$q_2 = \alpha_1 a_1 \cos \Phi_1 + \alpha_2 a_2 \cos \Phi_2$$

$$\dot{q}_1 = a_1 \omega_1 \cos \Phi_1 + a_2 \omega_2 \cos \Phi_2$$

$$\dot{q}_2 = -\alpha_1 a_1 \overline{\omega}_1 \sin \Phi_1 - \alpha_2 a_2 \overline{\omega}_2 \sin \Phi_2,$$

where

$$\Phi_1 = \overline{\omega}_1 t+\phi_1, \quad \Phi_2 = \overline{\omega}_2 t+\phi_2,$$

$$\alpha_1 = \frac{\overline{\omega}_1^2 - (\omega_1^2-\Omega^2)}{2\Omega\overline{\omega}_1} = \frac{2\Omega\overline{\omega}_1}{\overline{\omega}_1^2 - (\omega_2^2-\Omega^2)} .$$

$$\alpha_2 = \frac{\overline{\omega}_2^2 - (\omega_1^2-\Omega^2)}{2\Omega\overline{\omega}_2} = \frac{2\Omega\overline{\omega}_2}{\overline{\omega}_2^2 - (\omega_2^2-\Omega^2)} .$$

Proceeding now exactly as for non-gyroscopic systems and assuming the spectral density of f(t) to be $\varepsilon S(\omega)$, averaged equations may be set up for the squares of the amplitudes a_1, a_2 from which the following conditions for stability in the mean may be obtained:

For $\omega_o = 2\bar{\omega}_i$ (i = 1,2):

$$\beta > \frac{(\omega_1^2-\omega_2^2)^2}{16\Omega^2\bar{\omega}_i^2} S(2\bar{\omega}_i) \quad \text{if } \Omega < \omega_1$$

$$\beta > \frac{(\omega_1^2-\omega_2^2)^2}{2(\bar{\omega}_1^2-\bar{\omega}_2^2+4\Omega^2)\bar{\omega}_i^2} S(2\bar{\omega}_i) \quad \text{if } \Omega > \omega_2.$$

For $\omega_o = \bar{\omega}_1+\bar{\omega}_2$:

$$\beta > \frac{(\bar{\omega}_1\bar{\omega}_2+\omega_1^2-\Omega^2)^2}{4(\omega_1^2-\Omega^2)\bar{\omega}_1\bar{\omega}_2} S(\bar{\omega}_1+\bar{\omega}_2) \quad \text{if } \Omega < \omega_1$$

$$\beta > \frac{8(\bar{\omega}_1\bar{\omega}_2+\omega_1^2-\Omega^2)^2\Omega^4}{(\Omega^2-\omega_1^2)[16\Omega^4-(\bar{\omega}_1^2-\bar{\omega}_2^2)^2]\bar{\omega}_1\bar{\omega}_2} S(\bar{\omega}_1+\bar{\omega}_2) \quad \text{if } \Omega > \omega_2$$

For $\omega_o = \bar{\omega}_1-\bar{\omega}_2$:

$$\beta > \frac{8(\bar{\omega}_1\bar{\omega}_2-\omega_1^2+\Omega^2)^2\Omega^4}{(\omega_1^2-\Omega^2)[16\Omega^4-(\bar{\omega}_1^2-\bar{\omega}_2^2)^2]\bar{\omega}_1\bar{\omega}_2} S(\bar{\omega}_1-\bar{\omega}_2) \quad \text{if } \Omega < \omega_1$$

$$\beta > \frac{(\bar{\omega}_1\bar{\omega}_2-\omega_1^2+\Omega^2)^2}{4(\Omega^2-\omega_1^2)\bar{\omega}_1\bar{\omega}_2} S(\bar{\omega}_1-\bar{\omega}_2) \quad \text{if } \Omega > \omega_2.$$

Sufficient conditions for asymptotic stability w.p.1 may also be obtained as before using the Liapunov function $V(a) = a_1^2+a_2^2$ but are not presented here. Again, the qualitative similarity between the stochastic and the deterministic results may be noted. However, while in the case of sinusoidal excitation, instability near $\omega_o = \bar{\omega}_1+\bar{\omega}_2$ is present only when $\Omega < \omega_1$ and that near $\omega_o=\bar{\omega}_1-\bar{\omega}_2$ only when $\Omega>\omega_2$, see Mettler (1967), in the stochastic case there is no such distinction; both forms of instability can occur in both ranges of values of Ω.

CONCLUSIONS

This survey has dealt with the stability of coupled linear oscillatory systems under stochastic excitation of small intensity. The systems considered are typically encountered in the study of the dynamic stability of elastic structural and mechanical systems subjected to randomly fluctuating loads. Certain similarities between the stability conditions for the case of stochastic excitation and those for deterministic sinusoidal excitation have been emphasized.

The results presented here are obtained using the method of averaging which is valid to the first approximation only and hence correspond only to instabilities of the first order. To obtain conditions corresponding to higher order instabilities, higher approximations must be made in the solution of the governing equations in standard form. It may then be possible to investigate the influence, if any, of the values of the excitation spectrum at higher multiples and fractions of the natural frequencies and combination frequencies. Evidence that the spectrum at these frequencies may influence the stability is provided by some recent results of Bolotin (1971) for a single degree of freedom system excited parametrically by filtered Gaussian white noise. For such a system, differential equations governing the response moments were derived by Bogdanoff and Kozin (1962). Using a truncated linearised form of these equations, Bolotin has obtained approximate conditions for second moment stability and shown them graphically on a plot similar to the Strutt diagram for the Mathieu equation for various values of the damping parameter. The stability boundaries reveal a dip when the centre frequency of the excitation is close to the system natural frequency, in addition to the one near twice the natural frequency. While these results do indicate a qualitative trend, the validity of the approximations made in obtaining them seem somewhat drastic and questionable. It appears that Wedig (1972) has also obtained similar results in his paper at this Symposium.

Non-linear oscillatory systems under harmonic parametric excitation possess several complicated and interesting instability phenomena, Mettler (1965, 1967). A corresponding general investigation of such systems under stochastic excitation appears to be lacking.

ACKNOWLEDGMENTS

The research for this paper was supported (in part) by the National Research Council of Canada under Grant No. A-1815. The assistance of Mr. D.S.F. Tam and Mr. S.C. Oen in the preparation of the paper is gratefully acknowledged.

REFERENCES

Ariaratnam, S.T. and Graefe, P.W.U., 1965, Int. J. Control, 2, 161.

Ariaratnam, S.T., 1965, Proc. Int. Conf. on Dynamic Stability of Structures, Ed: G. Herrmann, Pergamon, 1966, 255.

Ariaratnam, S.T., 1967, Proc. Can. Cong. Appl. Mech., Vol. III, 163.

Ariaratnam, S.T., 1969, Proc. IUTAM Symp. on Instability of Continuous Systems, Ed: H. Leipholz, Springer-Verlag, 78.

Bogdanoff, J.L. and Kozin, F., 1962, J. Acous. Soc. Am., 34, 1063.

Bolotin, V.V., 1964, The Dynamic Stability of Elastic Systems, Holden-Day, Inc.

Bolotin, V.V., 1971, Study No. 6, Stability, Ed: H. Leipholz, Solid Mechanics Division, University of Waterloo, 385.

Caughey, T.K. and Dienes, J.K., 1962, J. Math. Phys., 40-41, 300.

Caughey, T.K. and Gray, A.H. Jr.,1965, J. Appl. Mech., 32, 365.

Chelpanov, I.B., 1962, PMM, 26, 762, English Translation, 1145.

Dimentberg, F., 1961, Flexural Vibration of Rotating Shafts, Butterworths.

Graefe, P.W.U., 1966, Ing. Arch., 35, 202.

Gray, A.H. Jr., 1967, J. Appl. Mech., 34, 1017.

Infante, E.T., 1968, J. Appl. Mech., 35.

Kozin, F., 1963, J. Math. Phys., 42, 59.

Kushner, H.J., 1967, Stochastic Stability and Control, Academic Press.

Man, F.T., 1970, J. Appl. Mech., 37, 541.

Mettler, E., 1965, Proc. Int. Conf. on Dynamic Stability of Structures, Ed: G. Herrmann, Pergamon, 1966, 169.

Mettler, E., 1967, Proc. Fourth Conf. on Nonlinear Oscillations, Prague, Academia, 1968; 51.

Samuels, J.C., 1960, J. Acoust. Soc. Am., 32, 594.

Samuels, J.C., 1961, J. Acoust. Soc. Am., 33, 1782.

Stratonovich, R.L., 1963, Topics in the Theory of Random Noise, Vol. I, Gordon and Breach.

Stratonovich, R.L., and Romanovskii, Yu.M., 1958, Nauchnye doklady vysshei shkoly, Fiziko-mat. nauk., 3, 221, Reprinted in Non-Linear Transformations of Random Processes, Ed: Kuznetsov, P.I., R.L. Stratonovich and V.I. Tikhonov, Pergamon, 1965, 332.

Wedig, W., 1972, Proc. IUTAM Symposium on Stability of Stochastic Dynamical Systems, Springer-Verlag.

Weidenhammer, F., 1964, Ing. Arch., 33, 404.

WAVES IN A ROTATING STRATIFIED FLUID WITH LATERALLY VARYING RANDOM INHOMOGENEITIES

By Lawrence A. Mysak

Department of Mathematics and Institute of Oceanography
University of British Columbia
Vancouver, B.C., Canada

ABSTRACT

We discuss the propagation and stability of internal waves in a rotating stratified unbounded fluid with randomly varying buoyancy frequency, N. The first order smoothing approximation is used to derive the mean wave dispersion relation when N is of the form $N^2 = N_0^2(1+\varepsilon\mu)$, where N_0 = constant, $0<\varepsilon^2<<1$ and μ is a centered stationary random function of the horizontal direction x. This form for μ represents a stochastic model of the lateral variations in the temperature and salinity microstructure in the ocean. From the complex dispersion relation, expressions are obtained for the phase speed change and spatial growth rate (§ 2); in particular, attention is focused on the asymptotic behaviour of these expressions for short and long correlation lengths (§ 3).

1. INTRODUCTION

It is now well established from observations that on the gross temperature and salinity depth profiles in the ocean, there are superimposed small step-like variations (e.g., see Gregg and Cox, 1972, and their references). That is, below the surface mixed layer it is apparent that in many regions of the ocean, the density stratification consists of thin, sharply-defined homogeneous layers with thicknesses varying from a few centimeters to several meters. Consequently, observed depth profiles of the Brunt-Väisälä or buoyancy frequency, which involves the vertical density gradient and which is of fundamental importance in the theory of internal waves, exhibit highly irregular fluctuations about their mean values (e.g., see Gregg and Cox, 1972; McGorman and Mysak, 1972, hereafter referrred to as I).

Although it is not yet fully understood how this so-called 'fine-structure' or 'microstructure' arises in practice, it is believed that both convective and diffusive processes are involved in its formation (Gregg and Cox, 1972; J.S. Turner and C.F. Chen, private communication). One very plausible mechanism is the so called salt finger instability which is caused by the different rates of diffusivity of heat and salt (Turner and Stommel, 1964). If a warm salty body of water overlies a cold, fresh and initially heavier body of water, then, because of the rapid diffusion of heat downward, the system becomes gravitationally unstable and long narrow convection cells (salt fingers) are formed. Under certain conditions the salt fingers become self limiting and a fairly regular step-like structure in the temperature and salinity profiles is formed. Fine-structure suggestive of salt fingering has been reported, for example, west of the Strait of Gibraltar in the depth range 1200-1500 m, which is just below the warm salty core of the Mediterranean water (Tait and Howe, 1968). The inverse situation, in which warm salty water overlies lighter, cold fresh water, also appears to lead to layering. In this case sharp interfaces separating layers in turbulent convective motion are formed. Such layering has been observed recently in the Arctic ocean in the depth range 200-500 m, which is just above the warm saline core of the Atlantic Intermediate Water (Neshyba et al, 1971). In an attempt to understand this inverse situation, Drs. Chen and Turner have recently shown in the laboratory that horizontal gradients in the diffusing components can lead to a fairly well defined (though changeable) layer structure.

A particularly interesting feature of the Chen-Turner experiments was the existence of horizontal variations in the microstructure. This is to say, the thickness of a particular layer varied considerably in the horizontal direction and often two layers merged into one. Similar remarks also apply to the horizontal behaviour of microstructure observations recently made in the Pacific off Vancouver Island (Nasmyth, 1973). Thus, in view of the considerable observational, experimental and theoretical interest in internal waves (see Turner, 1973), it appears worthwhile at this time to determine the effect of the horizontal variability in the microstructure on the propagation and stability of internal waves. Below, we shall discuss this problem via a study of the internal wave dispersion relation that is implied by a stochastic wave equation. The starting point of our analysis is the assumption that in the internal wave equation, the buoyancy frequency, N, has the form

$$N^2 = N_0^2 \left(1 + \varepsilon \mu(x)\right) \tag{1.1}$$

where $N_0 =$ constant, $0 < \varepsilon^2 \ll 1$, x is a horizontal coordinate and μ is a centered stationary random function of order unity. Eq. (1.1) represents the simplist possible model of the horizontal variability in the microstructure; we adopt a stochastic model since the observed horizontal variablity is rather irregular in character. It should be mentioned here, however, that this model is rather naive in that it does not include any basic velocity field associated with the convective motions; but work is in hand on this aspect of the problem.

Finally, we note here that this paper represents a continuation of the work initiated in I, where the effect of random depth or time dependent variations in N^2 on the dispersion of internal waves was discussed. In I the appropriate statistics associated with observed depth dependent fluctuations were incorporated into the theory, and it was found that the changes in both the phase speed and growth rate due to these fluctuations could be significant. Thus an important next step is the application of the results obtained below to ocean regions where the appropriate data is available.

2. DISPERSION RELATION

For a uniformly rotating, inviscid and incompressible fluid in $\mathbb{R}^3$ with axis of rotation in the z (vertically upward) direction and in which the motion is independent of one horizontal coordinate, the stream function, $\phi(x, z, t)$, of the perturbation velocity field satisfies the equation (see I)

$$\{\Delta \partial_t^2 + N^2 \partial_x^2 + f^2 \partial_z^2 - (N^2/g)(\partial_{ztt}^3 + f^2 \partial_z)\} \phi = 0 \tag{2.1}$$

where $\Delta = \partial_x^2 + \partial_z^2$, $N^2 = -g\rho_{oz}/\rho_o$, g is the acceleration of gravity, f is the Coriolis parameter, and ρ_o is the mean stable density field. For $N = N_o =$ constant, (2.1) admits plane wave solutions of the form

$$\phi = e^{dz/2 + i(kx + \ell z - \sigma t)} \tag{2.2}$$

with dispersion relation

$$\omega^2 k^2 - \ell^2 = d^2/4 \tag{2.3}$$

where $d = N_o^2/g > 0$ and $\omega^2 = (N_o^2 - \sigma^2)/(\sigma^2 - f^2) > 0$. The latter inequality holds for the frequency passbands $f^2 < \sigma^2 < N_o^2$ (PBI) and $N_o^2 < \sigma^2 < f^2$ (PBII); PBI is the usual oceanic situation. Since $\rho_o \propto e^{-dz}$ for $N = N_o$ and the wave kinetic energy is proportional to $\rho_o A^2$ ($A =$ wave amplitude), the factor $e^{dz/2}$ in (2.2) ensures that the kinetic energy is conserved. In the subsequent analysis we do not set $d = 0$ (the Boussinesq approximation), so that (2.2), which we shall call the deterministic solution, implies amplification for $\ell > 0$ (upward propagating wave phase) and attenuation for $\ell < 0$ (downward propagating wave phase).

In terms of the polar coordinates K_o, θ where $(k, \ell) = K_o(\cos\theta, \sin\theta)$, eq. (2.3) takes the form

$$K_o = d/2(\omega^2 \cos^2\theta - \sin^2\theta)^{1/2} \tag{2.4}$$

where $\theta \in (-\theta_A, \theta_A) \cup (\pi - \theta_A, \pi + \theta_A)$ and $0 \leq \theta_A = \tan^{-1}\omega \leq \pi/2$. The deterministic phase velocity thus has magnitude $c_o = \sigma/K_o$ and the group velocity, which is along the direction of energy propagation, is given by

$$\begin{aligned} \underset{\sim}{c}_{og} &\equiv (\sigma_k, \sigma_\ell) \\ &= k^2(N^2 - f^2)(\omega^2 k, -\ell)/\sigma(K_o^2 + d^2/4)^2. \end{aligned} \tag{2.5}$$

It is easy to show that the group velocity is always orthogonal to the local tangent of the $\sigma =$ constant curves as given by (2.3); for relatively short waves $(K_o \gg d/2\omega)$, $\underset{\sim}{c}_o$ and $\underset{\sim}{c}_{og}$ are essentially orthogonal.

For N^2 of the form (1.1), eq. (2.1) becomes a stochastic differential equation for $\psi(x,z)$, where $\phi(x,z,t) = \psi(x,z)e^{-i\sigma t}$ $(\sigma > 0)$:

$$(\mathcal{L} + M)\psi = 0 \tag{2.6}$$

where

$$\left.\begin{aligned} \mathcal{L} &= \partial_z^2 - \omega^2 \partial_x^2 - d\,\partial_z \\ M &= -\varepsilon\mu[N_0^2(\sigma^2 - f^2)^{-1}\partial_x^2 + d\,\partial_z] \end{aligned}\right\} \tag{2.7}$$

are respectively deterministic and stochastic differential operators, and $\langle M \rangle = 0$, where $\langle \cdot \rangle$ denotes ensemble average.

According to Keller and Veronis (1969), (2.6) admits plane wave solutions of the form $\langle \psi \rangle = e^{i(kx+\ell z)}$ provided that the following dispersion relation, which is correct to $O(\varepsilon^2)$, is satisfied:

$$D_0 + D_1 = 0 \tag{2.8}$$

where

$$D_0 = e^{-q}\mathcal{L}\, e^{q}$$

$$D_1 = -e^{-q}\langle M\mathcal{L}^{-1}M\rangle e^{q}$$

$$q = i(kx + \ell z)$$

and $\mathcal{L}^{-1}$ is the integral operator

$$\mathcal{L}^{-1} f(x,z) = \iint_{-\infty}^{\infty} G(x-x', z-z') f(x', z')\, dx'\, dz' \tag{2.9}$$

in which G is the following Green's function:

$$\mathcal{L}\, G(x,z) = \delta(x)\,\delta(z). \tag{2.10}$$

From (2.10) it is easy to show that

$$\left.\begin{aligned} \hat{G}(x,\ell) &\equiv \int_{-\infty}^{\infty} G(x,z)\, e^{-i\ell z}\, dz \\ &= (i/2\omega a)\, e^{ia|x|/\omega} \end{aligned}\right. \tag{2.11}$$

where $a = [\ell(\ell + id)]^{1/2}$, with positive square root implied. Eq. (2.8) represents the so called first order smoothing approximation (Frisch, 1968) or binary collision approximation (Howe, 1971) for the mean wave dispersion relation.

Using (2.9) and (2.11) in (2.8) and simplifying, we obtain

$$\left.\begin{aligned} &\omega^2 k^2 - a^2 \\ &+ \varepsilon^2 B\left\{\frac{N_0^2\Gamma(0)}{N_0^2 - \sigma^2} - \frac{i}{\omega a}\left[i\ell d - \frac{N_0^2 a^2}{N_0^2 - \sigma^2}\right]\int_0^{\infty}\Gamma(x)\cos kx\; e^{iax/\omega}\, dx\right\} = 0 \end{aligned}\right\} \tag{2.12}$$

where $B=-N_0^2k^2(\sigma^2-f^2)^{-1}+ild$ and $\Gamma(x)=\Gamma(-x)=\langle\mu(x')\mu(x'+x)\rangle$

is the autocovariance function of the fluctuations $\mu(x)$.

To analyze (2.12) we first introduce polar coordinates K,θ such that

$$k = K\cos\theta \;, \quad l = K\sin\theta - id/2$$

where θ, the angle of phase propagation, is real and K may be complex. Then, for given $\sigma, f, N_0, d, \Gamma$ and θ, (2.12) reduces to an equation for K which can be solved by iteration. On setting $\varepsilon=0$, one obtains $K=K_0$, the deterministic solution (2.4); then, upon substituting $K=K_0$ into the $O(\varepsilon^2)$ terms in (2.12) (now written in terms of K,θ), one obtains the first iteration solution, which in turn can be separated into real and imaginary parts to give

$$\frac{c_0/c-1}{\varepsilon^2/d^2} = AP - \frac{2Adl_0}{\omega^2|k_0|}\int_0^\infty \Gamma(x)(1+\cos 2k_0x)\,dx + \frac{A^2-d^2l_0^2}{\omega^2 k_0}\int_0^\infty \Gamma(x)\sin 2k_0x\,dx \tag{2.13}$$

$$\frac{-\mathrm{Im}\{K\}/K_0}{\varepsilon^2 d^2} = dl_0P + \frac{A^2-d^2l_0^2}{\omega^2|k_0|}\int_0^\infty \Gamma(x)(1+\cos 2k_0x)\,dx + \frac{2Adl_0}{\omega^2 k_0}\int_0^\infty \Gamma(x)\sin 2k_0x\,dx \tag{2.14}$$

where $c=\sigma/\mathrm{Re}\{K\}$, $A=N_0^2k_0^2/(\sigma^2-f^2)-d^2/2$, $P=2N_0^2\Gamma(0)/(N_0^2-\sigma^2)$ and $(k_0,l_0)=K_0(\cos\theta,\sin\theta)$. Eqs. (2.13) and (2.14) represent respectively the first correction to the relative change in phase speed and the spatial growth rate of the mean wave field due to the random fluctuations in N^2. If the right hand sides of (2.13) and (2.14) are both positive, for example, the mean wave travels slower and is amplified. It is interesting to note that both (2.13) and (2.14) are even functions of the horizontal wave number component, k_0, which is consistent with the lateral symmetry in the problem.

3. ASYMPTOTIC SOLUTIONS

Let L denote the correlation length of the process $\mu(x)$, i.e. a cut-off length such that for $|x|>L$, $\Gamma(x)$ is essentially zero. We now consider (2.13) and (2.14) in the limiting cases $K_0L\ll 1$ and $K_0L\gg 1$, corresponding respectively to correlation lengths that are short and long compared to the deterministic wavelength.

In (2.13) and (2.14) let $x=Lx'$ and set $\beta=K_0L$; also we assume that the integrals

$$\int_0^\infty \Gamma(Lx')x'^n\,dx' \simeq \int_0^1 \Gamma(Lx')x'^n\,dx' \;, \quad n=0,1,2,\ldots \tag{3.1}$$

are of order unity. Then, using the fact that $AL/d \sim \beta$ and assumption (3.1), it follows that for $\beta \ll 1$, short correlation lengths,

$$\left.\begin{aligned} c_0/c - 1 &= \varepsilon^2 AP/d^2 + O(\varepsilon^2\beta) \\ -\mathrm{Im}\{K\}/K_0 &= \varepsilon^2 P\ell_0/d + O(\varepsilon^2\beta) \end{aligned}\right\} \quad (3.2)$$

where it is assumed that $P = O(1)$. For $\beta \gg 1$, long correlation lengths, it follows that

$$\left.\begin{aligned} c_0/c - 1 &= -\varepsilon^2 A\ell_0\Gamma_0/\omega^2 d|k_0| + O(\varepsilon^2) \\ -\mathrm{Im}\{K\}/K_0 &= \varepsilon^2(A^2/d^2 - \ell_0^2)\Gamma_0/\omega^2|k_0| + O(\varepsilon^2) \end{aligned}\right\} \quad (3.3)$$

where $\Gamma_0 = \int_{-\infty}^{\infty}\Gamma(x)dx > 0$ and it is assumed that $\omega^2 = O(1)$. Note that the leading terms in (3.3) are $O(\varepsilon^2\beta)$; further, the expression for $c_0/c - 1$ as given by (2.13), being a correction to the relative change in phase speed, should be somewhat less than unity. Therefore, for (3.3) to be valid, we require that $\varepsilon^2\beta \lesssim 1$, which holds provided ε^2 is sufficiently small. By contrast, we note that the leading terms in (3.2) are $O(\varepsilon^2)$ and hence that for short correlation lengths, the amplitude of the random fluctuations does not have to be infinitesimally small for the first order smoothing approximation to be valid. These comments on the asymptotic validity of the first order smoothing approximation are in qualitative agreement with those made by Frisch (1968) in connection with the Helmholtz equation.

We note that $A < 0$ for PBII and that, provided $\sigma^2 \gg f^2$, $A > 0$ for PBI (See I, Figure 4). Thus, with the exception of the long correlation length growth rate, we can summarize the above asymptotic results quite simply - see Table I. For comparative purposes, the asymptotic results for the case $N^2 = N_0^2(1 + \varepsilon\mu(z))$, which was examined in I, are given in Table II.

TABLE I

Summary of results for short and long correlation lengths, as given by (3.2) and (3.3). $l_o > 0$ (<0) corresponds to upward (downward) propagating waves.

		$-Im\{K\}/K_o$ (relative growth rate)	c/c_o (relative phase speed)
$K_oL \ll 1$	PBI	$>0, \quad l_o > 0$ $<0, \quad l_o < 0$	< 1
	PBII	$<0, \quad l_o > 0$ $>0, \quad l_o < 0$	< 1
$K_cL \gg 1$	PBI	$-$	$>1, \quad l_c > 0$ $<1, \quad l_c < 0$
	PBII	$-$	$<1, \quad l_o > 0$ $>1, \quad l_o < 0$

TABLE II

Summary of asymptotic results for the case $M = M(z)$, as obtained in I. The sign of the growth rate for $K_oL \ll 1$ is independent of the passband.

		$-Im\{K\}/K_o$ (relative growth rate)	c/c_o (relative phase speed)
$K_oL \ll 1$	PBI	$\left\{ \begin{array}{l} >0, \quad l_o > 0 \\ <0, \quad l_o < 0 \end{array} \right\}$	< 1
	PBII		>1
$K_oL \gg 1$	PBI	$-$	<1
	PBII	$-$	>1

From the above tables we thus draw the following geophysically relevant conclusion: For PBI internal waves and for short correlation lengths (which is the case in practice for $\mu = \mu(z)$ (see I) and probably also the case for $\mu = \mu(x)$, although this point needs to be confirmed), vertical and lateral random fluctuations in N^2 have the same average effect on the propagation and stability of the waves. The random fluctuations make the waves travel slower and enhance the deterministic growth/decay behaviour.

Acknowledgements

This paper was partially supported by the National Research Council of Canada and was written while the author was a Senior Visitor during 1971-1972 in the Department of Applied Mathematics and Theoretical Physics, University of Cambridge.

REFERENCES

Frisch, U. Probabilistic Methods in Applied Mathematics, Vol 1. (ed. A.T. Bharucha-Reid), Academic Press, New York (1968)

Gregg, M. C., and Cox, C. S. Deep-Sea Research, 19, 355-376 (1972)

Keller, J. B., and Veronis, G. J. Geophys. Research, 74, 1941-1951 (1969)

McGorman, R. E., and Mysak, L. A. Geophys. Fluid Dynamics, in press (1972)

Nasmyth, P. In proceedings of 'Fourth Colloquium on Ocean Hydrodynamics', Liège University, in press (1973)

Neshyba, S., Neal, V. T., and Denner, W. J. Geophys. Research, 76, 8107-8120 (1971)

Tait, R. I., and Howe, M. R. Deep-Sea Research, 15, 275-280 (1968)

Turner, J. S. Buoyancy Effects in Fluids, Cambridge University Press, in press (1973)

Turner, J. S., and Stommel, H. Proc. Nat. Acad. Sci. USA, 52, 49-53 (1964)

THE STABILITY OF A SATELLITE WITH PARAMETRIC EXCITATION BY THE FLUCTUATIONS OF THE GEOMAGNETIC FIELD

Peter S. Sagirow
University of Stuttgart, Germany

The motion of a satellite stabilized with respect to the geomagnetic field by a satellite fixed magnetic rod is considered. The stability domains corresponding to different linearization models and to different definitions of the stochastic integral are compared.

1. Satellite in the Geomagnetic Field

Satellites with magnetic elements are influenced by the geomagnetic field in two different ways [1]. The stabilizing magnetic rod interacts with the field directly causing a restoring moment proportional to the intensity H of the field. A second moment is caused by the interaction between the field of the earth and the eddy currents induced in the shell of the satellite by the field. As these currents are proportional to H the second moment is proportional to H^2. It is a damping moment depending on the angular velocity of the satellite with respect to the geomagnetic field.

In the most simple case of a satellite in a circular equatorial orbit the yaw oscillations of the satellite caused by the moments mentioned above are described by equation [6]

(1) $$B\Psi'' + kH^2\Psi' + HI\sin\Psi = 0 \quad .$$

Here, Ψ is the yaw angle, B is the moment of inertia with respect to the yaw axis, k is a positive constant depending on the magnetic properties of the shell, and I is the moment of the magnetic rod fixed in the pitch axis of the satellite. The intensity H can be described by

(2) $$H = H_0\,(1+\eta)$$

where H_0 = const and η denotes the fluctuations of the geomagnetic field. Transforming the time and introducing the constant

(3) $$c = k(H_0)^{3/2}/(BI)^{1/2}$$

we obtain the normed equation

(4) $$\ddot{\Psi} + c(1+\eta)^2\,\dot{\Psi} + (1+\eta)\sin\Psi = 0$$

where the dot denotes the derivative with respect to the transformed time.

2. Linearization of the Nonlinear Noise Term

Suppose the fluctuations $\eta(t)$ to be Gaussian with $E\eta = 0$, $E\eta^2 = \sigma^2$. The nonlinear term $v = (1+\eta)^2 = 1+2\eta+\eta^2$ can be linearized in different ways. Assuming η to be small we can neglect η^2 and replace v by $v_1 = 1+2\eta$. A more sophisticated method is the statistical linearization proposed by Kazakov [3]. We replace v by $v_i = \alpha_i + \beta_i\eta$ and determine the constants α_i, β_i either by the requirements

$$Ev_2 = Ev \quad , \quad E(v_2 - Ev_2)^2 = E(v - Ev)^2$$

or by the requirement

$$E(v-v_3)^2 \stackrel{!}{=} \min.$$

In the first case we obtain $\alpha_2 = 1+\sigma^2$, $\beta_2 = \sqrt{4+2\sigma^2}$ and in the second case $\alpha_3 = 1+\sigma^2$, $\beta_3 = 2$. A further possibility [3] is $\alpha_4 = 1+\sigma^2$, $\beta_4 = (\beta_2 + \beta_3)/2$. Thus, four linearized models are obtained:

(5) $$\ddot{\Psi} + c(\alpha_i + \beta_i\eta)\dot{\Psi} + (1+\eta)\sin\Psi = 0 .$$

The coefficients α_i, β_i are given below:

model	α_i	β_i
1	1	2
2	$1+\sigma^2$	$\sqrt{4+2\sigma^2}$
3	$1+\sigma^2$	2
4	$1+\sigma^2$	$1+\sqrt{1+\sigma^2/2}$

3. Introduction of the White Noise

The fluctuations $\eta(t)$ of the geomagnetic field have rapidly changing frequencies and amplitudes [2] and can be approximated by white noise

$$\eta(t) = \sigma\dot{w}(t)$$

where $w(t)$ is a normed Wiener process with $Ew(t) = 0$, $Ew^2(t) = t$.

Then, equation (5) reads as

(6) $$\ddot{\Psi} + c(\alpha_i + \beta_i\sigma\dot{w})\dot{\Psi} + (1+\sigma\dot{w})\sin\Psi = 0$$

or written correctly as a stochastic system

$$\text{(7)} \qquad dx = \begin{pmatrix} x_2 \\ -\sin x_1 - c\alpha_i x_2 \end{pmatrix} dt + \begin{pmatrix} 0 \\ -\sigma \sin x_1 - c\beta_i \sigma x_2 \end{pmatrix} dw(t)$$

where $x = (x_1, x_2)^T = (\psi, \dot{\psi})^T$ and T denotes transposition.

For small oscillations the eqs. (6) and (7) can be linearized:

$$\text{(8)} \qquad \ddot{\psi} + c(\alpha_i + \beta_i \sigma \dot{w})\dot{\psi} + (1 + \sigma\dot{w})\psi = 0$$

or

$$\text{(9)} \qquad dx = \begin{pmatrix} 0 & 1 \\ -1 & -c\alpha_i \end{pmatrix} x\,dt + \begin{pmatrix} 0 & 0 \\ -\sigma & -c\beta_i\sigma \end{pmatrix} x\,dw(t) .$$

4. Stability in the Mean Square

The mean square stability can be checked by the criterion of Nevel'son-Khasminski [5]. Interpreting eq. (8) in the sense of Itô we directly obtain the stability condition

$$\text{(10)} \qquad 2c\alpha_i > c^2\beta_i^2\sigma^2 + \sigma^2 .$$

To the linearized models derived above correspond the three explicit stability conditions

$$\sigma_{I1}^2 < 2c/(1+4c^2)$$

$$\sigma_{I2}^2 < \left[-(1-2c+4c^2) + \sqrt{(1-2c+4c^2)^2 + 16c^3}\,\right]/4c^2$$

$$\sigma_{I3}^2 < 2c/(1-2c+4c^2)$$

and in the fourth case the condition

$$c^4\sigma_{I4}^8 + 4(1-2c)c^2\sigma_{I4}^6 + 4(1-4c+8c^2-10c^3)\sigma_{I4}^4 +$$

$$+ 16c(2c-2c^2-1)\sigma_{I4}^2 + 16c^2 > 0 .$$

The index I stands for Itô. Fig. 1 shows the stability domains in the (c, σ^2)-plane.

Interpreting eq. (8) in the sense of Stratonovich [7], first we have to derive the equivalent Itô equation. This equation is

$$\text{(11)} \qquad \ddot{\psi} + (c\alpha_i - \tfrac{1}{2}\sigma^2c^2\beta_i^2 + \sigma c\beta_i\dot{w})\dot{\psi} + (1 - \tfrac{1}{2}\sigma^2c\beta_i + \sigma\dot{w})\psi = 0$$

and leads to the Nevel'son-Khasminski stability condition

$$\text{(12)} \qquad 2(c\alpha_i - \tfrac{1}{2}\sigma^2c^2\beta_i^2)(1-\tfrac{1}{2}\sigma^2c\beta_i) > c^2\beta_i^2\sigma^2(1-\tfrac{1}{2}\sigma^2c\beta_i) + \sigma^2 .$$

The stability domains corresponding to the different linearization models are shown in Fig. 2.

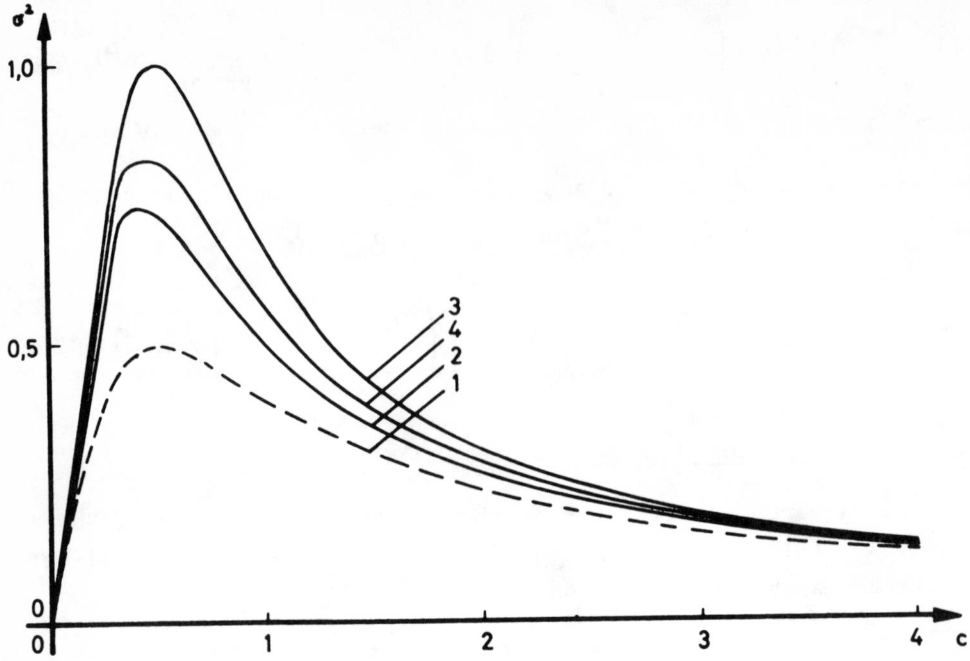

Fig. 1: Domains of stability in mean square for eq. (8) interpreted in the sense of Itô

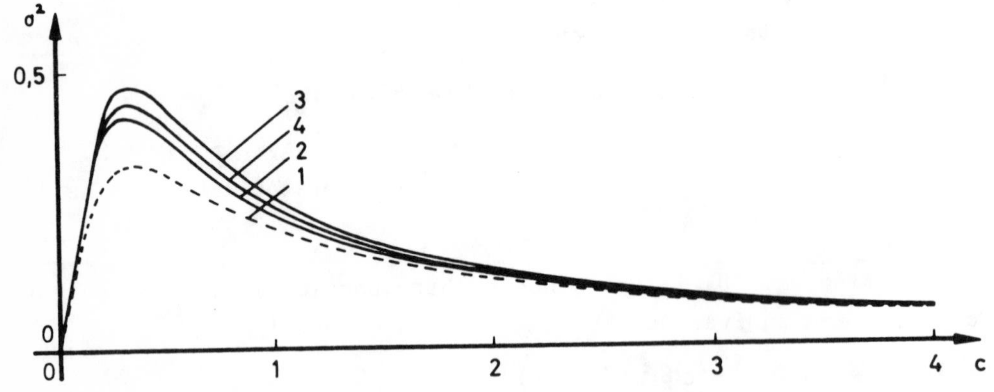

Fig. 2: Domains of stability in mean square for eq. (8) interpreted in the sense of Stratonovich

5. Stability in Probability

The necessary and sufficient condition for stability in probability can be obtained introducing the polar coordinates $r = (x_1^2 + x_2^2)^{1/2}$, $\varphi = \tan^{-1}(x_1/x_2)$ and considering the processes $\varphi = \varphi(t)$ on $r = 1$ and $\rho = \ln r(t)$. This method is proposed in [4] and leads to a stability condition in integral form. However, due to the numerical difficulties evaluating the occuring integrals this condition turns out to be non effective.

Sufficient stability conditions can be obtained by stochastic Liapunov techniques. As Liapunov function for the linear system (9) can serve the quadratic form

(13) $$v(x) = A x_1^2 + B x_1 x_2 + C x_2^2 \quad .$$

The stability condition depends on the constants A,B,C and is derived applying the L-operator to (13) and requiring the positive definiteness of the functions $v(x)$ and

(14) $$-Lv(x) = (B - C\sigma^2)x_1^2 + (Bc\alpha_i - 2A + 2C - 2C\sigma^2 c\beta_i)\, x_1 x_2 + $$
$$+ (2Cc\alpha_i - Cc^2\beta_i^2\sigma^2 - B)\, x_2^2 \quad .$$

However, the best possible choice of A,B,C leads to the old condition (10).

Considering the stability of the nonlinear system (7) and using the Liapunov function

(15) $$v(x) = Ax_1^2 + Bx_1 x_2 + Cx_2^2 + D\sin^2\frac{x_1}{2}$$

with

$$A = c\alpha_i/2 \quad , \quad B = 1 \quad , \quad C = (1 + c^2\beta_i^2)/2c\alpha_i \quad , \quad D = 4C(1 - \sigma^2 c\beta_i)$$

the stability in probability can be shown to hold for σ satisfying once more condition (10). This result follows also directly from the stability of the linearized system.

6. Conclusions

All results obtained agree qualitatively with the general insight in the considered motion. All stability bounds $\sigma^2(c)$ tend to zero for $c \to 0$ and $c \to \infty$, e.g. for $k \to 0$ (vanishing damping) and for $I \to 0$ (vanishing magnetic moment of the stabilizing rod). The maximum of $\sigma^2(c)$ which must be expected between the mentioned minima occurs for all models in the relatively small intervall $0{,}3 < c < 0{,}5$. Given B and k the value $c = 0{,}4$ can serve for the choice of the optimal magnetic moment I for the stabilizing rod.

The linearization model 1 was obtained neglecting a positive part of the damping. Thus, the curves σ^2_{I1} and σ^2_{S1} indicate sure lower bounds of the stability domains. For small and great values of c the stability bounds for the models 2,3 and 4 differ very slightly. In the neighborhood of the maxima these differences are still below 15 %.

The difference between the Itô and Stratonovich models is considerable greater. Roughly speaking the stability region of the Stratonovich equation is two times smaller than the stability region of Itô. Thus, the different definitions of the stochastic integral turn out to have a much stronger influence on the results than the linearization.

The use of conventional Liapunov functions does not enlarge the stability regions obtained by mean square techniques. The known necessary and sufficient conditions turn out to be non effective for applications.

References

[1] Beletski, V.V., Motion of an Artificial Satellite About its Centre of Mass, Jerusalem, 1966

[2] Handbook of Geophysics and Space Environments, McGraw Hill, 1965

[3] Kazakov, I.E., Automatika i Telemekhanika, XVII, Nr. 5, 1956, 385 - 409

[4] Khas'minski, R.Z., Stability of Systems of Differential Equations with Random Perturbation of Parameters, Nauka, Moscow, 1969

[5] Nevel'son, M.B., Khas'minski, R.Z., On the Stability of a Linear System with Random Perturbated Parameters, Prikl. Mat. i Mekh. 30, 2, 1966, 404 - 409

[6] Sagirov, P.S., Stochastic Methods in the Dynamics of Satellites, CISM Lecture Notes Nr. 57, Udine, 1970

[7] Stratonovich, R.L., Conditional Markov Processes and Their Application to the Theory of Optimal Control, Elsevier, New York, 1968

APPLICATION OF AVERAGING PRINCIPLE IN NONLINEAR OSCILLATORY STOCHASTIC SYSTEMS

V.G.Kolomietz

U.S.S.R., Kiev, Institute of Mathematics of Academy of Sciences of the Ukr.S.S.R.

The study of the influence of random forces on nonlinear systems is a very practical problem. The problems of this type are of great importance in many branches of science and engineering bound up with improvement of noise stability of various devices, with the stability of systems under random effects, etc.

Increased interest to stochastic systems has stimulated not only mathematical works [4], [5], [6], [10] but the investigations having nature of application [1], [7], [8], [9], etc. The effective method for investigation of random processes in nonlinear oscillatory stochastic systems is the successive application of Krylov-Bogoliubov-Mitropolsky asymptotic method of nonlinear mechanics and the method of Kolmogorov-Fokker-Plank (KFP) equations, though the latter yield to analytical solution with difficulty in large number of cases except the case of linear systems.

The application of Bogoliubov-Mitropolsky averaging principle allows us to obtain interesting and important results also for quasi-linear systems containing a small parameter. The method of KFP equations gives in this case satisfactory results if the initial equations under consideration can be reduced to the standard form [2].

Averaging may be carried out either in the standard equations themselves which then can be readily analyzed by KFP equations, or in the constructed for them KFP equation which in this case is also of standard form.

We shall consider an autonomous oscillating system with one degree of freedom which is under effect of stationary Gaussian "white noise" and is described by the following second order differential equation

$$\frac{d^2x}{dt^2} + \omega^2 x = \varepsilon f_1\left(x, \frac{dx}{dt}\right) + \sqrt{\varepsilon}\,\sigma f_2\left(x, \frac{dx}{dt}\right) \dot{\xi}(t), \qquad (1)$$

where ε is a small parameter, ω, σ are some constants, f_1 and f_2 are nonlinear functions satisfying all necessary conditions which can depend periodically upon time, $\dot{\xi}(t)$ is "white noise" process being generalized derivative of the Wiener process $\xi(t)$.

From mathematical point of view the equation (1) is quasilinear differential stochastic equation of two-dimensional diffusion Markov process.

On account of smallness of the parameter ε the application of Krylov-Bogoliubov-Mitropolsky asymptotic methods [2] is possible. While using this method it is reasonable to pass from the equation (1) to the system of the first order equations in terms of amplitude and phase of random oscillations. For this, as a rule, the change of variables will be done by

$$x = a\cos\psi, \qquad (2)$$
$$\frac{dx}{dt} = -a\omega\sin\psi$$
$$(\psi = \omega t + \theta).$$

Then, for new variables a and θ being amplitude and phase, we shall obtain the following system of stochastic differential equations

$$\frac{da}{dt} = -\frac{\varepsilon}{\omega} f_1(a\cos\psi, -a\omega\sin\psi)\sin\psi -$$
$$-\frac{\sqrt{\varepsilon}\,\sigma}{\omega} f_2(a\cos\psi, -a\omega\sin\psi)\sin\psi\, \dot{\xi}(t), \qquad (3)$$

$$\frac{d\theta}{dt} = -\frac{\varepsilon}{a\omega} f_1(a\cos\psi, -a\omega\sin\psi)\cos\psi -$$

$$-\frac{\sqrt{\varepsilon}\,\sigma}{a\omega} f_2(a\cos\psi, -a\omega\sin\psi)\cos\psi\, \dot{\xi}(t).$$

The equations obtained represent also the system of stochastic difretential equations in standard form for two-dimensional diffusion Markov process.

In the works by I.L.Berstein [3], R.L.Stratonovicz [8] and by some others, "incomplete" averaging has been carried out - only non-random terms in the right-hand side of the standard equation (3) were averaged, then fluctuation terms were averaged in the constructed for them KFP equation, that is the averaging has been done in two steps. These averagings involving two steps can be combined into one "complete" averaging if under the averaged system of stochastic differential equations, corresponding to system (3) one understands the system

$$\frac{d\bar{a}}{dt} = \overline{f_1^{(1)}}(\bar{a}) + \overline{f_2^{(1)}}(\bar{a})\,\dot{\xi}(t),$$

$$\frac{d\bar{\theta}}{dt} = \overline{f_1^{(2)}}(\bar{a}) + \overline{f_2^{(2)}}(\bar{a})\,\dot{\xi}(t), \tag{4}$$

where

$$\overline{f_1^{(1)}}(\bar{a}) = M_t\left\{-\frac{\varepsilon}{\omega} f_1(\bar{a}\cos\bar{\psi}, -\bar{a}\omega\sin\bar{\psi})\sin\bar{\psi}\right\},$$

$$\overline{f_2^{(1)}}(\bar{a}) = \sqrt{M_t\left\{\frac{\varepsilon\sigma^2}{\omega^2} f_2^2(\bar{a}\cos\bar{\psi}, -\bar{a}\omega\sin\bar{\psi})\sin^2\bar{\psi}\right\}}$$

$$\overline{f_1^{(2)}}(\bar{a}) = M_t\left\{-\frac{\varepsilon}{\bar{a}\omega} f_1(\bar{a}\cos\bar{\psi}, -\bar{a}\omega\sin\bar{\psi})\cos\bar{\psi}\right\},$$

$$\overline{f_2^{(2)}}(\bar{a}) = \sqrt{M_t\left\{\frac{\varepsilon\sigma^2}{\bar{a}^2\omega^2} f_2^2(\bar{a}\cos\bar{\psi}, -\bar{a}\omega\sin\bar{\psi})\cos^2\psi\right\}},$$

M_t is an operator of averaging in time which is explicitly presented. Under certain conditions the solution of system (3′) is reduced to that of system (4) in different senses, say, in the sense of root-mean-square convergence of stochastic variables, et al.

To the system of equation (4) the following KFP equation

$$\frac{\partial W}{\partial t} + \frac{\partial}{\partial \bar{a}}[K_{\bar{a}} W] + \frac{\partial}{\partial \bar{\theta}}[K_{\bar{\theta}} W] =$$

$$= \frac{1}{2}\left\{\frac{\partial^2}{\partial \bar{a}^2}[\mathcal{D}_{\bar{a}} W] + 2\frac{\partial^2}{\partial \bar{a}\partial\bar{\theta}}[\mathcal{D}_{\bar{a}\bar{\theta}} W] + \right. \qquad (5)$$

$$\left. + \frac{\partial^2}{\partial \bar{\theta}^2}[\mathcal{D}_{\bar{\theta}} W]\right\},$$

is put in correspondence where $W = W(\bar{a}, \bar{\theta}, t \mid \bar{a}_0, \bar{\theta}_0, t_0)$ - the joint amplitude and phase density function, K_a, K_θ - amplitude and phase transfer coefficients, $\mathcal{D}_{\bar{a}}, \mathcal{D}_{\bar{\theta}}$ - amplitude and phase diffusion coefficients respectively, $\mathcal{D}_{\bar{a}\bar{\theta}}$ - amplitude and phase mixed diffusion coefficient are of the form

$$K_{\bar{a}}(\bar{a}) = \overline{f_1^{(1)}}(\bar{a}),$$

$$K_{\bar{\theta}}(\bar{a}) = \overline{f_1^{(2)}}(\bar{a}),$$

$$\mathcal{D}_{\bar{a}}(\bar{a}) = \left[\overline{f_2^{(1)}}(\bar{a})\right]^2,$$

$$\mathcal{D}_{\bar{\theta}}(\bar{a}) = \left[\bar{f}_2^{(2)}(\bar{a})\right]^2,$$

$$\mathcal{D}_{\bar{a}\bar{\theta}}(\bar{a}) = \left[\bar{f}_2^{(1)}(\bar{a})\right]\left[\bar{f}_2^{(2)}(\bar{a})\right].$$

In the case of nonautonomous stochastic systems the resonance case and nonresonance case are distinguished. The nonresonance case is investigated similarly to the above mentioned. In the resonance case with the transfer and diffusion coefficients depending on the phase θ the investigation becomes essentially difficult.

For the analysis of oscillating systems with random effects the stationary amplitude density function is of great importance inasmuch as stationary points of this density (if they exist) correspond to stable and unstable states of the initial oscillating system depending upon maximum or minimum is achieved at every of these points.

Consider the familiar "radio engineering" system which is described by Van der Pole equation and is subjected to the "white noise"

$$\frac{d^2x}{dt^2} + \left(1 + \sqrt{\varepsilon}\,\sigma\,\dot{\xi}(t)\right)x = \varepsilon\left(1 - x^2\right)\frac{dx}{dt}. \qquad (6)$$

The averaged stable KFP equation for stationary amplitude density function $W_{st}(\bar{a})$ in this case takes the form

$$\frac{d}{d\bar{a}}\left[\frac{\varepsilon\bar{a}}{2}\left(1 - \frac{\bar{a}^2}{4}\right)W_{st}(\bar{a})\right] - \frac{1}{2}\frac{d^2}{d\bar{a}^2}\left[\frac{\varepsilon\sigma^2\bar{a}^2}{8}W_{st}(\bar{a})\right] = 0. \qquad (7)$$

Taking into account the boundary conditions $W_{st}(\bar{a}) \to 0$ and $\frac{dW_{st}(\bar{a})}{d\bar{a}} \to 0$ with $a \to \infty$, the equation (7) is readily integrated

$$W_{st}(\bar{a}) = C a^{2\left(\frac{4}{\sigma^2}-1\right)} e^{-\frac{a^2}{\sigma^2}}, \qquad (8)$$

where the constant C is found from the condition of normaliza- tion function $W_{st}(\bar{a})$. From the analysis of (8) it is evident that it has the only maximum at

$$\bar{a} = \sqrt{4 - \sigma^2} \qquad (9)$$

For $\sigma \to 0$, the amplitude of random oscillations coincides with the one of determinate oscillations of the system where random dis- turbance is absent. In the case when $\sigma = 2$, no oscillations are present. When $\sigma < 2$, stationary random oscillations are stable and possess of the amplitude (9). The algorithm quoted can be gene- ralized to investigate the stochastic oscillations in the systems with delay if the equations of oscillations can be reduced to the standard form (3).

References

1. Barret J.F., Application of Kolmogorov equations for the investigation of automatic control systems with random disturbances, Proc. 1 Intern. Congress IFAC, Ed. USSR Academy of Sciences, v.III, 84-97, 1961. In Russian.

2. Bogoliubov N.N., Mitropolsky Yu.A., Asymptotic methods in the theory of nonlinear oscillations, "Nauka", 1963. In Russian.

3. Berstein I.L., Lamp generator amplitude and phase fluctuations, Izv. USSR Academy of Sciences, ser. Phys., v. 14, N 2, 145-173, 1950. In Russian.

4. Gihman I.I., Skorohod A.V., Stochastic differential equations, "Naukova Dumka", 1968. In Russian.

5. Mitropolsky Yu.A., Averaging method in nonlinear mechanics, "Naukova Dumka", 1971. In Russian.

6. Mitropolsky Yu.A., Kolomietz V.G., Averagings in stochastic systems, UMZ., v.23, N 3, 318-345, 1971. In Russian.

7. Rytov S.M., Introduction to statical radiophysics, "Nauka", 1966, In Russian.

8. Stratonovicz R.L., Selected problems of fluctuation theory in radioengineering, "Sovetskoe Radio", 1961. In Russian.

9. Tihonov B.I., Statical radioengineering, "Sovetskoe Radio", 1966. In Russian.

10. Hasminsky R.Z., Stability of the systems of differential equations under random disturbances of their parameters, "Nauka", 1969. In Russian.

11. Crandall S.H., Some heuristic procedures for analizing random vibration of nonlinear oscillators, Proc. V International Conference on Nonlinear Oscillations, Kiev, v. 3, 21-36, 1970.

OPTIMIZATION OF MULTI-DIMENSIONAL STOCHASTIC SYSTEMS AND STABILITY OF SOLUTIONS

V.V. Solodovnikov and V.F. Biriukov

Bauman Institute U.S.S.R.

ABSTRACT

Two important cases of the problem of optimization of multidimensional stochastic systems are considered. The first case is related to the regulator problem for a stochastic system which describes a regulating nonstochastic and nonstationary device. The second one arises from problems in which some moments must have specified properties. It is shown that the above problems are among the non-well-posed-ones. Hence it is difficult to use experimental data on moments of random processes for the study of an optimization process. The method of synthesis of multi-dimensional systems is proposed and a projection method for the solutions of the exact equation with respect to the impulse response functions is obtained. Theorems on evaluation of the errors of approximate solutions are formulated.

ОПТИМИЗАЦИЯ МНОГОМЕРНЫХ СТОХАСТИЧЕСКИХ СИСТЕМ И ПРОБЛЕМА УСТОЙЧИВОСТИ РЕШЕНИЙ

В.В.Солодовников, В.Ф.Бирюков

СССР, МВТУ им. Н.Э.Баумана

Проблемам, связанным с оптимизацией многомерных стохастических систем при воздействии стохастических процессов уделяется еще недостаточно внимания несмотря на их большую актуальность. Часто задачу оптимизации многомерной стохастической системы целесообразно трактовать как задачу оптимизации многомерной линейной нестационарной системы со стохастическими воздействиями. К такой задаче сводится задача оптимизации многомерного корректирующего устройства в классе линейных нестационарных устройств,если объект стохастический, и некоторые другие аналогичные задачи, а также оптимизация многомерной стохастической системы, когда требуется, чтобы средние характеристики обеспечивали нужное качество системы. Ниже рассматриваются некоторые вопросы, естественным образом возникающие при рассмотрении подобных задач, но еще недостаточно исследованные.

Известно, что многомерную линейную систему можно описывать с помощью матричной функции-матрицы $\tilde{K}(t,\tau)$, составленной из импульсных переходных функций размером $m \times n$

$$\tilde{K}(t,\tau) = \left(K_{ij}(t,\tau)\right), \qquad i = 1,\dots,m, \; j = 1,\dots,n., \tag{I}$$

где $K_{ij}(t,\tau)$ - импульсная переходная функция между j-м входом и i-м выходом; $t_0 \leq t < \infty$, в частности $t_0 = -\infty$ либо $t_0 = 0$; $K_{ij}(t,\tau) \equiv 0$ для $t < \tau$.

Задача оптимизации такой многомерной системы по критерию минимума математического ожидания квадрата евклидовой нормы вектора ошибки приводит к необходимости решения уравнений

$$R_{z_j \varphi}(t,\tau) = \int_{t_0}^{t} R_{\varphi\varphi}(\tau,\lambda) K_j^T(t,\lambda)\, d\lambda . \tag{2}$$

$$t_0 \leq \tau \leq t \;;\quad t \in [t_0,\infty) \;;\; j = 1, 2, \dots, m.$$

Если у векторного случайного процесса размерности n на входе системы $\varphi(t) = M(t) + N(t)$ компоненты имеют нулевые математические ожидания, компоненты вектора желаемого случайного процесса $Z(t)$ на выходе системы размерности m также имеют нулевые математические ожидания, тогда в (2) $R_{\varphi\varphi}(\tau,\lambda)$ - корреляционная $n\times n$ матрица-функция процесса $\varphi(t)$;

$R_{z_j\varphi}(t,\tau)$ - взаимная корреляционная вектор-функция между j -ой компонентой $z(t)$ и $\varphi(t)$;

$K_j^T(t,\lambda)$ - j -я транспонированная строка матрицы-функции (I);

Уравнение (2) при каждом j можно рассматривать как операторное уравнение в линейном пространстве n -мерных вектор-функций двух переменных, определенных в области $\Delta=\{(t,\tau):t,\tau\in[t_0,\infty),\tau\leqslant t\}$ Однако, при каждом фиксированном $t=T$ из области допустимых значений t уравнение (2) может также рассматриваться как операторное в пространстве n -мерных вектор-функций одной переменной, определенных для $t_0\leqslant\tau\leqslant T$. Таким образом, в первом случае имеем m операторных уравнений

$$R_{z_j\varphi}(t,\tau)=A\,K_j^T(t,\tau)\quad;\quad j=1,\dots,m\,, \tag{3}$$

а во втором случае m операторных уравнений

$$R_{z_j\varphi}(T,\tau)=A_T\,K_j^T(T,\tau) \tag{4}$$

$$j=1,\dots,m\,,\qquad t_0\leqslant\tau\leqslant T\,.$$

Прежде чем переходить к исследованию и приближенному решению операторных уравнений (3), (4), отметим некоторые важные практические обстоятельства, поясняющие эти исследования.

I. Отсутствие в общем случае убедительных аналитических процедур, приводящих к построению точного решения в случае его существования в выбранном множестве функций. Среди предлагаемых ранее методов решения уравнений (3),(4) таких как: метод канонических разложений случайных функций, метод интегральных канонических представлений и различные его варианты, метод Шинброта, достаточной общностью вероятно обладает метод канонических разложений. Однако, этим методом в общем случае строится приближенное решение и остаются открытыми вопросы об оценке погрешности приближенного решения по сравнению с точным, а также о характере сходимости частичных сумм ряда к точному решению. Кроме этого, все перечисленные методы не учитывают некорректности этих уравнений.

2. Необходимость получения корреляционных матриц-функций, определяющих операторы A , A_T , а также взаимных корреляционных вектор-функций $R_{z_j\varphi}(t,\tau)$. Информация об этих функциях в настоящее время получается следующими способами. Или исследуется математическая модель случайного векторного процесса, или обрабатываются экспериментально его реализации. Эти пути неизбежно связаны с погрешностями, например, из-за неадэкватности процесса и его модели. Эти погреш-

ности могут вызвать значительные изменения в $k_j^T(t,\tau)$.

3. Необходимость учета возможностей практической реализации, получаемых в результате оптимизации операторов, учет неизменяемых частей системы.

Из сказанного следует, что алгоритмы решения уравнение (3),(4) должны быть основаны на приближенных численных методах. При построении этих алгоритмов особое внимание должно быть уделено следующим вопросам:

I) устойчивость алгоритмов;
2) адэкватность получаемой информации об оптимальном операторе в результате использования численного метода способу реализации этого оператора (например, возможность простого перехода от импульсной переходной функции к дифференциальному уравнению);
3) возможность оценки погрешности между точным и приближенным решением задачи оптимизации;
4) минимальность требований к вычислительным машинам в смысле быстродействия и памяти.

В свете сказанного, перейдем к исследованию операторных уравнений (3), (4) и построению численных алгоритмов решения таких и аналогичных уравнений, способных удовлетворять перечисленным требованиям.

Рассмотрим операторное уравнение (3) в линейном нормированном пространстве C_Δ, состоящем из непрерывных ограниченных по модулю n-мерных вектор-функций двух переменных, определенных на Δ с нормой

$$\| k_j(t,\tau)\|_{C_\Delta} = \sup\left\{\sup_\Delta \left|k_{j1}(t,\tau)\right|, \dots \sup_\Delta \left|k_{jn}(t,\tau)\right|\right\} \tag{5}$$

Пусть матрица-функция $R_{\varphi\varphi}(t,\tau)$ непрерывная квадратично суммируемая функция. Тогда оператор A действует из пространства C_Δ в пространство C_Δ. Оператор A является при этом непрерывным оператором. Оставляя в стороне вопросы существования обратного (обобщенного обратного) оператора A^{-1} для оператора A. Исследуем вопрос об устойчивости решений относительно возмущений $R_{z j\varphi}(t,\tau)$. Покажем, что если обратный оператор для оператора A существует, то он не ограничен. Для этого рассмотрим в Δ последовательность вектор-функций, сходящейся в норме (5) к нулю за исключением множества нулевой лебеговой меры, а образ этой последовательности посредством оператора A сходится к нулю в норме (5). Поскольку такие последовательности существуют, то оператор A^{-1} неограничен. Следовательно, задача решения операторного уравнения (3) в C_Δ некорректно поставлена по Адамару. Некорректность уравнений (4) при каждом j следует из теорем о вполне непрерывных операторах, поскольку при сделанных предположениях о $R_{\varphi\varphi}(t,\tau)$ и множестве функций, где отыскивается решение, оператор Δ_T является вполне непрерывным в множестве непрерывных вектор-функций.

То обстоятельство, что небольшие возмущения левых частей (3),(4) дают значительные отклонения от точного решения, а значит значительные изменения в структуре оптимального оператора, может приводить к ухудшению обработки случайных процессов системой.

Предполагая, что функциональное пространство матриц-функций (I) состоит из функций двух переменных, определенных на Δ и суммируемых с квадратом, норму элемента $K_j(t,\tau)$ можно также задать следующим образом

$$\|K_j(t,\tau)\|_{L_{2\Delta}} = \left(\sum_{i=1}^{n} \iint_{\Delta} K_{ji}^2(t,\tau)\, dt\, d\tau \right)^{\frac{1}{2}} . \tag{6}$$

Если кроме этого функции $K_{ji}(t,\tau)$ являются непрерывно дифференцируемыми функциями τ до порядка q, а получающиеся после дифференцирования функции суммируемы с квадратом на Δ, то норму можно определить, исходя из выражения

$$\|K_j(t,\tau)\|_{L_{2\Delta}^q} = \left(\sum_{i=1}^{n} \sum_{\nu=1}^{q} \iint_{\Delta} \left[\frac{\partial^\nu K_{ji}(t,\tau)}{\partial \tau^\nu} \right]^2 dt\, d\tau \right)^{\frac{1}{2}} . \tag{7}$$

При соответствующих предположениях, норму в пространстве функций для операторного уравнения (4) можно задать следующими способами:

$$\|K_j(T,\tau)\|_{L_2} = \left(\sum_{i=1}^{n} \int_{t_0}^{t=T} K_{ji}^2(T,\tau)\, d\tau \right)^{\frac{1}{2}} , \tag{8}$$

$$\|K_j(t,\tau)\|_{L_2^q} = \left(\sum_{i=1}^{n} \sum_{\nu=1}^{q} \int_{t_0}^{t=T} \left[\frac{\partial^\nu K_{ji}(T,\tau)}{\partial \tau^\nu} \right]^2 d\tau \right)^{\frac{1}{2}} . \tag{9}$$

Очевидно, что если функции, удовлетворяющие операторным уравнениям (3),(4) не удовлетворяют условиям, при которых было возможно писать нормы (6),(7),(8),(9), то эти нормы могут рассматриваться как ограничения на класс систем.

Выше была показана неограниченность обратного оператора A^{-1} для нормы, определяемой выражением (5). Аналогичное исследование может быть проведено и для норм (6),(7),(8),(9). Поскольку, приведенная задача оптимизации многомерной системы не учитывает практических возможностей реализации получаемых решений или корректирующих устройств, то для обеспечения этого учета необходимо вводить ограничение на множество оптимизируемых систем.

Устойчивость решения относительно возмущений исходных данных, реализуемость систем и корректирующих устройств можно обеспечить, используя постановку задачи оптимизации многомерной системы по принципу минимальной сложности. Формулировка этого принципа имеется в статьях [1], [2] В.Солодовникова, В.Бирюкова. Если в качестве функционала сложности взять норму, определяемую соотношением (8) при каждом t, то вместо уравнений (2) нужно будет решать уравнения

$$R_{z_j\varphi}(t,\tau)=\int_{t_0}^{t} R_{\varphi\varphi}(\tau,\lambda)\,k_j^T(t,\lambda)\,d\lambda+\mu(t)\,k_j^T(t,\tau). \qquad (10)$$
$$t_0\le\tau\le t;\quad t\in[t_0,\infty);\quad j=1,2,\dots,m.$$

В (10) $\mu(t)$ – неопределенный множитель Лагранжа, зависящий от t. $\mu(t)$ определяется путем подстановки решения уравнения (10) в критерий качества и должен обеспечивать требуемую точность системы.

Для приближенного численного решения уравнения (10) при каждом $t=T$ предлагается использовать методику, которая будет излагаться далее ради простоты обозначений для уравнения Винера-Хопфа второго рода на полупрямой

$$\theta k(t)+\int_0^{\infty} R_1(t-\tau)\,k(\tau)\,d\tau=R_2(t),\ t\in[0,\infty), \qquad (11)$$

где θ – вещественное число, $\theta>0$;

$R_1(t)$ – автокорреляционная функция входного скалярного случайного процесса системы;

$R_2(t)$ – взаимная корреляционная скалярная функция между входным случайным процессом и его желаемым преобразованием;

$k(t)$ – импульсная переходная функция одномерной системы.

Пусть $R_1(t)\in L_1(-\infty,+\infty)$, $R_2(t)\in L_2[0,\infty)$ для $t\geq 0$, у уравнения (11) существует единственное решение

$$k(t)\in L_2[0,\infty).$$

С данным уравнением связан ограниченный в L_2 самосопряженный оператор

$$Bk\equiv\int_0^{\infty} R_1(t-\tau)\,k(\tau)\,d\tau.$$

Пусть $\{\varphi_i(t)\}$, $i=1,2,\dots$ произвольная счетная система линейно-независимых функций из $L_2[0,\infty)$. Для любой функции $k(t)\in L_2[0,\infty)$ определен оператор Φ_n, действующий по правилу

$$\Phi_n[k(t)]=\sum_{i=1}^{n} c_i\,\varphi_i(t),$$

где $c_i=\int_0^{\infty} k(t)\,\varphi_i(t)\,dt=(k,\varphi_i)$ – компоненты функции $k(t)$ по системе $\{\varphi_i(t)\}$, $i=1,2,\dots$

В случае функции двух переменных $F(t,\tau)$ оператор Φ_n рассматривается действующим на $F(t,\tau)$ как функцию t, а τ при этом рассматривается как параметр,

$$\Phi_n\left[F(t,\tau)\right]=\sum_{i=1}^{n} b_i(\tau)\varphi_i(t),$$

где $b_i(\tau)=\int_0^\infty F(t,\tau)\varphi_i(t)\,dt$.

В качестве $\{\varphi_i(t)\}$, $i=1,2,\ldots$ может быть взята, например, некоторая ортогональная или ортонормированная система функций из $L_2[0,\infty)$.

Уравнению (II) ставится в соответствие уравнение

$$\theta k_n(t)+\int_0^\infty \Phi_n\left[R_1(t-\tau)\right]k_n(\tau)\,d\tau=\Phi_n\left[R_2(t)\right],\quad t\in[0,\infty). \tag{I2}$$

Так как $\Phi_n[R_2(t)]$, $\Phi_n[R_1(t-\tau)]$ представляют собой обобщенные полиномы порядка n, то решение $k_n(t)$ будет обобщенным полиномом порядка n. Коэффициенты этого полинома могут быть найдены из системы линейных алгебраических уравнений, полученных из (2) путем умножения обеих частей (2) последовательно на $\varphi_i(t)$, $i=1,2,\ldots,n$ и интегрированием на $[0,\infty)$. Система линейных алгебраических уравнений будет иметь наиболее простой вид, если система функций $\{\varphi_i(t)\}$, $i=1,2,\ldots$ ортонормированная. Пусть эта система для данного θ и данного n имеет единственное решение. Требуется построить оценки уклонения решения уравнения (I2) от решения уравнения (II) через оценки уклонения решения $k(t)$ уравнения (II), если бы оно было нам известно от $\Phi_n[k(t)]$. Последние оценки могут эффективно вычисляться при помощи аппарата теории приближений.

Имеет место <u>лемма</u>:

$$\Phi_n\left[\int_0^\infty R_1(t-\tau)k(\tau)\,d\tau\right]=\int_0^\infty \Phi_n\left[R_1(t-\tau)\right]k(\tau)\,d\tau$$

для любой $k(t)\in L_2[0,\infty)$.

Наряду с нормой в гильбертовом пространстве $L_2[0,\infty)$

$$\|f\|_{L_2}=\left(\int_0^\infty f^2(t)\,dt\right)^{\frac{1}{2}}$$

вводим норму $\|f\|_{L_{2B}}=\left(\int_0^\infty f^2(t)e^{-ct}\,dt\right)^{\frac{1}{2}},\quad c>0.$

Между квадратами норм $\|\cdot\|^2_{L_{2B}}$ и $\|\cdot\|^2_{L_2}$ имеет место неравенство

$$\|f\|^2_{L_{2B}}\leq\|f\|^2_{L_2},\qquad f\in L_2$$

Нам потребуется еще одно соотношение между этими нормами, а именно:

$$\|f\|^2_{L_2}\leq M(f)\cdot\|f\|^2_{L_{2B}},\quad f\in L_2.$$

С учетом леммы имеет место <u>теорема I</u>

Пусть в уравнении (II) оператор B является самосопряженным оператором, действующим в вещественном гильбертовом пространстве $L_2[0,\infty)$ с ядром $R_1(t)\in L_1(-\infty;+\infty)$. Далее, пусть у уравнения (II) существует резольвента $R(t,\tau)$, удовлетворяющая условию

$$\int_0^\infty e^{-ct}\left(\int_0^\infty \left|R(t,\tau)\right|^2 d\tau\right) dt = A_1 < +\infty , \qquad (13)$$

тогда для любого оператора Φ_n при замене уравнения (II) уравнением (I2) оценка погрешности приближения точного решения имеет вид

$$\left.\begin{array}{l} \left\|K(t)-K_n(t)\right\|^2_{L_{2B}} \leqslant \dfrac{2\,(1+B_2)}{1-2B_1(n)M_2}\left\|K(t)-\Phi_n\left[K(t)\right]\right\|^2_{L_2}, \\ \text{если} \quad 2B_1(n)\cdot M_2 < 1 \\ \left\|K(t)-K_n(t)\right\|^2_{L_{2B}} < +\infty , \\ \text{если} \quad 2B_1(n)\,M_2 \geqslant 1 . \end{array}\right\} \qquad (14)$$

В (I4) обозначения имеют следующий смысл:

$$B_1(n) = 4\lambda^2 C_1(n)\cdot(1+A_1M_1\lambda^2);$$

$$B_2 = 4\lambda^2 C_2(1+A_1M_1\lambda^2); \quad \lambda = -\frac{1}{\theta};$$

$$C_1(n) = \int_0^\infty \left\|\Phi_n\left[R_1(t-\tau)\right]-R_1(t-\tau)\right\|^2_{L_2} e^{-ct}\,dt;$$

$$C_2 = \int_0^\infty e^{-ct}\left\|R_1(t-\tau)\right\|^2_{L_2} dt .$$

Также имеет место <u>теорема 2</u>.

Пусть дано уравнение

$$\theta K(t) + \int_0^\infty R_1(t,\tau)\,K(\tau)\,d\tau = R_2(t), \qquad t \in [0,\infty), \qquad (15)$$

где $R_1(t,\tau)$ - функция двух переменных, не являющая функцией, зависящей от разности аргументов;

θ - вещественное число, оператор $\mathcal{D}$;

$\mathcal{D}K \equiv \int_0^\infty R_1(t,\tau)K(\tau)d\tau$ - самосопряженный, действующий в вещественном гильбертовом пространстве $L_2[0,\infty)$. Пусть ядро оператора $\mathcal{D}$ удовлетворяет условию

$$\int_0^\infty e^{-ct}\left(\int_0^\infty \left|R_1(t,\tau)\right|^2 d\tau\right) dt < +\infty$$

у уравнения (I5) существует резольвента $R(t,\tau)$, удовлетворяющая условию (I3). Тогда для любого оператора Φ_n при замене уравнения (I5) уравнением типа (I2) оценка погрешности приближения точного решения имеет вид (I4) (если заменить $R_1(t-\tau)$ на $R_1(t,\tau)$)

<u>Замечание</u>.

Аналогичные результаты имеют место, если вместо вещественных гильбертовых пространств рассматривать комплексные.

Переход от уравнений (2) к уравнениям (I0) в общем случае будет обеспечивать слабую регуляризацию (непрерывную зависимость решения (10) от вариации исходных данных в среднем). Часто необходимо обеспечивать непрерывную зависимость в более хороших метриках. Это обстоятельство иногда требует решения краевых задач для интегро-дифференциальных уравнений вместо решение интегральных уравнений второго рода. Предлагаемый метод решения и оценку погрешности можно применить и в этом случае, так как, например, можно с помощью функции Грина для краевой задачи перейти от интегро-дифференциального уравнения к интегральному уравнению второго рода.

Цитированная литература

I. Солодовников В.В., Бирюков В.Ф.
Оптимальная обработка случайных сигналов в измерительных и информационных системах при наличии помех.
Труды У Международного конгресса ИМЕКО. Франция.Версаль.I970.

2. Solodownikow, W.W., Birjukow, V.F.
Über die Anwendung von Digitalrechnern zur Stochastischen Optimierung und Identifikation automatischer Systeme.
Berlin, VEB Verlag Technik, "Messen, Steuern, Regeln", 1971, Nr. 9.

Lecture Notes in Mathematics

Comprehensive leaflet on request

Vol. 111: K. H. Mayer, Relationen zwischen charakteristischen Zahlen. III, 99 Seiten. 1969. DM 16,–

Vol. 112: Colloquium on Methods of Optimization. Edited by N. N. Moiseev. IV, 293 pages. 1970. DM 18,–

Vol. 113: R. Wille, Kongruenzklassengeometrien. III, 99 Seiten. 1970. DM 16,–

Vol. 114: H. Jacquet and R. P. Langlands, Automorphic Forms on GL (2). VII, 548 pages. 1970. DM 24,–

Vol. 115: K. H. Roggenkamp and V. Huber-Dyson, Lattices over Orders I. XIX, 290 pages. 1970. DM 18,–

Vol. 116: Sèminaire Pierre Lelong (Analyse) Année 1969. IV, 195 pages. 1970. DM 16,–

Vol. 117: Y. Meyer, Nombres de Pisot, Nombres de Salem et Analyse Harmonique. 63 pages. 1970. DM 16,–

Vol. 118: Proceedings of the 15th Scandinavian Congress, Oslo 1968. Edited by K. E. Aubert and W. Ljunggren. IV, 162 pages. 1970. DM 16,–

Vol. 119: M. Raynaud, Faisceaux amples sur les schémas en groupes et les espaces homogènes. III, 219 pages. 1970. DM 16,–

Vol. 120: D. Siefkes, Büchi's Monadic Second Order Successor Arithmetic. XII, 130 Seiten. 1970. DM 16,–

Vol. 121: H. S. Bear, Lectures on Gleason Parts. III, 47 pages. 1970. DM 16,–

Vol. 122: H. Zieschang, E. Vogt und H.-D. Coldewey, Flächen und ebene diskontinuierliche Gruppen. VIII, 203 Seiten. 1970. DM 16,–

Vol. 123: A. V. Jategaonkar, Left Principal Ideal Rings. VI, 145 pages. 1970. DM 16,–

Vol. 124: Sèminare de Probabilités IV. Edited by P. A. Meyer. IV, 282 pages. 1970. DM 20,–

Vol. 125: Symposium on Automatic Demonstration. V, 310 pages. 1970. DM 20,–

Vol. 126: P. Schapira, Théorie des Hyperfonctions. XI, 157 pages. 1970. DM 16,–

Vol. 127: I. Stewart, Lie Algebras. IV, 97 pages. 1970. DM 16,–

Vol. 128: M. Takesaki, Tomita's Theory of Modular Hilbert Algebras and its Applications. II, 123 pages. 1970. DM 16,–

Vol. 129: K. H. Hofmann, The Duality of Compact Semigroups and C*- Bigebras. XII, 142 pages. 1970. DM 16,–

Vol. 130: F. Lorenz, Quadratische Formen über Körpern. II, 77 Seiten. 1970. DM 16,–

Vol. 131: A Borel et al., Seminar on Algebraic Groups and Related Finite Groups. VII, 321 pages. 1970. DM 22,–

Vol. 132: Symposium on Optimization. III, 348 pages. 1970. DM 22,–

Vol. 133: F. Topsøe, Topology and Measure. XIV, 79 pages. 1970. DM 16,–

Vol. 134: L. Smith, Lectures on the Eilenberg-Moore Spectral Sequence. VII, 142 pages. 1970. DM 16,–

Vol. 135: W. Stoll, Value Distribution of Holomorphic Maps into Compact Complex Manifolds. II, 267 pages. 1970. DM 18,–

Vol. 136: M. Karoubi et al., Séminaire Heidelberg-Saarbrücken-Strasbuorg sur la K-Théorie. IV, 264 pages. 1970. DM 18,–

Vol. 137: Reports of the Midwest Category Seminar IV. Edited by S. MacLane. III, 139 pages. 1970. DM 16,–

Vol. 138: D. Foata et M. Schützenberger, Théorie Géométrique des Polynômes Eulériens. V, 94 pages. 1970. DM 16,–

Vol. 139: A. Badrikian, Séminaire sur les Fonctions Aléatoires Linéaires et les Mesures Cylindriques. VII, 221 pages. 1970. DM 18,–

Vol. 140: Lectures in Modern Analysis and Applications II. Edited by C. T. Taam. VI, 119 pages. 1970. DM 16,–

Vol. 141: G. Jameson, Ordered Linear Spaces. XV, 194 pages. 1970. DM 16,–

Vol. 142: K. W. Roggenkamp, Lattices over Orders II. V, 388 pages. 1970. DM 22,–

Vol. 143: K. W. Gruenberg, Cohomological Topics in Group Theory. XIV, 275 pages. 1970. DM 20,–

Vol. 144: Seminar on Differential Equations and Dynamical Systems, II. Edited by J. A. Yorke. VIII, 268 pages. 1970. DM 20,–

Vol. 145: E. J. Dubuc, Kan Extensions in Enriched Category Theory. XVI, 173 pages. 1970. DM 16,–

Vol. 146: A. B. Altman and S. Kleiman, Introduction to Grothendieck Duality Theory. II, 192 pages. 1970. DM 18,–

Vol. 147: D. E. Dobbs, Cech Cohomological Dimensions for Commutative Rings. VI, 176 pages. 1970. DM 16,–

Vol. 148: R. Azencott, Espaces de Poisson des Groupes Localement Compacts. IX, 141 pages. 1970. DM 16,–

Vol. 149: R. G. Swan and E. G. Evans, K-Theory of Finite Groups and Orders. IV, 237 pages. 1970. DM 20,–

Vol. 150: Heyer, Dualität lokalkompakter Gruppen. XIII, 372 Seiten. 1970. DM 20,–

Vol. 151: M. Demazure et A. Grothendieck, Schémas en Groupes I. (SGA 3). XV, 562 pages. 1970. DM 24,–

Vol. 152: M. Demazure et A. Grothendieck, Schémas en Groupes II. (SGA 3). IX, 654 pages. 1970. DM 24,–

Vol. 153: M. Demazure et A. Grothendieck, Schémas en Groupes III. (SGA 3). VIII, 529 pages. 1970. DM 24,–

Vol. 154: A. Lascoux et M. Berger, Variétés Kähleriennes Compactes. VII, 83 pages. 1970. DM 16,–

Vol. 155: Several Complex Variables I, Maryland 1970. Edited by J. Horváth. IV, 214 pages. 1970. DM 18,–

Vol. 156: R. Hartshorne, Ample Subvarieties of Algebraic Varieties. XIV, 256 pages. 1970. DM 20,–

Vol. 157: T. tom Dieck, K. H. Kamps und D. Puppe, Homotopietheorie. VI, 265 Seiten. 1970. DM 20,–

Vol. 158: T. G. Ostrom, Finite Translation Planes. IV. 112 pages. 1970. DM 16,–

Vol. 159: R. Ansorge und R. Hass. Konvergenz von Differenzenverfahren für lineare und nichtlineare Anfangswertaufgaben. VIII, 145 Seiten. 1970. DM 16,–

Vol. 160: L. Sucheston, Constributions to Ergodic Theory and Probability. VII, 277 pages. 1970. DM 20,–

Vol. 161: J. Stasheff, H-Spaces from a Homotopy Point of View. VI, 95 pages. 1970. DM 16,–

Vol. 162: Harish-Chandra and van Dijk, Harmonic Analysis on Reductive p-adic Groups. IV, 125 pages. 1970. DM 16,–

Vol. 163: P. Deligne, Equations Différentielles à Points Singuliers Reguliers. III, 133 pages. 1970. DM 16,–

Vol. 164: J. P. Ferrier, Seminaire sur les Algebres Complétes. II, 69 pages. 1970. DM 16,–

Vol. 165: J. M. Cohen, Stable Homotopy. V, 194 pages. 1970. DM 16,–

Vol. 166: A. J. Silberger, PGL_2 over the p-adics: its Representations, Spherical Functions, and Fourier Analysis. VII, 202 pages. 1970. DM 18,–

Vol. 167: Lavrentiev, Romanov and Vasiliev, Multidimensional Inverse Problems for Differential Equations. V, 59 pages. 1970. DM 16,–

Vol. 168: F. P. Peterson, The Steenrod Algebra and its Applications: A conference to Celebrate N. E. Steenrod's Sixtieth Birthday. VII, 317 pages. 1970. DM 22,–

Vol. 169: M. Raynaud, Anneaux Locaux Henséliens. V, 129 pages. 1970. DM 16,–

Vol. 170: Lectures in Modern Analysis and Applications III. Edited by C. T. Taam. VI, 213 pages. 1970. DM 18,–

Vol. 171: Set-Valued Mappings, Selections and Topological Properties of 2^X. Edited by W. M. Fleischman. X, 110 pages. 1970. DM 16,–

Vol. 172: Y.-T. Siu and G. Trautmann, Gap-Sheaves and Extension of Coherent Analytic Subsheaves. V, 172 pages. 1971. DM 16,–

Vol. 173: J. N. Mordeson and B. Vinograde, Structure of Arbitrary Purely Inseparable Extension Fields. IV, 138 pages. 1970. DM 16,–

Vol. 174: B. Iversen, Linear Determinants with Applications to the Picard Scheme of a Family of Algebraic Curves. VI, 69 pages. 1970. DM 16,–

Vol. 175: M. Brelot, On Topologies and Boundaries in Potential Theory. VI, 176 pages. 1971. DM 18,–

Vol. 176: H. Popp, Fundamentalgruppen algebraischer Mannigfaltigkeiten. IV, 154 Seiten. 1970. DM 16,–

Vol. 177: J. Lambek, Torsion Theories, Additive Semantics and Rings of Quotients. VI, 94 pages. 1971. DM 16,–

Please turn over

Vol. 178: Th. Bröcker und T. tom Dieck, Kobordismentheorie. XVI, 191 Seiten. 1970. DM 18,–

Vol. 179: Seminaire Bourbaki – vol. 1968/69. Exposés 347-363. IV. 295 pages. 1971. DM 22,–

Vol. 180: Séminaire Bourbaki – vol. 1969/70. Exposés 364-381. IV, 310 pages. 1971. DM 22,–

Vol. 181: F. DeMeyer and E. Ingraham, Separable Algebras over Commutative Rings. V, 157 pages. 1971. DM 16.–

Vol. 182: L. D. Baumert. Cyclic Difference Sets. VI, 166 pages. 1971. DM 16,–

Vol. 183: Analytic Theory of Differential Equations. Edited by P. F. Hsieh and A. W. J. Stoddart. VI, 225 pages. 1971. DM 20,–

Vol. 184: Symposium on Several Complex Variables, Park City, Utah, 1970. Edited by R. M. Brooks. V, 234 pages. 1971. DM 20,–

Vol. 185: Several Complex Variables II, Maryland 1970. Edited by J. Horváth. III, 287 pages. 1971. DM 24,–

Vol. 186: Recent Trends in Graph Theory. Edited by M. Capobianco/ J. B. Frechen/M. Krolik. VI, 219 pages. 1971. DM 18.–

Vol. 187: H. S. Shapiro, Topics in Approximation Theory. VIII, 275 pages. 1971. DM 22,–

Vol. 188: Symposium on Semantics of Algorithmic Languages. Edited by E. Engeler. VI, 372 pages. 1971. DM 26,–

Vol. 189: A. Weil, Dirichlet Series and Automorphic Forms. V, 164 pages. 1971. DM 16,–

Vol. 190: Martingales. A Report on a Meeting at Oberwolfach, May 17-23, 1970. Edited by H. Dinges. V, 75 pages. 1971. DM 16,–

Vol. 191: Séminaire de Probabilités V. Edited by P. A. Meyer. IV, 372 pages. 1971. DM 26,–

Vol. 192: Proceedings of Liverpool Singularities – Symposium I. Edited by C. T. C. Wall. V, 319 pages. 1971. DM 24,–

Vol. 193: Symposium on the Theory of Numerical Analysis. Edited by J. Ll. Morris. VI, 152 pages. 1971. DM 16,–

Vol. 194: M. Berger, P. Gauduchon et E. Mazet. Le Spectre d'une Variété Riemannienne. VII, 251 pages. 1971. DM 22,–

Vol. 195: Reports of the Midwest Category Seminar V. Edited by J.W. Gray and S. Mac Lane.III, 255 pages. 1971. DM 22,–

Vol. 196: H-spaces – Neuchâtel (Suisse)- Août 1970. Edited by F. Sigrist, V, 156 pages. 1971. DM 16,–

Vol. 197: Manifolds – Amsterdam 1970. Edited by N. H. Kuiper. V, 231 pages. 1971. DM 20,–

Vol. 198: M. Hervé, Analytic and Plurisubharmonic Functions in Finite and Infinite Dimensional Spaces. VI, 90 pages. 1971. DM 16.–

Vol. 199: Ch. J. Mozzochi, On the Pointwise Convergence of Fourier Series. VII, 87 pages. 1971. DM 16,–

Vol. 200: U. Neri, Singular Integrals. VII, 272 pages. 1971. DM 22,–

Vol. 201: J. H. van Lint, Coding Theory. VII, 136 pages. 1971. DM 16,–

Vol. 202: J. Benedetto, Harmonic Analysis on Totally Disconnected Sets. VIII, 261 pages. 1971. DM 22,–

Vol. 203: D. Knutson, Algebraic Spaces. VI, 261 pages. 1971. DM 22,–

Vol. 204: A. Zygmund, Intégrales Singulières. IV, 53 pages. 1971. DM 16,–

Vol. 205: Séminaire Pierre Lelong (Analyse) Année 1970. VI, 243 pages. 1971. DM 20,–

Vol. 206: Symposium on Differential Equations and Dynamical Systems. Edited by D. Chillingworth. XI, 173 pages. 1971. DM 16,–

Vol. 207: L. Bernstein, The Jacobi-Perron Algorithm – Its Theory and Application. IV, 161 pages. 1971. DM 16,–

Vol. 208: A. Grothendieck and J. P. Murre, The Tame Fundamental Group of a Formal Neighbourhood of a Divisor with Normal Crossings on a Scheme. VIII, 133 pages. 1971. DM 16,–

Vol. 209: Proceedings of Liverpool Singularities Symposium II. Edited by C. T. C. Wall. V, 280 pages. 1971. DM 22,–

Vol. 210: M. Eichler, Projective Varieties and Modular Forms. III, 118 pages. 1971. DM 16,–

Vol. 211: Théorie des Matroïdes. Edité par C. P. Bruter. III, 108 pages. 1971. DM 16,–

Vol. 212: B. Scarpellini, Proof Theory and Intuitionistic Systems. VII, 291 pages. 1971. DM 24,–

Vol. 213: H. Hogbe-Nlend, Théorie des Bornologies et Applications. V, 168 pages. 1971. DM 18,–

Vol. 214: M. Smorodinsky, Ergodic Theory, Entropy. V, 64 pages. 1971. DM 16,–

Vol. 215: P. Antonelli, D. Burghelea and P. J. Kahn, The Concordance-Homotopy Groups of Geometric Automorphism Groups. X, 140 pages. 1971. DM 16,–

Vol. 216: H. Maaß, Siegel's Modular Forms and Dirichlet Series. VII, 328 pages. 1971. DM 20,–

Vol. 217: T. J. Jech, Lectures in Set Theory with Particular Emphasis on the Method of Forcing. V, 137 pages. 1971. DM 16,–

Vol. 218: C. P. Schnorr, Zufälligkeit und Wahrscheinlichkeit. IV, 212 Seiten 1971. DM 20,–

Vol. 219: N. L. Alling and N. Greenleaf, Foundations of the Theory of Klein Surfaces. IX, 117 pages. 1971. DM 16,–

Vol. 220: W. A. Coppel, Disconjugacy. V, 148 pages. 1971. DM 16,–

Vol. 221: P. Gabriel und F. Ulmer, Lokal präsentierbare Kategorien. V, 200 Seiten. 1971. DM 18,–

Vol. 222: C. Meghea, Compactification des Espaces Harmoniques. III, 108 pages. 1971. DM 16,–

Vol. 223: U. Felgner, Models of ZF-Set Theory. VI, 173 pages. 1971. DM 16,–

Vol. 224: Revètements Etales et Groupe Fondamental. (SGA 1). Dirigé par A. Grothendieck XXII, 447 pages. 1971. DM 30,–

Vol. 225: Théorie des Intersections et Théorème de Riemann-Roch. (SGA 6). Dirigé par P. Berthelot, A. Grothendieck et L. Illusie. XII, 700 pages. 1971. DM 40,–

Vol. 226: Seminar on Potential Theory, II. Edited by H. Bauer. IV, 170 pages. 1971. DM 18,–

Vol. 227: H. L. Montgomery, Topics in Multiplicative Number Theory. IX, 178 pages. 1971. DM 18,–

Vol. 228: Conference on Applications of Numerical Analysis. Edited by J. Ll. Morris. X, 358 pages. 1971. DM 26,–

Vol. 229: J. Väisälä, Lectures on n-Dimensional Quasiconformal Mappings. XIV, 144 pages. 1971. DM 16,–

Vol. 230: L. Waelbroeck, Topological Vector Spaces and Algebras. VII, 158 pages. 1971. DM 16,–

Vol. 231: H. Reiter, L^1-Algebras and Segal Algebras. XI, 113 pages. 1971. DM 16,–

Vol. 232: T. H. Ganelius, Tauberian Remainder Theorems. VI, 75 pages. 1971. DM 16,–

Vol. 233: C. P. Tsokos and W. J. Padgett. Random Integral Equations with Applications to Stochastic Systems. VII, 174 pages. 1971. DM 18,–

Vol. 234: A. Andreotti and W. Stoll. Analytic and Algebraic Dependence of Meromorphic Functions. III, 390 pages. 1971. DM 26,–

Vol. 235: Global Differentiable Dynamics. Edited by O. Hájek, A. J. Lohwater, and R. McCann. X, 140 pages. 1971. DM 16,–

Vol. 236: M. Barr, P. A. Grillet, and D. H. van Osdol. Exact Categories and Categories of Sheaves. VII, 239 pages. 1971, DM 20,–

Vol. 237: B. Stenström. Rings and Modules of Quotients. VII, 136 pages. 1971. DM 16,–

Vol. 238: Der kanonische Modul eines Cohen-Macaulay-Rings. Herausgegeben von Jürgen Herzog und Ernst Kunz. VI, 103 Seiten. 1971. DM 16,–

Vol. 239: L. Illusie, Complexe Cotangent et Déformations I. XV, 355 pages. 1971. DM 26,–

Vol. 240: A. Kerber, Representations of Permutation Groups I. VII, 192 pages. 1971. DM 18,–

Vol. 241: S. Kaneyuki, Homogeneous Bounded Domains and Siegel Domains. V, 89 pages. 1971. DM 16,–

Vol. 242: R. R. Coifman et G. Weiss, Analyse Harmonique Non-Commutative sur Certains Espaces. V, 160 pages. 1971. DM 16,–

Vol. 243: Japan-United States Seminar on Ordinary Differential and Functional Equations. Edited by M. Urabe. VIII, 332 pages. 1971. DM 26,–

Vol. 244: Séminaire Bourbaki – vol. 1970/71. Exposés 382–399. IV, 356 pages. 1971. DM 26,–

Vol. 245: D. E. Cohen, Groups of Cohomological Dimension One. V, 99 pages. 1972. DM 16,–